Waste Treatment Processes in Environmental Engineering

The Editors

Dr. S.N. Kaul has obtained his bachelor and master degrees from IIT, Kanpur and was awarded Ph.D. degree from the University of New Castle Upon Tyne, UK. Dr. Kaul joined CSIR-National Environmental Engineering Research Institute [NEERI] Nagpur in 1970 and worked for over three decades. Dr. Kaul superannuated as Acting Director and Director Grade Scientist from NEERI and was also associated with MIT College of Engineering, Pune as its Principal.

Dr.Kaul has published over 350 Papers in National and International journals of repute beside authorship of over 30 Books on Environmental Engineering and Science, 10 Patents and 150 conference papers. Dr. Kaul has designed and operated several waste treatment plants in India. Dr. Kaul has supervised over a dozen of Ph.D. students and 40 M Tech students. In addition, Dr. Kaul was responsible for organizing several National and International Conferences and has traveled extensively in India and aboard concerning various R and D programs. Dr. Kaul has received several National and International awards and is expert member in importants National committees.

Dr. D.R. Saini obtained his B Tech and M Tech from HBTI,Kanpur and PhD [Chem.Engg.] from IIT-Delhi. Presently working as Principal Scientist at CSIR-NCL, Pune. Dr. Saini is co-author of 8 books, six chapters and published seventy papers in international journals. He has guided three PhD students and few M.Tech. students.

Dr. B.N. Rai is presently working as Professor in the Deapartment of Chemical Engineering and Technology, IIT (BHU), Varanasi. Dr. Rai has done B.E. (Chem. Eng.) from Bangalore University and M.Tech. (Chem) and Ph.D. from Banaras Hindu University. He has more than 20 years of teaching, research and administrative experiences in various capacities. He has more than 40 publications in international journals, book chapters, conferences, etc. and delivered invited key note lectures in various conferences. He has guided more than 20 M.Tech. and two Ph.D. students. Dr. Rai was awarded by the Ministry of Non-Conventional Energy, Govt of India for writing a book in Hindi on fundamental topics. His research interests include Environmental Biotechnology, Industrial Pollution and Control. He has also organized international conferences, refresher courses and workshops on waste management. Besides above, he has served the Universitiy in various capacities such as President, IT Gymkhana, Coordinator, National Youth Festival, etc.

Er. Prateek Kaul has obtained M.Tech. from V.NIT, Nagpur. Er. Kaul travelled respectively in India and abroad. Er. Kaul has several research publications, books and patents to his credit. Er. Kaul has worked in several multinational industries and is presently associated with Symantec (USA), Pune.

Waste Treatment Processes in Environmental Engineering

Volume 4

Editors

Prof. (Dr.) S.N. Kaul
Principal
MIT College of Engineering, Pune
Former Acting Director and Director-Grade
Scientist, NEERI [CSIR], Nagpur, M.S.

Dr. D.R. Saini
Scientist
National Chemical Laboratory [CSIR],
Pune, M.S.

Dr. B.N. Rai
Professor, Department of Chemical Engineering & Technology,
Indian Institute of Technology (BHU),
Banaras Hindu University, Varanasi, U.P.

Er. Prateek Kaul
Sr. Manager,
Symantec Corporation (USA), Pune

2016

Daya Publishing House®

A Division of

Astral International Pvt. Ltd.

New Delhi – 110 002

Waste Treatment Processes in Environmental Engineering
(4 Volumes Set)
Volume 1: Page 0001-0306
Volume 2: Page 0307-0628
Volume 3: Page 0629-0908
Volume 4: Page 0909-1184

Cataloging in Publication Data—DK
Courtesy: D.K. Agencies (P) Ltd. <docinfo@dkagencies.com>

Waste treatment processes in environmental engineering / editors, Prof. (Dr.) S.N. Kaul, Dr. D.R. Saini, Dr. B.N. Rai, Er. Prateek Kaul.
4 volumes cm
Includes indexes.

ISBN 9789351309147 (Vol. 4)
ISBN 978-93-5124-372-4 (International Edition)

1. Sewage—Purification—India. 2. Sewage—Purification. 3. Environmental engineering—India. 4. Environmental engineering. I. Kaul, S. N., editor.

DDC 628.30954 23

Published by : **Daya Publishing House®**
A Division of
Astral International Pvt. Ltd.
– ISO 9001:2008 Certified Company –
4760-61/23, Ansari Road, Darya Ganj
New Delhi-110 002
Ph. 011-43549197, 23278134
E-mail: info@astralint.com
Website: www.astralint.com

Laser Typesetting : **Classic Computer Services**, Delhi - 110 035

Printed at : **Thomson Press India Limited**

Dedicated to the Sacred Memory of

Sri Sri Sarda Maa

&

Swami Vivekananda

Preface

The Editors of the present edition on Waste Treatment Processes in Environmental Engineering, have been associated with various educational institutes, research organizations, universities etc. It was felt that a comprehensive selective topics in Environmental Engineering concerning India with Indian Experiences from various part of the India with Indian data from various Indian experts must be collected to disseminate the authentic information to our readers.

Text books and reference books are already available in the market but most often they relate to information and experiences of the western world, which is not relevant to Indian conditions and economics. The editors have been contacting various experts throughout the country in order to present the Indian experiences on which the contributing authors have been working. Often the good and invaluable work gets published in foreign journals which are expansive and, therefore,beyond the reach of most of the readers to acquire. To reduce this gaps the editors have made some efforts in this direction, and the editors feel that readers will appreciate this and try to benefits from this compilation. It is also primarily with this view to make aware Indian readers about the work of Indian experts and their experiences.

Waste Treatment Processes in Environmental Engineering is a compendium of knowledge concerning Environmental Issues from several experts who are associated with prestigious institutes in India, *viz.*: IITs, CSIR, NITS, Central Universities, Pollution Control Board. This book endeavors to explore various important topics of immediate concern to all our stake holders, especially students, researches, practitioners, teachers, NGOs etc. This book also introduces the reader about areas of research of the experts in India.

Economic progress is directly connected with environment and ecology. Material progress goes linear while environmental degradation follows geometric progression. Therefore, environment and ecology is the rate determining step in the progress of the country. This aspect is extensively exhibited in this book.

Emphasis has been laid on environmental pollution and control and the editors feel that these this specific book will be immensely useful to all concerned with Environmental and Ecology encompassing all the three component of Environment, *viz.*: Air, Water and Solid Waste besides EIAs, cleaner technology, problem associated with implementation including environmental education etc.Finally, the editors would like to thanks all the contributing authors for accepting our invitation to contribute their valuable research and development work for our readers and editors would also like to thanks them for the time and efforts devoted while compiling their work.

The trust reposed by the Astral International (P) Ltd., New Delhi is thankfully acknowledged.

Prof. (Dr.) S.N. Kaul

Dr. D.R. Saini

Prof. (Dr.) B.N. Rai

Er. Prateek Kaul

Contents

Volume 2

Volume 3

Volume 4

Chapter 56

Assessment of Groundwater Contamination due to Improper Functioning of Sewage Treatment Plants: A Case Study

P. Nema, C.G. Mehta and S.N. Kaul

CSIR-NEERI, Nagpur

ABSTRACT

A detailed investigation into the various aspects of the groundwater contamination reportedly caused by the improper functioning of two sewage treatment plants was carried out. A number of groundwater sources inside as well as outside areas of the STP were identified for water quality monitoring. Seasonal variations in groundwater quality with specific reference to physico-chemical and bacteriological quality were studied. The paper presents the data, analyses various causes and also suggests remedial measures, both preventive as well as curative, to arrest further contamination of groundwater.

Keywords: *Sewage treatment plant, Groundwater, Contamination, Total coliforms, Faecal coliforms, Eschertia coli., Faecal streptococci.*

Introduction

Two sewage treatment plants of 75 MLD and 180 MLD capacities have been in operation in a metro city in western India since 1995-96. These STPs are located in the close proximity to a National Highway (hereafter referred to STP-1 and STP-2). The treatment schemes of the STPs comprise of pretreatment (screen and grit removal) followed by biological treatment (a series combination of anaerobic lagoon followed by aerated lagoon cum settling zones).

Residents in the area near the sewage treatment plant expressed, concern on several occasions, over the potential threat of groundwater contamination caused by the malfunctioning of the STP and the consequential adverse health impacts. Their appeal sought a detailed scientific investigation of the problem to help concerned authority to consider effective measures for the better functioning of the STPs that would ensure prevention of groundwater pollution.

Methodology

In order to assess the groundwater contamination due to possible infiltration of sewage, a water quality-monitoring programme was formulated. Sampling locations were fixed within 1 to 1.5 kms distance of the STPs taking into consideration the distance that the enteric bacteria can traverse. Another criteria followed for selection was upstream and downstream areas of the STPs with respect to their discharge points in river so as to find out change, if any, in groundwater quality due to infiltration of sewage in downstream and upstream areas of STPs. The layouts of the respective study area including sampling locations are shown in Figures 56.1 and 56.2.

Sampling procedure was followed as per WHO guidelines (Guidelines For Drinking Water Quality, Vol. 3, WHO, 1985). The samples were analyzed for the relevant physico-chemical and bacteriological parameters following the standard procedures (Standard Methods for Water and Wastewater analysis, 18th edition, 1992; D.D.Mara, 1968).

Results and Discussion

STP-1

Physico-chemical Quality of Groundwater

Samples of groundwater were collected from different sources in upstream and downstream areas of STP-1. Details of these sources are given in Table 56.1.

Physico-chemical analysis of the water samples from the sources upstream of STP-1 showed that pH, sulphate and nitrate were within IS 10500-1995 limits but Mg hardness, chlorides and TDS exceeded these limits at all locations during post monsoon, winter and summer seasons. Fluoride concentration was also more than the Drinking Water Specifications IS 10500- 1991 (First Revision, 2nd Reprint 1995) during all seasons in all the samples except the handpump at sampling point 5. However, Fluoride concentration in the samples from handpump at sampling point 3 was found within IS limits in winter.

A comparison of water quality of these samples with those from suburb areas of the city (Pre-and Post-monsoon Potable Water Quality Assessment in Some Major Cities in India, NEERI Report, 2000) indicated that the total hardness in the samples from handpump at sampling point 2 and sampling point 4 was higher during all seasons whereas Ca-hardness in the samples from the borewell at sampling point 4

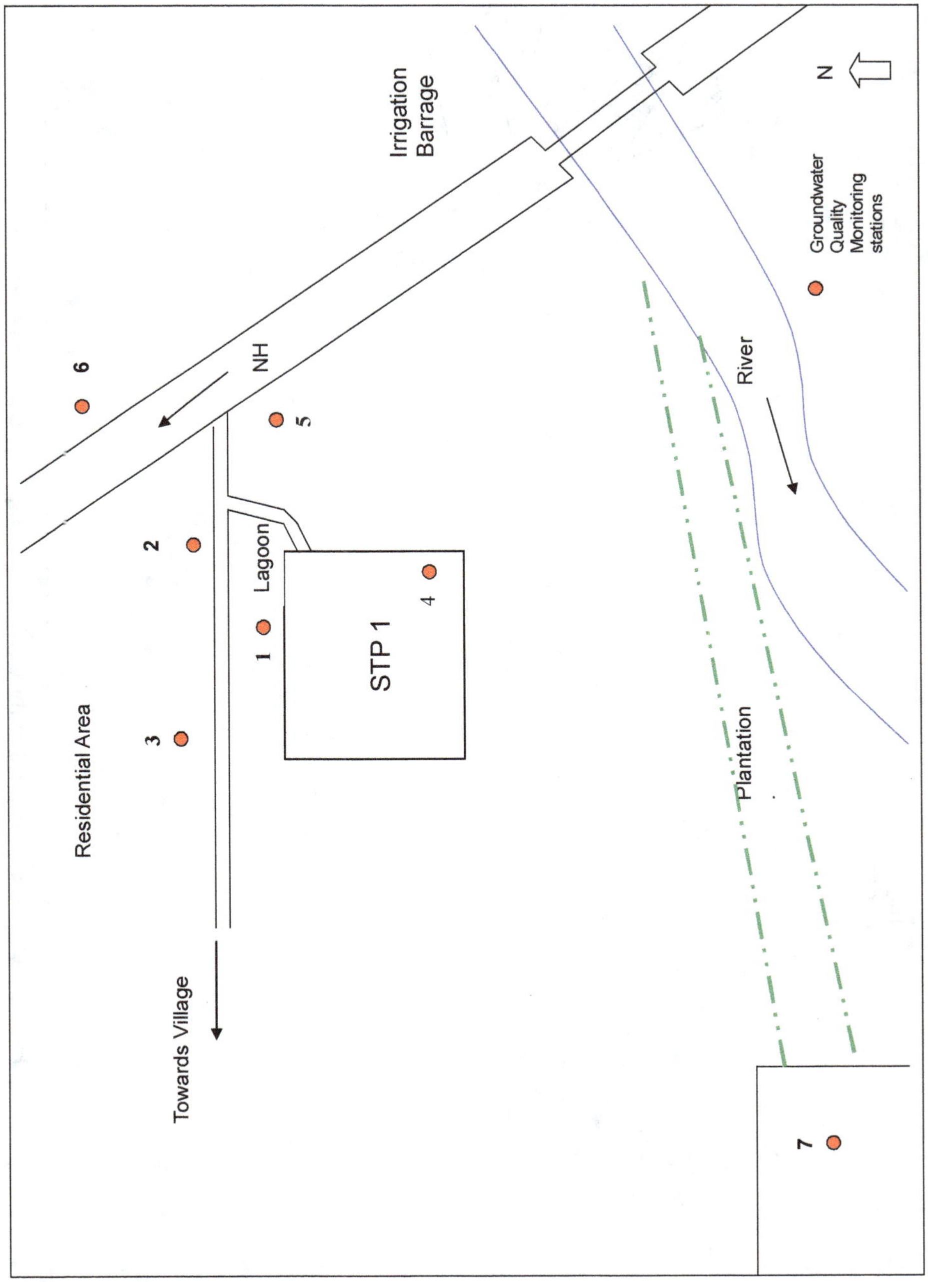

Figure 56.1: Groundwater Quality Monitoring Sampling Locations at STP–1.

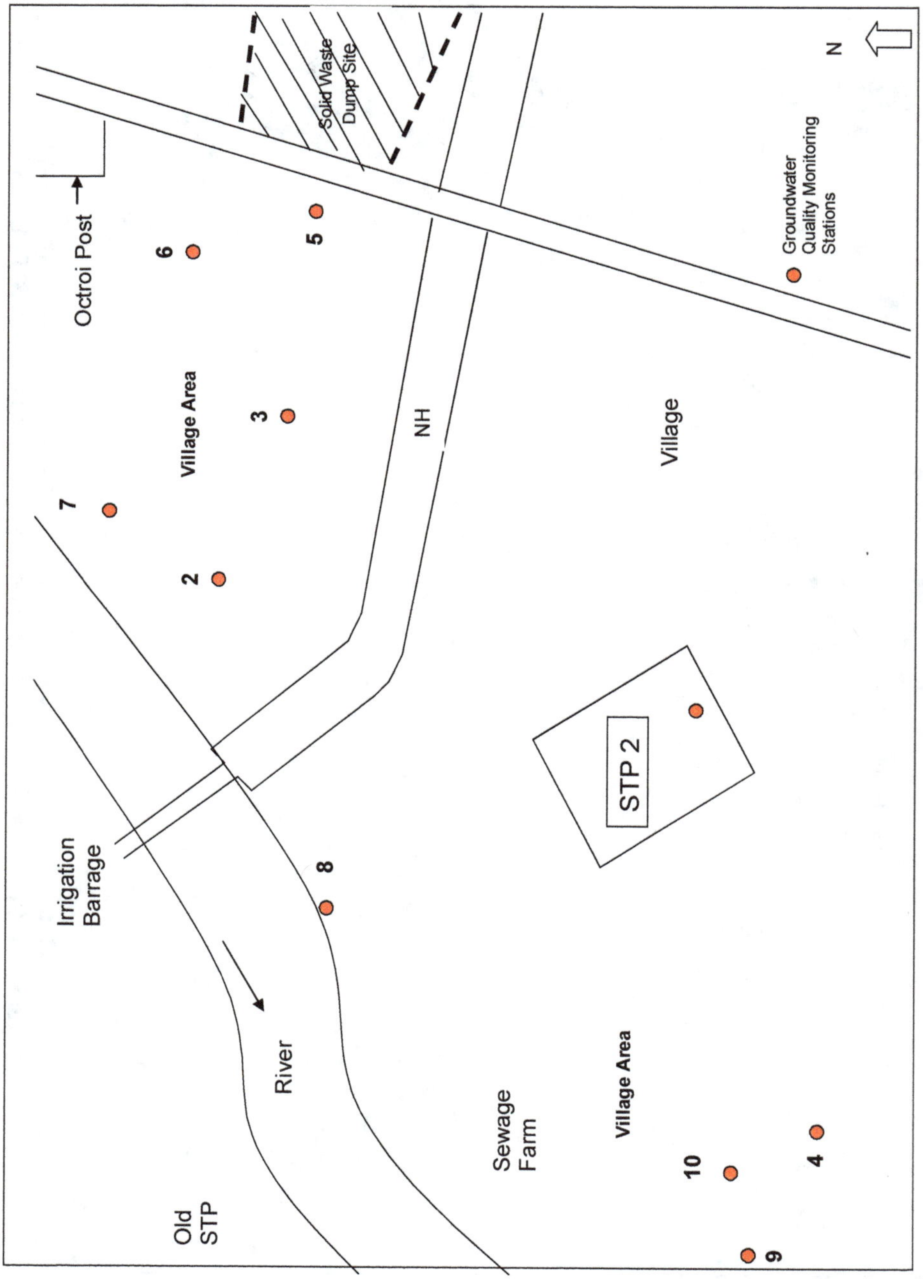

Figure 56.2: Groundwater Quality Monitoring Sampling Locations at STP–2.

Table 56.1: Groundwater Sources: In and Around STP-1

*Source**	*1*	*2*	*3*	*4*	*5*	*6*	*7*
Distance from STP (m)	5	80	100	150	300	500	1500
Direction from STP	North	West	West	East	North	Northwest	Southwest
Type of Source	Borewell	Handpump	Handpump	Borewell	Handpump	Borewell	Borewell
Depth (m)	200	20	8.5	100	10	200	—
Year of Commissioning	1998	1996	1998	—**	1995	1997	—
Population served	—	50	10	—	70	250	10
Water Consumption (KLD)	—	1.8	1	—	11	30	10
Present Usage of Water	Domestic	Washing	Washing	Domestic	Domestic	Domestic	Gardening
Existence of soak pit (Distance/Direction from the source)	80 m Northwest	8 m South	15 m South	No soak pit	No soak pit	—	—

** Data Not Available.

1-7 are the sampling points.

and Mg-hardness in all the samples from all locations were higher during all seasons (Tables 56.2A, 56.2B).

Physico-chemical quality of groundwater in the downstream area of STP-1 discharge point in the river showed pH, Ca-Hardness, Mg-Hardness, Chlorides, Fluoride and TDS exceeding the IS: 10500-1991 (First Revision, 2nd Reprint 1995) Standards for drinking water during summer season. When compared with the water quality of suburb area of the city, all parameters except pH and Alkalinity were comparable with city water quality (Table 56.2B).

A comparison of water quality data of upstream and downstream areas of STP-1 revealed that the concentration of Chlorides, Sulphate, TDS and COD were more in the samples from downstream sources than those from the sources upstream (Figure 56.3). The increase in concentration of Chlorides, Sulphate, TDS and COD were found as 19 per cent, 46 per cent, 12 per cent, and 62 per cent respectively. Higher COD, Chlorides and Sulphate in samples from downstream area as compared to upstream area indicated presence of organic matter, which could be attributed to continuous discharge of treated sewage in the river and its possible seepage into the groundwater (Figure 56.3).

Bacteriological Analysis

- ✰ Water samples from the borewell located at a distance of about 5m North of lagoon towards residential area (sampling point 1) were analyzed. Six out of seven samples showed presence of total coliforms. Faecal coliforms were detected in five samples, whereas *E.coli* was found in two samples (Table 3A). However, all the samples were found free from *Faecal streptococci*. The presence of faecal coliforms and *E.coli* indicated faecal contamination (D.D.Mara: 1968).
- ✰ Water samples from a handpump located at 80 m West from lagoon (sampling point 2) were analyzed. The sample showed presence of faecal coliforms in four out of seven samples, *E.coli* was detected in three samples (Table 56.3B). However all the samples were free from *Faecal streptococci*. The presence of faecal coliforms and *E.coli* indicated faecal contamination.
- ✰ Water samples collected from a handpump located at a distance of 100m in the West direction from the lagoon (sampling point 3) showed presence of total coliforms in five out of seven samples. The presence of faecal coliforms and E.coli was detected in four samples (Table 56.3C) which indicated faecal contamination. However all the samples were free from *Faecal streptococci*.
- ✰ Water samples collected from the borewell in STP-1 treatment plant premises located at a distance of 150 m east from lagoon (sampling point 4) showed presence of total coliforms in six out of seven samples. However faecal coliforms, *E.coli* and *Faecal streptococci* were detected in five, four and two samples respectively (Table 56.3D). The presence of *Faecal streptococci* is indicative of recent contamination. The results showed bacterial contamination of the source.

Table 56.2A: Physico-chemical Quality of Groundwater: STP-1

Parameter	Source*												IS: 10500-1995 Standards/ WHO Guidelines	Water Quality of Suburban Areas of the City
	1			2			3			4				
	PM	W	S	PM	W	S	PM	W	S	PM	W	S		
Alkalinity ($CaCO_3$)	410	435	460	500	455	475	515	545	485	440	415	630	—**	330 – 600
Total Hardness ($CaCO_3$)	227	230	350	432	455	500	402	410	390	595	565	460	300	110 – 420
Ca-Hardness ($CaCO_3$)	125	115	200	225	260	365	217	210	270	452	430	280	75	70 – 330
Mg-Hardness ($CaCO_3$)	102	115	150	207	195	135	185	200	120	142	135	180	30	40 – 100
Chlorides	380	410	415	398	395	375	395	425	390	366	390	380	250	240 – 600
Sulphate (SO_4)	74	78	63	57	59	58	56	52	80	51	63	51	150	37 – 322
Nitrate (NO_3)	12	8	9	5.7	5	4	5.7	6	7	19	12	6	45	—
Fluoride	5.0	1.7	2.5	1.2	3.0	3.2	3.2	0.9	2.4	1.3	3.2	3.2	0.6-1.2	—
Sodium	287	290	350	282	300	339	321	300	306	246	251	266	—	—
Potassium	0.9	0.9	1.1	0.9	0.7	0.8	0.4	0.6	0.6	0.1	0.6	0.7	—	—
TDS	1236	1220	1126	1210	1201	1135	1319	1284	1144	1245	1221	1055	500	1042 – 2468
COD	10	14	10	23	18	16	8	4	7	10	11	10	—	—

All parameters are in mg/L.

*Data: average of six sets.

PM: Post Monsoon; W: Winter; S: Summer.

Table 56.2B: Physico-chemical Quality of Groundwater: STP-1

Parameter	Source*							IS: 10500-1995 Standards/WHO Guidelines	Water Quality of Suburban Areas of the City
	5			6			7		
	PM	W	S	PM	W	S	S		
Alkalinity ($CaCO_3$)	435	450	410	430	340	400	636	—**	330 – 600
Total Hardness ($CaCO_3$)	365	365	285	305	290	285	265	300	110 – 420
Ca-Hardness ($CaCO_3$)	237	240	175	185	160	170	175	75	70 – 330
Mg-Hardness ($CaCO_3$)	127	115	110	120	130	115	73	30	40 – 100
Chlorides	591	600	395	434	420	395	533	250	240 – 600
Sulphate (SO_4)	82	88	105	117	93	94	159	150	37 – 322
Nitrate (NO_3)	14	15	15	17	18	8	2.3	45	—
Fluoride	0.9	0.8	1.5	2.6	2.4	2.1	2.5	0.6-1.2	—
Sodium	331	330	353	338	330	356	442	—	—
Potassium	0.9	0.5	0.8	0.9	0.7	1.1	12	—	—
TDS	1563	1559	1236	1504	1502	1164	1526	500	1042 – 2468
COD	11	11	13	7	4	7	29	—	—

All parameters are in mg/L.; * Data: average of six sets.

PM: Post Monsoon; W: Winter; S: Summer.

Points	Distance from the STP (m)	Points	Distance from the STP (m)
1	5	2	80
3	100	4	150
5	300	6	500
7	1500		

- Water samples collected from a handpump located towards north of the lagoon (sampling point 5) were analyzed. All the five samples showed presence of total coliforms. However faecal coliforms, *E.coli* and *Faecal streptococci* were absent (Table 56.3E). From the presence of only total coliforms, it may be inferred that the bacteria might be of soil origin.
- Water samples from the borewell located at 500 m NW of the lagoon (sampling point 6) were analyzed. Total coliforms were detected in six out of seven samples. Faecal coliforms, *E.coli* and *Faecal streptococci* were not detected in any of the samples (Table 56.3F). This indicated that the bacteria present in the water might be of soil origin.
- Water samples from the borewell located at a distance of about 1500m S E from the STP (sampling point 7) showed presence of total coliforms, faecal coliforms *and E.coli* (Table 56.3G). However *Faecal streptococci* were not detected in any of the samples. The presence of faecal coliforms and *E.coli* indicated faecal contamination.

Table 56.3: Bacteriological Quality of Groundwater: STP-1

Parameter	*Post Monsoon*		*Winter*			*Summer*	
A: Borewell (Sampling Point 1)							
Total coliforms*	2	20	1	4	Nil	12	18
Faecal coliforms*	2	12	2	Nil	Nil	8	6
*E. coli**	Nil	Nil	Nil	Nil	Nil	3	4
*Faecal streptococci**	Nil	Nil	Nil	Nil	Nil	Nil	6
B: Handpump (Sampling Point 2)							
Total coliforms	TNC	18	Nil	Nil	Nil	22	30
Faecal coliforms	TNC	11	Nil	Nil	Nil	15	14
E. coli	27	Nil	Nil	Nil	Nil	5	6
Faecal streptococci	Nil	Nil	Nil	Nil	Nil	Nil	Nil
C: Handpump (Sampling Point 3)							
Total coliforms	7	1	1	Nil	Nil	4	6
Faecal coliforms	2	1	Nil	Nil	Nil	1	2
E. coli	Nil	Nil	Nil	Nil	Nil	1	1
Faecal streptococci	Nil	Nil	Nil	Nil	Nil	Nil	Nil
D: Borewell (Sampling Point 4)							
Total coliforms	15	54	TNC	1	Nil	48	28
Faecal coliforms	3	3	3	Nil	Nil	19	12
E. coli	3	3	Nil	Nil	Nil	5	8
Faecal streptococci	3	Nil	Nil	Nil	Nil	1	Nil

Contd...

Table 56.3–*Contd...*

Parameter	*Post Monsoon*		*Winter*			*Summer*	
E: Handpump (Sampling Point 5)							
Total coliforms	TNC	11	TNC	8	18	9	11
Faecal coliforms	13	Nil	Nil	Nil	Nil	Nil	Nil
E. coli	Nil	Nil	Nil	Nil	Nil	Nil	Nil
Faecal streptococci	Nil	Nil	Nil	Nil	Nil	Nil	Nil
F: Borewell (Sampling Point 6)							
Total coliforms	81	Nil	4	TNC	2	2	4
Faecal coliforms	Nil	Nil	Nil	Nil	Nil	Nil	Nil
E. coli	Nil	Nil	Nil	Nil	Nil	Nil	Nil
Faecal streptococci	Nil	Nil	Nil	Nil	Nil	Nil	Nil
Parameter				*Summer*			
G: Borewell (Sampling Point 7)							
Total coliforms	65	58	68	TNC	50	38	
Faecal coliforms	42	33	40	45	38	21	
E. coli	12	20	18	32	10	5	
Faecal streptococci				—Nil—-			

TNC: Too Numerous to Count

* CFU per 100ml

STP-2

Physico-Chemical Quality of Groundwater

Samples of groundwater were collected from different sources located upstream and downstream of STP-2. Details of the sources are given in Table 56.4.

The data on various physico-chemical parameters are presented in Table 56.5A and 56.5B. As can be seen, physico-chemical quality of groundwater in upstream area of STP-2 showed that pH and nitrate were within IS 10500–1991 (First Revision, 2nd Reprint 1995) standards during postmonsoon, winter and summer seasons. Sulphate exceeded the IS 10500–1995 standards in the samples from sampling point 2, 6 and 7. Ca and Mg hardness and Chlorides also exceeded IS 10500–1991 (First Revision, 2nd Reprint 1995) standards at all locations during all the seasons. Total hardness exceeded IS 10500–1991 (First Revision, 2nd Reprint 1995) standards at all locations in all the seasons barring the sample from sampling point 1. The observed fluoride concentration was always more than the IS 10500–1991 (First Revision, 2nd Reprint 1995) standards in all the samples during all the seasons.

A comparison of water quality of the groundwater sources sampled during the study period with that of suburbs areas of the city indicated that alkalinity at sampling points 1 and 2 were higher in all seasons. Total hardness was also more in the

Table 56.4: Groundwater Sources: In and Around STP-2

*Source**	*1*	*2*	*3*	*4*	*5*	*6*	*7*	*8*	*9*	*10*
Distance from STP (m)	200	1000	1100	1100	1500	1500	1500	500	1000	1100
Direction from STP	East	North	North	South	East	North	North	North	Southwest	Southwest
Source Type	Borewell	Handpump	Handpump	Borewell	Borewell	Handpump	Handpump	Handpump	Handpump	Handpump
Depth (m)	130	23	23	167	—	23	23	26	—	15
Year of Commission	—		1970		—**		1975	1985	—	1992
Population served	—		30	2000	—		100	100	Occasio-nally used	Occasio-nally used
Water Consumption (KLD)	—		3	160	—		8	5	—	—
Usage of Water	Domestic	Domestic	Domestic	Domestic	Farming	Domestic	Domestic	Drinking	Drinking	Domestic
Existence of soakpit (Distance/Direction from the source)	——			No soak pit			——	—	—	—

** Data Not Available

* Location (Figure 56.1)

1 to 10 are the sampling points.

Table 56.5A: Physico-chemical Quality of Groundwater: STP-2

Parameter	*Source*															*IS: 10500-1995 Standards/ WHO Guidelines*	*Water Quality of Suburban Areas of the City*
	1			*2*			*3*			*4*			*5*				
	PM	*W*	*S*	*PM*	*W*	*S*	*PM*	*W*	*S*	*PM*	*W*	*S*	*PM*	*W*	*S*		
Alkalinity ($CaCO_3$)	355	365	465	715	690	790	665	665	745	385	375	490	565	600	510	—**	330 – 600
Total Hardness ($CaCO_3$)	290	315	335	712	700	710	557	515	595	542	535	535	429	405	410	300	110 – 420
Ca-Hardness ($CaCO_3$)	200	230	220	432	430	400	295	345	335	327	340	330	425	225	290	75	70 – 330
Mg-Hardness ($CaCO_3$)	90	80	115	280	270	310	262	220	260	210	195	205	204	180	120	30	40 – 100
Chlorides	350	349	345	941	920	895	530	520	515	448	450	440	395	390	380	250	240 – 600
Sulphate (SO_4)	106	119	116	248	255	293	127	121	236	86	97	104	129	139	150	150	37 – 322
Nitrate (NO_3)	9	13	11	11	12	15	18	18	7	10	14	9	13	8	12	45	—
Fluoride	2.2	2.2	2.3	3.2	3.4	3.5	2.6	2.2	2.4	2.3	2.5	2.5	2.5	2.6	2.6	0.6-1.2	—
Sodium	298	284	272	588	556	530	398	370	385	245	244	230	306	303	348	—	—
Potassium	0.9	1.3	1.2	0.7	1.2	0.7	1.2	1.7	1.3	1.3	1.1	1.3	1.7	1.5	1.7	—	—
TDS	1353	1388	1272	2695	2654	2513	1864	1815	1726	1326	1302	1217	1384	1308	1510	500	1042 – 2468
COD	6	7		8	10	7	7	6	7	11	9	6	—	—	—	—	—

All parameters are in mg/L.

*Data: average of six sets.

PM: Post monsoon; W: Winter; S: Summer.

Table 56.5B: Physico-chemical Quality of Groundwater: STP-2

Parameter	Source									IS: 10500-1995 Standards/ WHO Guidelines	Water Quality of Suburban areas of the City
	6			7			8	9	10		
	PM	W	S	PM	W	S	S	S	S		
Alkalinity ($CaCO_3$)	600	655	690	580	550	**635**	950	868	865	—**	**330 – 600**
Total Hardness ($CaCO_3$)	394	425	475	505	475	610	383	556	490	**300**	**110 – 420**
Ca-Hardness ($CaCO_3$)	275	250	270	290	280	335	290	448	400	**75**	**70 – 330**
Mg-Hardness ($CaCO_3$)	123	175	205	215	195	275	93	108	80	30	40 – 100
Chlorides	312	320	305	902	910	855	600	820	890	250	240 – 600
Sulphate (SO_4)	163	162	187	226	226	305	360	440	496	150	37 – 322
Nitrate (NO_3)	17	17	11	15	19	15	8.1	29.5	7.7	45	—
Fluoride	2.9	2.8	3.1	3.3	3.6	3.7	3.2	1.6	2.2	0.6-1.2	—
Sodium	431	385	350	577	560	510	667	830	898	—	—
Potassium	1.0	0.7	1.5	1.0	2.0	1.2	22	27	30	—	—
TDS	1968	1981	1727	2564	2517	2594	2457	3102	3338	500	1042 – 2468
COD	8	11	12	5	4	9	48	64	23.5	—	—

All parameters are in mg/L.; * Data: average of six sets.

PM: Post Monsoon; W: Winter; S: Summer.

Points	Distance from the STP (m)	Points	Distance from the STP (m)
1	200	2	1000
3	1100	4	1100
5	1500	6	1500
7	1500	8	500
9	1000	10	1100

samples collected from sampling point 2, 3, 4 and 7 during all seasons. Ca-hardness at sampling point 2 whereas Mg-hardness showed higher values at all locations except sampling point 1 during all seasons. Chlorides and TDS, however, showed lower value at all locations except sampling points 2 and 7 during all seasons (Tables 565A and 565B).

Physico-chemical quality of water samples from the sources in the downstream surrounding areas of STP-2 showed that the Total Hardness, Mg-Hardness, Chlorides, Sulphate, Fluoride and TDS exceeded the limits prescribed under IS: 10500- 1991 (First Revision, 2nd Reprint 1995) during summer (Table 565B).

A comparison of water quality of the above-mentioned groundwater sources with that of suburb area of the city indicated that the sulphate and TDS concentration were more at all locations. However, only two sources (sampling point 9 and 10) showed Ca-hardness and Chlorides in higher concentration.

The groundwater samples from the downstream area of STP-2 showed Chlorides, Sulphate, TDS and COD in higher concentrations as compared to those from upstream area (Figure 56.4). The increase observed in Chlorides, Sulphate, TDS and COD concentrations were 28 per cent, 64 per cent, 37 per cent, and 82 per cent respectively. Higher concentration of COD, Chlorides and Sulphate in samples from downstream are vis-a-vis upstream area was indicative of addition of organic matter which could be attributed to the continuous discharge of treated sewage in the river and possibly its seepage into the groundwater.

Bacteriological Analysis

- ☆ Water samples from the borewell located at a distance of about 200m from the lagoon towards East (sampling point 1) were analyzed. Total coliforms, faecal coliforms and *E.coli* were detected in two of the five samples (Table 56.6A). However, all the samples were found to be free from *Faecal streptococci*. The presence of faecal coliforms and *E.coli* indicated faecal contamination.
- ☆ Water samples from the new handpump located at a distance of 1 km North of lagoon (sampling point 2) were analyzed. None of the sample showed presence of Total coliforms, faecal coliforms, *E. coli* and *Faecal streptococci*. (Table 56.6B). However, four of the samples collected from an old handpump located at a distance of 1.1 km in the same area/direction (sampling point 3), showed presence of total coliforms but faecal coliforms, *E.coli* and *Faecal streptococci* were absent (Table 56.6C). The presence of only total coliform indicated that they might be of soil or vegetation origin.
- ☆ Water samples from the borewell located at a distance of 1.1 km in South from the lagoon (sampling point 10) were analyzed. All the seven samples showed the presence of total coliforms. Three samples showed presence of faecal coliforms whereas only one sample showed the presence of *E.coli* (Table 56.6D). However all the samples were free from *Faecal streptococci*. The results indicated faecal contamination of the water source.
- ☆ Water samples from the borewell located at a distance of 1.5 km towards East of the lagoon (sampling point 5) were analyzed. The samples showed

absence of all indicator organisms *i.e.* total coliforms, faecal coliforms, *E.coli* and *Faecal streptococci* indicating that the water source was free from contamination (Table 56.6E).

Table 56.6: Bacteriological Quality of Groundwater: STP-2

Parameter	*Post Monsoon*			*Winter*		*Summer*	
A: Borewell (Sampling Point 1)							
Total coliforms*	TNC	TNC	Nil	Nil	Nil	Nil	Nil
Faecal coliforms*	100	100	Nil	Nil	Nil	Nil	Nil
*E. coli**	100	47	***Nil***	Nil	Nil	Nil	Nil
*Faecal streptococci**	Nil	Nil	Nil	Nil	Nil	Nil	Nil
B: New handpump (Sampling Point 2)							
Total coliforms*	Nil	Nil	Nil	Nil	Nil	Nil	Nil
Faecal coliforms*	Nil	Nil	Nil	Nil	Nil	Nil	Nil
*E. coli**	Nil	Nil	***Nil***	Nil	Nil	Nil	Nil
*Faecal streptococci**	Nil	Nil	Nil	Nil	Nil	Nil	Nil
C: Old handpump (Sampling Point 3)							
Total coliforms*	2	4	7	Nil	Nil	Nil	1
Faecal coliforms*	Nil	Nil	Nil	Nil	Nil	Nil	Nil
*E. coli**	Nil	Nil	***Nil***	Nil	Nil	Nil	Nil
*Faecal streptococci**	Nil	Nil	Nil	Nil	Nil	Nil	Nil
D: Borewell (Sampling Point 4)							
Total coliforms*	33	11	3	4	2	6	12
Faecal coliforms*	25	Nil	Nil	Nil	2	1	Nil
*E. coli**	Nil	Nil	***Nil***	Nil	Nil	Nil	Nil
*Faecal streptococci**	Nil	Nil	Nil	Nil	Nil	Nil	Nil
E: Borewell (Sampling Point 5)							
Total coliforms*	Nil	Nil	Nil	Nil	Nil	Nil	Nil
Faecal coliforms*	Nil	Nil	Nil	Nil	Nil	Nil	Nil
*E. coli**	1	Nil	***Nil***	Nil	1	***Nil***	Nil
*Faecal streptococci**	1	Nil	Nil	Nil	Nil	Nil	Nil
F: Borewell (Sampling Point 6)							
Total coliforms*	13	1	1	Nil	Nil	10	15
Faecal coliforms*	5	Nil	Nil	Nil	Nil	5	5
*E. coli**	Nil	Nil	***Nil***	Nil	Nil	***Nil***	2
*Faecal streptococci**	Nil	Nil	Nil	Nil	Nil	Nil	Nil

Contd...

Table 56.6–*Contd...*

Parameter	*Post Monsoon*			*Winter*		*Summer*	
G: Handpump (Sampling Point 7)							
Total coliforms*	25	Nil	2	Nil	Nil	Nil	Nil
Faecal coliforms*	18	Nil	Nil	Nil	Nil	Nil	Nil
*E. coli**	15	Nil	***Nil***	Nil	Nil	Nil	Nil
*Faecal streptococci**	1	Nil	Nil	Nil	Nil	Nil	Nil
Parameter				*Summer*			
H: Handpump (Sampling Point 8)							
Total coliforms*	78	83	TNC	TNC	TNC	TNC	
Faecal coliforms*	65	59	TNC	TNC	TNC	TNC	
*E. coli**	10	22	TNC	TNC	TNC	Nil	
*Faecal streptococci**				—Nil—-			
I: Handpump (Sampling Point 9)							
Total coliforms*	57	60	54	TNC	71	TNC	
Faecal coliforms*	32	28	30	32	45	45	
*E. coli**	24	18	20	25	21	15	
*Faecal streptococci**				—Nil—-			
J: Handpump (Sampling Point 10)							
Total coliforms*	TNC	TNC					
Faecal coliforms*	32	42					
*E. coli**	14	8					
*Faecal streptococci**	—Nil—-						

TNC: Too Numerous to count.

CFU per 100ml.

- ☆ Water samples from the handpump located at a distance of 1.5 km towards North of lagoon (sampling point 6) showed presence of total coliforms in five out of seven samples. The faecal coliforms were detected in three samples whereas *E.coli* was found in one sample. All the samples were however, free from *Faecal streptococci*. The presence of faecal coliforms indicated that the source was contaminated (Table 56.6F).
- ☆ Water samples from the handpump located at a distance of 1.5 km towards North of lagoon (sampling point 7) were analyzed. The results showed presence of total coliforms, faecal coliforms, *E.coli* and *Faecal streptococci* in only one of the seven samples. However, the other six samples did not show presence of these indicator organisms (Table 56.6G). The source may be considered as not contaminated.

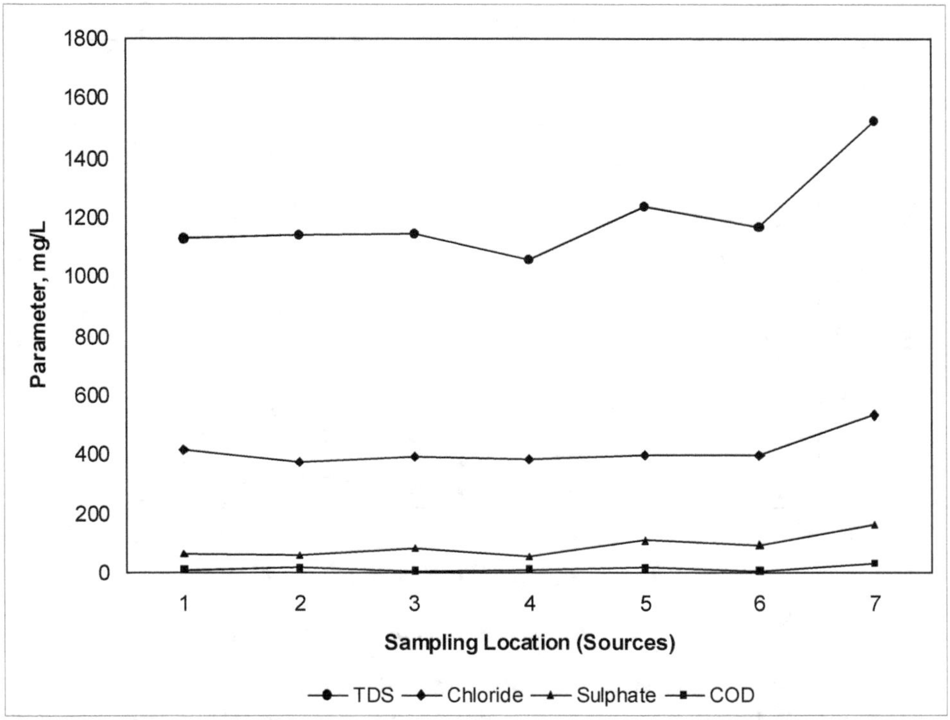

Source No: 1 -6 Upstream, 7 Downstream

Figure 56.3: Variation in Groundwater Quality: Sources at Upstream and Downstream of STP-1 Effluent Discharge Point in River.

- Water samples from the handpump located in the fields at a distance of 500m from the STP-2 (sampling point 8) showed presence of total coliforms, faecal coliforms, *E.coli* in five out of six samples (Table 56.6H). However, all the samples were free from *Faecal streptococci*. The presence of faecal coliform and *E.coli* indicated faecal contamination of the source.
- Water samples from the handpump located at a distance of 1.1 km from STP-2 (sampling point 10) showed presence of total coliforms, faecal coliforms and *E.coli* (Table 56.6I). However, all the samples were free from *Faecal streptococci*. The presence of faecal coliform and *E.coli* indicated faecal contamination of the source.
- Water samples from the handpump located at a distance of 1 km from STP-2 (sampling point 9) showed presence of total coliforms, faecal coliforms, *E.coli* in two samples (Table 56.6J). However, all the samples were free from *Faecal streptococci*. The presence of faecal coliform and *E.coli* indicated faecal contamination of the source.

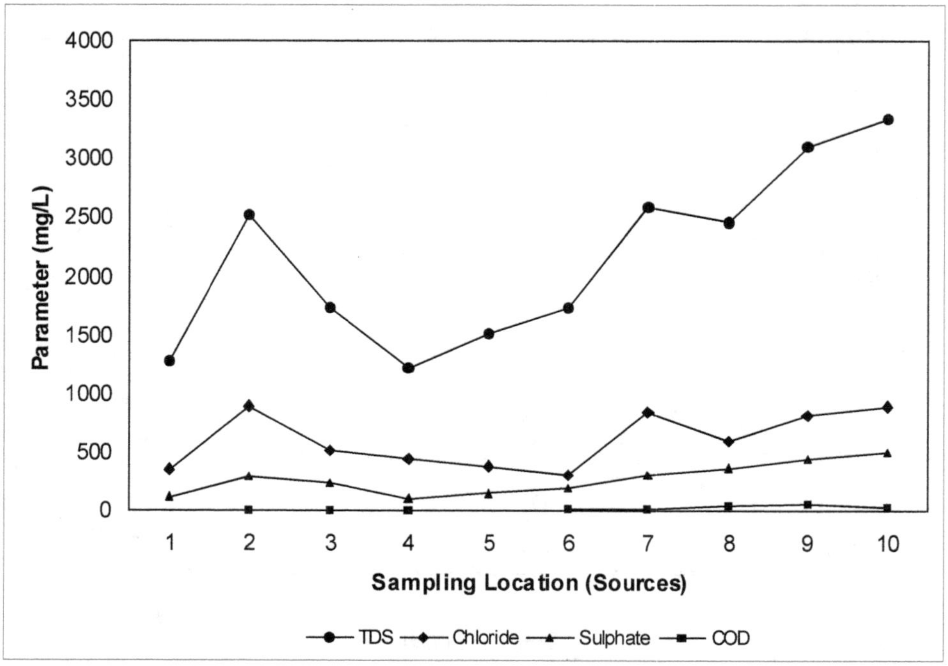

Source No: 1 -7 Upstream, 8-10 Downstream

Figure 56.4: Variation in Groundwater Quality: Sources at Upstream and Downstream of STP-2 Effluent Discharge Point in River.

Salient Findings

Water samples from the groundwater sources in the upstream of STP-1 discharge point in the river showed bacterial contamination at four out of six points. In addition, physico-chemical quality (hardness, chlorides, sulphate, fluoride and TDS) did not fully conform to IS 10500–1991 (First Revision, 2nd Reprint, 1995) standards for drinking water. Out of these four points, two points are located within the plant premises where contamination could be caused by seepage of sewage from the lagoons. The third and fourth points are located at a distance of 80m -100m from the lagoon and approximately 8m–15m from a soak pit. Hence the exact source of contamination could not be ascertained.

Water samples collected from the downstream area of STP-1 showed bacterial contamination. Moreover, physico-chemical quality in respect of hardness, chlorides, sulphate, fluoride and TDS exceeded IS 10500–1991 (First Revision, 2nd Reprint, 1995) limits. Continuous discharge of treated sewage into the river Sabarmati appeared to be the possible cause of groundwater contamination in this area.

Water samples from the groundwater sources in the upstream of STP-2 discharge point in the river showed bacterial contamination at two points. Further, physico-chemical quality of water (hardness, chlorides, sulphate, fluoride and TDS) did not conform to IS 10500–1991 (First Revision, 2nd Reprint, 1995) standards for drinking water. The contamination of borewell situated within the premises of STP-2 (sampling point 1) appeared to be due to seepage of sewage from the lagoons. The other source, namely the handpump (sampling point 6) located at a distance of 1.5 km from the STP also showed bacterial contamination. Here also the physico-chemical quality of the water (hardness, chlorides, sulphate, fluoride and TDS) did not conform to IS 10500–1991 (First Revision, 2nd Reprint, 1995) standards for drinking water. Use of untreated sewage for irrigation and its infiltration into the groundwater appeared to be the causative factor.

Water samples collected from the downstream of STP-2 discharge point in the river showed bacterial contamination. The physico-chemical quality of water with reference to hardness, chlorides, sulphate, fluoride and TDS also exceeded the IS 10500–1991 (First Revision, 2nd Reprint, 1995) limits. Discharge of treated sewage into the river Sabarmati was considered as the possible cause of groundwater contamination in this region.

Recommendations

1. Sources of contaminated water used for potable purposes as identified by the study should be red-marked and sealed. Alternate means of safe water supply to these areas must be found. In order to arrest seepage of sewage into the groundwater, the bottom and the sides of the lagoons should be made impervious employing suitable means *viz.* clay consolidation, HDPE sheets and/or chemical sprays etc.
2. A water quality-monitoring programme should be carried out on regular basis and necessary preventive measures should be taken to ensure that the residents get safe water. As an additional safeguard, periodic health check-ups of the residents should be carried out with a view to taking adequate measures both preventive and remedial to prevent water borne diseases.
3. The affected population in the vicinity of the treatment plants should be made aware in local language through direct interactive programmes about the risk associated with the use of contaminated water. Information should also be imparted to the residents about the simple methods of disinfecting the water supply using easily available disinfectants like chlorine (tablet/liquid form). However, the main responsibility of providing potable water would rest with the concerned civic body.

The findings of the study warrant a detailed hydrogeological study of the area with recourse to predictive modeling for assessment of water quality. The hydrogeological studies would comprise of quantitative/qualitative assessment of water reserves, its pathways (flow directions) and the identification of potential sources of contamination. The predictive modeling would also provide the future scenario on the basis of the present data and this will help concerned authorities in planning their future strategies.

References

1. Investigation of Environmental Problems and Evaluation Sewage Treatment Plants, *NEERI Report*, 2000.
2. *Guidelines for Drinking Water Quality*, Vol. 3, WHO, 1985.
3. D.D.Mara:, *Bacteriology for Sanitary Engineers*, John Wiley and Sons, New York, 1968.
4. *Standard Methods for Examination of Water and Wastewater*, 18th ed., American Public Health Association (APHA, AWWA, WEF), Washington, D.C., 1992.
5. Imhoff, K., *Handbook of Urban Drainage and Wastewater Disposal*, John Wiley and Sons, New–York, 1982.
6. Metcalf and Eddy; *Wastewater Engineering, Treatment, Disposal, and Reuse*, Tata McGraw-Hill, Inc., New Delhi, 1991.
7. Todd D. K.: *Groundwater Hydrology*, John Wiley and Sons, New York, 1980.
8. Pre-and Post-monsoon Potable Water Quality Assessment in Some Major Cities in India, *NEERI Report*, 2000.

Chapter 57

A Coupled Photocatalytic-Biological Process for Degradation of 1-amino-8-naphthol-3,6-disulphonic acid (H-acid)

S. Mohanty, N.N. Rao, P. Khare and S.N. Kaul

CSIR-NEERI, Nagpur

ABSTRACT

There has been growing emphasis on the development of coupled treatment systems (*e.g.*, advanced oxidation-biological) for treating poorly biodegradable wastewater. An attempt has been made in the present study to couple photocatalytic (TiO_2/UV) pretreatment with conventional activated sludge process to achieve improvement in biodegradation of H-acid. The combination of titanium dioxide and UV light has been known to generate strong oxidants that degrade several water bound organic pollutants into carbon dioxide via formation of some intermediates. The intermediates formed may undergo biodegradation readily. Accordingly, photodegradation experiments were carried out initially at an optimized TiO_2 dose and the minimum pretreatment time that required for transforming H-acid identified. For this purpose, UV-vis spectrophotometry and High Performance Liquid Chromatography (HPLC) were extensively used. Subsequently, it was attempted to biodegrade untreated and pretreated H-acid using activated sludge from textile industry acclimatized to H-acid. It was found that a photocatalytic pretreatment of H-acid for 30 min during which period approximately 8-10 per cent chemical oxygen demand (COD) removal occurred can be coupled to second stage biological treatment for achieving enhanced biodegradation of H-acid.

Keywords: *Photocatalysis, Pretreatment, Biodegradation, 1-amino-8-naphthol-3,6-disulphonic acid (H-acid).*

Introduction

1-amino-8–naphthol–3,6–disulphonic acid (H-acid) is an important intermediate for the preparation of several direct, acid and reactive dyes (Colour Index, 1971). It is also used in the production of chromotrophic acid, a dye stuff intermediate. It is also used as a corrosion inhibitor and in photography. Due to such diverse use, H-acid along with its derivatives may be present in textile/dyeing wastewater. H-acid has been a tough molecule to biodegrade (Chen and Pignatello, 1997). Presence of sulphonic acid groups in the structure of H-acid makes it highly soluble in water. Further the presence of fused ring system in H-acid molecule makes it difficult for treatment by any physico-chemical method (Wanpeng *et al.,* 1996). H-acid effluent is currently treated through evaporation-incineration technology (Kaul *et al.,* 1998). However, the energy and cost intensiveness of this technology makes it unfavorable. *In-situ* biodegradation of H-acid is very difficult and there are no known microbial cultures for meeting this requirement. Therefore, a technological approach that would lead to a solution for this existing problem is desired.

An improved system would be a physico-chemical process in tandem with biological process that would not only achieve degradation of H-acid molecules but also economize the energy and cost requirements (Xiao *et al.,* 2002; Mc Martin *et al.,* 2003; Li *et al.,* 1997). Hence, an attempt has been made in this study to treat H-acid by short duration photocatalytic pretreatment to generate intermediates, which were further taken care off by a second stage biological process. The current interest in photocatalysis is mainly due to its ability to destroy a variety of organic pollutants present in water or air (Serpone and Pelizzetti, 1989; Fujishima *et al.,* 2000; Legrini *et al.,* 1993; Herrmann *et al.,* 2002).

Experimental Section

Chemicals

H-acid was procured from M/s Rohan Dyes, Ahmedabad (India). Titanium Dioxide (predominantly anatase, surface area 50± 15 m^2g^{-1}, and mean particle size 21 nm) was used as a catalyst (Degussa AG, Germany). All the other chemicals used for the determination of COD and biochemical oxygen demand (BOD) were high purity analytical reagents. High performance liquid chromatography (HPLC) grade acetonitrile solvent was used for collecting HPLC data.

Photocatalytic Oxidation of H-acid

Photocatalysis of H-acid was performed in two phases. Initially the photocatalytic degradation on H-acid was assessed. Later, it was used as a pretreatment prior to biodegradation in second phase. For the initial phase, H-acid solutions (concentration ranging from 50-150 mg l^{-1}) containing 1g TiO_2 catalyst were stirred on the magnetic stirrer before commencing UV irradiation. A medium pressure mercury vapor lamp (400 W, peak emission at 365 nm, 5.0 ´ 10^{19} photons s^{-1}, Scientific Aids and Instruments Corporation, Chennai, India) was used for irradiating the test solutions. All photocatalytic experiments were carried out in an immersion well photoreactor as described previously (Rao *et al.,* 2003). The reacted samples were withdrawn after the time interval of 1,2,3,4 and 5h. After treatment, the solutions were filtered through 0.22 µm membrane and their absorbance was measured at 358 nm using UV-vis spectrophotometer. The test samples were analyzed for COD and BOD. Simultaneously, high performance liquid chromatograms (HPLC) and UV-vis Spectra were also recorded.

For biodegradation study under second phase, photocatalytic pretreatment experiments were carried out with H-acid solutions of different concentrations (500 and 1000 mg l^{-1}) for 30 and 60 minutes using the optimized TiO_2 dose. The pH prior to and after photocatalytic pretreatment was 6.0 and 5.5, respectively. The initial pH was further low at higher H-acid concentration. These pretreated samples were filtered and subsequently subjected to biodegradation. The toxicity of the pretreated effluent was analyzed as per the agar diffusion method (Blaise, 1991).

Nutrient Media

Biodegradation experiments were carried out in Minimal Salt Basal medium. The medium was composed of three different solutions: Solution A [Na_2HPO_4 (5.36 g l^{-1}), KH_2PO_4 (2.72 g l^{-1}), pH – 7.25 with KOH]; Solution B [Nitriloacetic acid (10 g l^{-1}), $MgSO_4$.$7H_2O$ (14.45 g l^{-1}), $CaCl_2$.$2H_2O$ (3.33 g l^{-1}), $(NH_4)_6Mo_7O_{24}$.$24H_2O$ (9.25 mg l^{-1}), $FeSO_4$.$7H_2O$ (99 mg l^{-1}), Metal 44 solution (50 ml l^{-1}), pH- 6.6-6.8] and Solution C [$(NH_4)_2SO_4$ (1 g l^{-1})]. The Metal 44 solution used in Solution B consists of Na_2EDTA (25 mg l^{-1}), $ZnSO_4$.$7H_2O$ (0.1095 mg l^{-1}), $FeSO_4$.$7H_2O$ (50 mg l^{-1}), $MnSO_4$.$7H_2O$ (15.4 mg l^{-1}), $CuSO_4$ (3.92 mg l^{-1}), Co $(NO_3)_2$.$6H_2O$ (2.48 mg l^{-1}), $Na_2B_4O_7$. $10H_2O$ (1.77 mg l^{-1}). The three solutions were mixed in the following proportion: Solution A (40 ml l^{-1}) + Solution B (40 ml l^{-1}) + Solution C (10 ml l^{-1}).

Biomass Acclimatization

Microbial biomass used for the biodegradability testing was activated sludge collected from a functional ETP catered for treating textile effluents. Acclimatization of the biomass to H-acid was done in mineral salts medium for a period of 12 days. After every 3 days, the biomass was pelleted and transferred to a fresh medium.

Biodegradation of H-acid

Biodegradation experiments were carried out in 250 ml Erlenmeyer flasks consisting of 100 ml of 50 mg l^{-1} H-acid solutions. The experiments were carried out in batch mode. The flasks were incubated at 30°C in static incubation condition for 28 days to study the utilization of H – acid by the co-culture. A control flask containing only bacterial culture was also incubated simultaneously. Samples were drawn on the fifth day and collected after filtration through 0.45 µm membrane filter. The samples were analyzed for their absorbance on the UV–vis Spectrophotometer, COD and BOD. HPL Chromatograms were also recorded for these samples.

Pilot Scale Experiments

In this case, 20 liters of 500 mg l^{-1} H-acid solution was subjected to photocatalytic pretreatment as described previously. The combined photocatalytic- biological reactor set up is shown in Figure 57.1. Influent from tank 1 was pumped into photoreactor 3

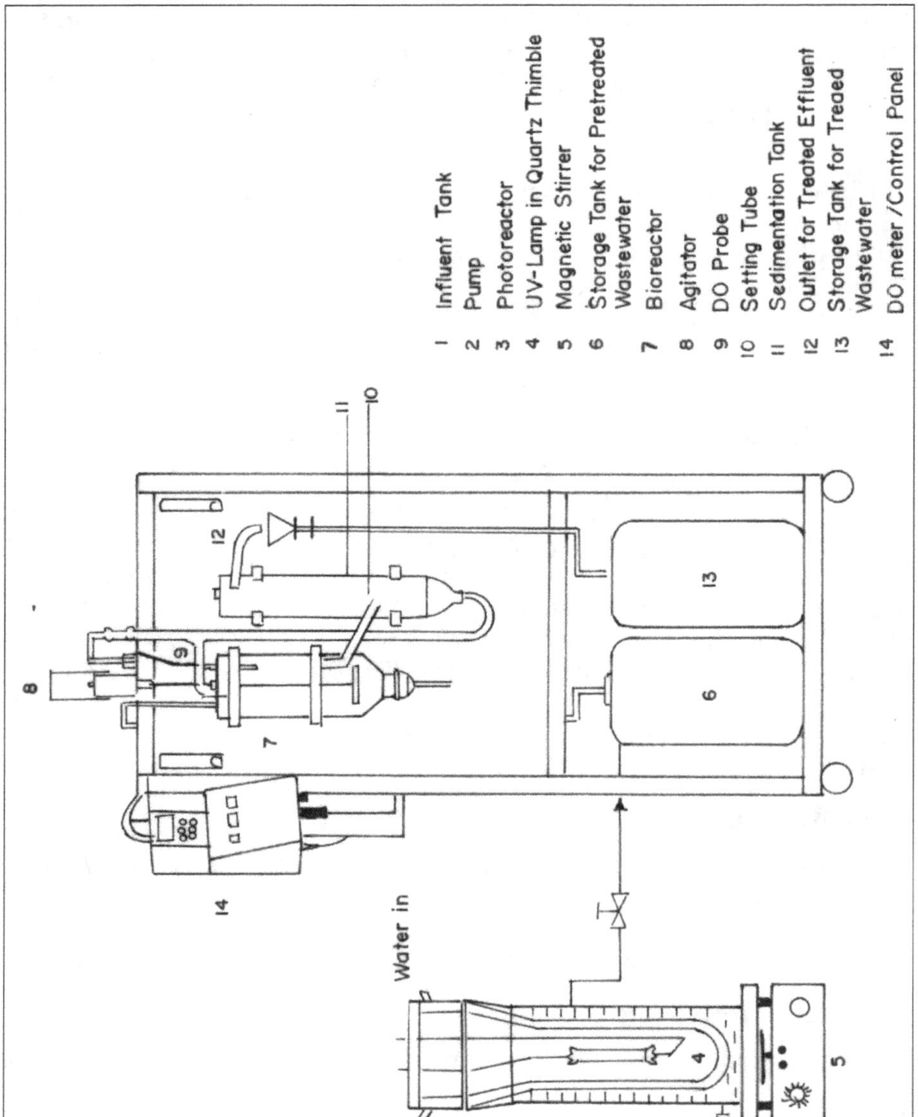

Figure 57.1: Schematic Diagram Coupled of Photoreactor-Bioreactor Set-up.

and pretreatment was given. The pretreated effluent from photoreactor 3 was collected in tank 6. Most of the TiO_2 slurry was retained in reactor 3 only; however, a small amount of TiO_2 was carried over to tank 6 whereby it settled to the bottom. Then the pretreated effluent from tank 6 was carried into bioreactor 7 (4.2 L capacity; feed rate = 2 ml min^{-1}; HRT = 35 h; Mixed Liquor Suspended Solids (MLSS) = 1500 mg l^{-1}; Dissolved Oxygen (DO) = 2.5-3.0 mg l^{-1}; pH =5.6) using panel mounted peristaltic pump (situated at rear side in the Figure 57.2). Agitation in bioreactor was controlled through agitator 8, and concentration of DO was maintained through DO probe/DO meter Control Panel 9 and 14, respectively. The overflow from bioreactor 7 was collected into sedimentation tank 11 and the settled biomass is recirculated into bioreactor 7 using sludge recycle pump situated on the rear of the unit. The clear effluent from sedimentation tank 11 was collected into final effluent collection tank 13. As there was no significant pH change due to short duration (30 min. or 1h) pretreatment step, the pH of the pretreated effluent was not adjusted. Unlike flask studies, there was no nutrient supplementation during the pilot scale experiments. Routine monitoring for maintenance of DO and MLSS. Samples were collected at different detention times and were analyzed for their COD and BOD.

Analysis

The test samples were analyzed for changes in UV absorption on UV – Vis Spectrophotometer (Lambda 900, Perkin Elmer). The COD was estimated by closed reflux method (APHA, 1989) using Hach COD reactor and Hach Spectrophotometer (DR 2010). BOD was measured by Oxitop Respirometers (WTW, Germany). The seed used for BOD determination was the same acclimatized biomass used for biodegradation experiments. These samples were also analyzed for degradation of H-acid and intermediate development using HPLC. HPL chromatograms were recorded using Model 481, Millipore – Waters make Lambda- Max LC spectrophotometer. The stationary phase was RP 18 (Lichrosorb, 2mm diameter, 250mm length, 5μm particle diameter; Germany). The mobile phase was acetonitrile: water solution (70: 30) at a flow rate of 1.2 ml min^{-1}. A 10 μL sample was injected and the eluent was detected at 358 nm. Alternatively, HPLC analysis was done using stationary phase Bondapak C-18 column (2 mm diameter, 250 mm length, 5 μm particle diameter). The mobile phase was acetonitrile: water solution (60: 40) at a flow rate of 0.9 ml min^{-1}. A 10 μL sample was injected and the eluent was detected at 234 nm. HPLC detection at 358 nm facilitated monitoring of H-acid molecule, while at 234 nm both H-acid and its intermediate were detected. The corresponding retention times were also slightly different due to use of two different HPLC columns.

Results and Discussion

Photocatalytic Degradation of H-acid

This catalyst dose was optimized previously (Khare *et al.*, 2003). H-acid displayed two UV absorption bands at wavelengths 358 and 234 nm. The band at 358 nm completely disappeared within 15 min treatment, whereas the intensity of band at 234 nm decreased significantly during 2h.

The photocatalytic degradation of H-acid also resulted in the reduction of chemical oxygen demand. More than 70-90 per cent reduction in COD occurred during 5h. It is found that the COD removal reaction is diffusion controlled, at least at lower COD concentration. The reaction is found to follow typical first order kinetics in H-acid concentration range 50-150 mg l^{-1}. The rate constants (k) for COD removal were calculated from the slopes of -log [$(COD)_t$/ $(COD)_o$] versus time plots for each concentration of H-acid (Table 57.1). The rate constant (k) varied from 1.30 ´ 10^{-3} min^{-1} to 3.43 ´ 10^{-3} min^{-1}. Further, H-acid wastewater having an initial COD of 1400 mgl^{-1} was also subjected to photocatalytic degradation under similar experimental conditions and a substantial decrease in the COD to 384 mg l^{-1} within 5 h was observed.

Table 57.1: Variation of Rate Constant for the Degradation of H-acid at different Concentrations of H-acid

Concentration of H-acid (mg/L)	*Per cent COD Removal*	*Rate Constant*k, h^{-1}*
50	73	0.0784
75	83	0.128
100	84	0.1549
150	91	0.2055

* Deduced from –log $[COD]_t/[COD]_o$ versus time plots.

Figure 57.2 shows the HPL chromatograms of photocatalytically treated H-acid solutions at different time intervals. The chromatogram of the initial H-acid sample showed a peak at a retention time of 1.3 min. Subsequently, this peak disappeared within 10-min only and a new peak at 1.47 min retention time appeared. The new peak may be attributed to formation of an intermediate. This intermediate was found to persist in all the subsequent samples collected up to 5-h photocatalytic treatment. The peak at 1.3 min due to H-acid was seen as a shoulder in this case indicating the presence of some unreacted H-acid also.

Biodegradation of H-acid

Figure 57.3 shows the UV-vis spectra of H-acid samples treated under different pretreatment and biodegradation conditions. The acclimatized biomass was capable of utilizing H-acid as evidenced from the UV-Vis spectra Comparison of the spectra labeled as 'initial' and 'BIO' suggest that the concentration of H-acid was reduced substantially without formation of intermediates. In contrast, photocatalysis pretreatment brought about transformation of the H-acid molecule (see spectra labeled as 'initial' and 'PC'). Moreover, the photocatalytic pretreatment when coupled with biodegradation (spectrum labeled as 'PC+BIO') showed better substrate reduction when compared to only photocatalytic treatment ('PC') or biodegradation ('BIO'). The HPLC data obtained using RP 18 column (λ for detection was 358 nm) also confirmed the enhancement of biodegradation after photocatalytic treatment (Figure 57.4). In these chromatograms, H-acid was eluted at a retention time of 1.6 min. The effectiveness of the coupled system (PC + BIO) is evident from the substantial decrease found in the peak height as compared to only photocatalytic treatment ('PC') or biodegradation ('BIO').

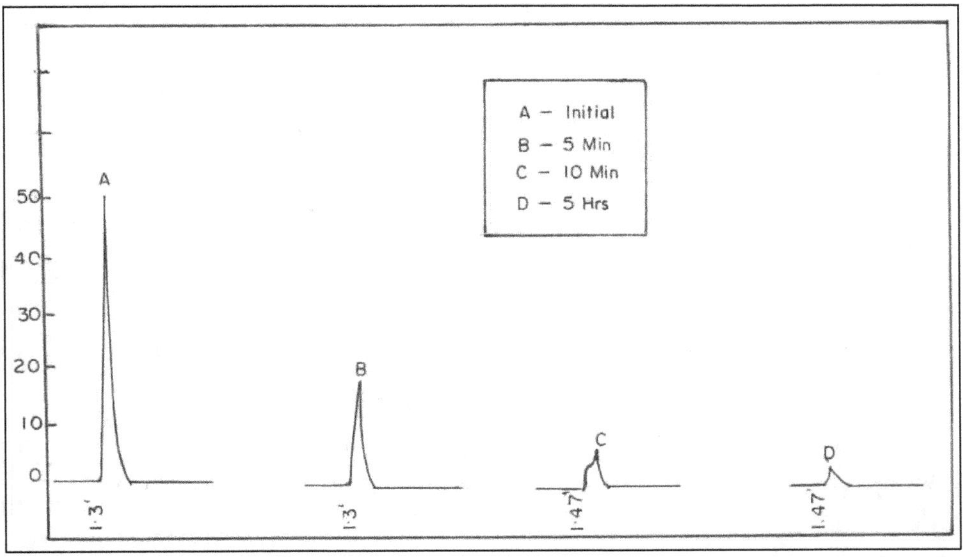

Figure 57.2: HPLC Chromatogram of Photocatalytically Treated H-acid Solutions (Bodapak C_{18}, 2 mm Diameter, 250 mm Length, 5 μm, 60: 40 Acetonitrile: Water, Flow Rate = 0.9 ml min^{-1}; Detection Wavelength, 234 nm).

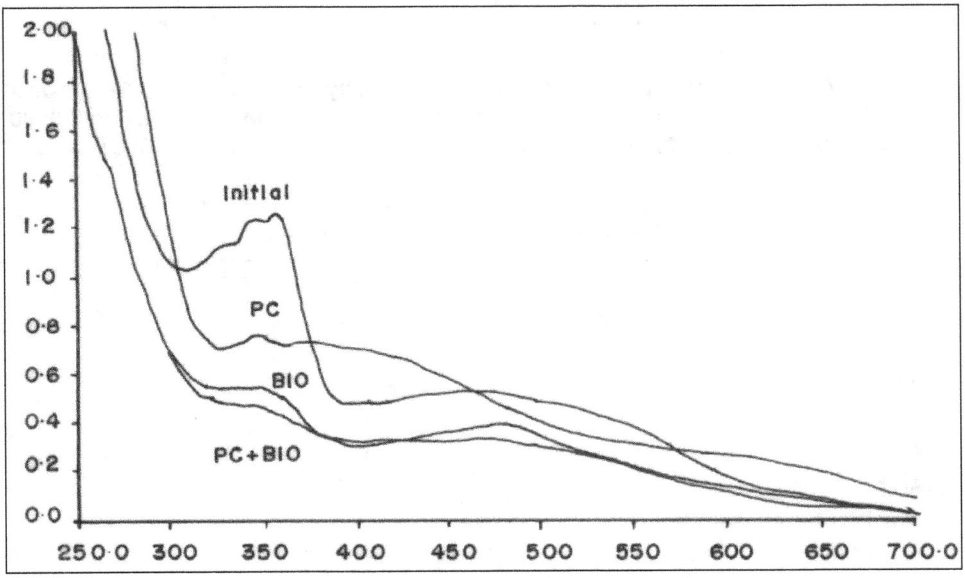

Figure 57.3: UV-vis Spectra of H-acid Solutions after Photocatalytic Pretreatment and Biodegradation. [H-acid] = 50 mg l^{-1}.

The results of biodegradation experiments using 500 and 1000 mg l^{-1} H-acid solutions are given in Table 57.2. The concentrations of COD, BOD and MLSS as well as per cent COD removals in each case are included in this table. Only biodegradation of H-acid resulted in 9.5 and 3.5 per cent COD reduction respectively for 500 and

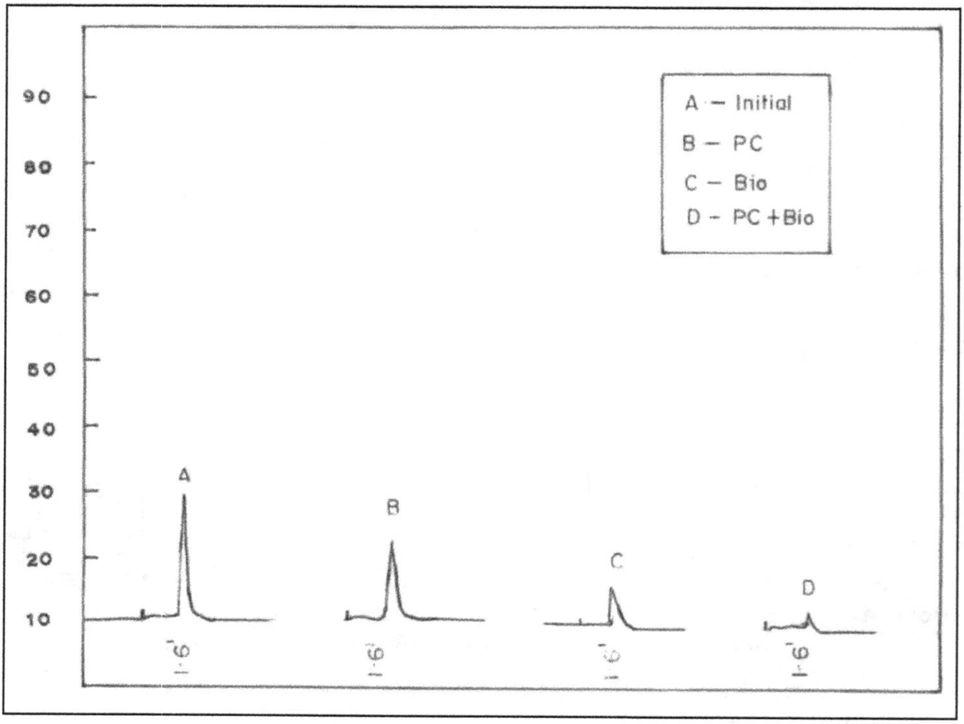

Figure 57.4: HPLC Chromatogram of Photocatalytically Pretreated H-acid Followed by Biodegradation. (Lichrosorb RP 18, 2 mm diameter, 250 mm Length, 5 μm, 70: 30 acetonitrile: water, flow rate = 0.9 ml min^{-1}; detection wavelength, 358 nm.

Table 47.2: Biodegradability Testing Parameters Before and After Photocatalytic Pretreatment of H-acid

Sample	*COD*, mg/L*	*% COD reduction*	*MLSS, mg/L*	*BOD, mg/L*
H-acid (500 mg/L)				
Initial	815		220	50
After Biodegradation	737	9.5	228	40
After PC (30 min)	737	9.5	298	60
After PC (30 min) + Biodegradation	617	24.3	–	40
After PC (60 min)	659	19.1	303	40
After PC (60 min) + Biodegradation	477	41.5	–	20
H-acid (1000 mg/L)				
Initial	1125		220	60
After Biodegradation	1086	3.5	231	40
After PC (30 min)	970	13.7	332	70
After PC (30 min) + Biodegradation	607	46.0	–	30
After PC (60 min)	815	27.5	345	70
After PC (60 min) + Biodegradation	435	61.3	–	20

* includes blank contribution from medium.

1000 mg l^{-1} H-acid solutions. The final pH of the effluent after biodegradation was 7.0 ± 0.1. The reduction in COD of 500 and 1000 g l^{-1} H-acid solutions was enhanced to 24.3 and 46 per cent respectively after 30 minutes photocatalytic pretreatment. This was further increased to 41.5 and 61.3 per cent respectively after 60 minutes of photocatalytic treatment (Table 57.2). The pretreatment also resulted in increase of BOD. The data suggests that the photocatalytic pretreatment is effective and enhances efficiency of subsequent biodegradation process. Moreover, there was substantial increase (20-25 per cent) in the quantity of MLSS after photocatalytic pretreatment. The concentration of MLSS further rises with increase in pretreatment time. This increase in concentration of MLSS indicates that the intermediate (s) formed during pretreatment stage are more biodegradable.

Table 57.3: Results of Pilot Scale Experiments*

Sample/Treatment	*COD (mg/L)*	*Per cent COD Reduction*	*BOD (mg/L)*
Initial	595	–	25
PC (30 min.)	547	8	32
PC (30 min.) + Bio (36 hrs)	488	17.9	40
PC (30 min.)+ Bio (72 hrs)	486	18.5	42
PC (30 min.)+ Bio (108 hrs)	482	19.0	42

* H-acid solution = 20 L of 500 mg/L

Biological reactor (4.2L capacity; feed rate = 2 mL min^{-1-}. HRT = 36-108h;

MLSS = 1500 mg/L; D. O. = 2.5-3.0 mg/L).

In the present case, there is a possibility for the formation of small quantities of hydrogen peroxide in the pretreatment stage (Legrini *et al.*, 1993; Fujishima *et al.*, 2000) that might be detrimental to bacteria in the second stage. However, as the pretreatment was given only for a short duration, the formation of peroxide may be very limited. This is also evident from the increase in MLSS concentration that was observed subsequent to photocatalytic pretreatment, which indicates the reduced toxicity in the pretreated effluent. The toxicity of the pretreated effluent as analyzed by agar diffusion method (Blaise, 1991) also indicated substantial low toxicity to textile activated sludge.

Pilot Experiments

The results of pilot experiments performed in the coupled photocatalytic – bioreactor setup (Figure 57.1) using 20 litres of 500 mg l^{-1} H-acid solution are given in Table 57.3. The COD after 30 minutes PC pretreatment was 547 mg l^{-1}. The per cent COD removal after 36, 72 and 108 h of detention periods under second stage biological reaction conditions was 17.9 per cent, 18.3 per cent and 18.6 per cent respectively. Correspondingly, there was also a certain increase in the BOD of these samples. It may be noted that no nutrients were added to biological reactor during these pilot experiments. This may be the reason for less effective biological treatment under pilot experiments. Nevertheless, it is clearly indicated that photocatalytic treatment was a useful technique for generating intermediates that were more biodegradable than the original substrate (H-acid).

Conclusion

The conclusions drawn from this study can be summarized as follows:

H-acid can be degraded in a TiO_2/UV photoreactor whereby 70-90 per cent COD reduction is attainable in 5h under the experimental conditions maintained in this paper. The degradation reaction obeys psuedo first-order rate law. During photodegradation of H-acid an intermediate builds up indicating transformation of H-acid molecule. Depending upon the duration of pretreatment, the reactor can be used to transform H-acid into intermediates or degrade to achieve high COD removals.

Photocatalytic pretreatment of H-acid for short duration ensures enhanced biodegradation of effluents bearing H-acid. A two-stage treatment comprising initial photocatalytic pretreatment followed by conventional activated sludge process can ensure enhanced biodegradation of H-acid bearing wastewater. For this purpose, a photocatalytic pretreatment to bring down COD by $\leq$ 8-10 per cent is adequate to achieve enhanced biodegradation of H-acid bearing wastewater in second stage biological treatment.

References

1. APHA *Standard Methods for Examination of Water and Wastewaters*, 17th ed. Washington, DC: American Public Health Association, 1989.
2. Blaise C. (1991). Microbiotests in aquatic ecotoxicology: characteristics, utility and prospects. *Toxic. Assess.* 6, 145-155.
3. Chen R. and Pignatello J.J. (1997). Role of quinone intermediates as electron shuttles in Fenton and photo-assisted Fenton oxidations of aromatic compounds. *Environ. Sci. and Technol.*, 31, 2399-2406.
4. *Colour Index* (1971). *Society of Dyers and Colorists*. Bradford, 4, 3rd edn.
5. Fujishima A., Rao T. N. and Tryk D. A. (2000). Titanium dioxide photocatalysis, *J. Photochem. Photobiol. C: Photochemistry Reviews* 1, 1-27.
6. Herrmann J-M., Guillard C., Disdier J., Lehaut C., Malato S. and Blanco J. (2002). New industrial titania photocatalysts for the solar detoxification of water containing various pollutants. *Appl. Catal. B: Environmental* 35, 281-294.
7. Kaul S.N., Nandy T., Deshpande C.V., Srivastava A., Sastry S. and Szpyrkowicz L. (1998). Application of full-scale evaporation-incineration technology for hazardous wastewater. *Wat. Sci. and Tech.*, 38 (4-5), 363- 372.
8. Khare P., Mohanty S., Rao N.N. and kaul S.N. (2003). TiO_2/UV catalyzed degradation of 1-amino-8-napthol-3,6- disulphonic acid,.*Proceedings of 16th National Symposium and 1st Indo-German Conference on "Catalysis – A Cross Disciplinary Vision"*, conducted by the Catalytic Society of India, at IICT, Hyderabad.

9. Legrini O., Oliveros E., and Braun A.M. (1993). Photochemical processes for water treatment. *Chem. Rev.*, 93, 671-698.

10. Li X.Z. and Zhao Y.G. (1997). On- site treatment of dyeing wastewater by a bio-photoreactor system. *Wat. Sci. and Technol.* 36 (2-3), 165–172.

11. Mc Martin D.W., Headley J.V., Neu Thomas, R., Friesen D.A. (2003). Photolysis and biodegradation of selected resin acids in river Saale water, Germany. *J. of Envt. Sci. and Health, Part A*, A38 (12), 2727- 2747.

12. Rao N.N., Dubey A., Mohanty S., Khare P., Jain R. and Kaul S.N (2003). Photocatalytic degradation of 2- Chlorophenol: a study of kinetic intermediates and biodegradability. *Journal of Hazardous Materials (B)*, 101 (2), 301-314.

13. Serpone N. and Pellizetti E. (Eds.). (1989). *Photocatalysis: Fundamentals and Applications*, Wiley/Interscience, New York.

14. Wanpeng Z., Yang Z. and Wang Li (1996). Application of ferrous hydrogen peroxide for the treatment of H-Acid manufacturing process wastewater. *Wat. Res.*, 30 (12), 2949-2954.

15. Xiao Y., Wang G., Liu H., Zhao H., Zhang J., Sun C. and Wu M. (2002). Treatment of H-acid wastewater by photo-Fenton reagent combined with a biotreatment process: A study on optimum conditions of Pretreatment by a Photo-Fenton Process. *Bull. Environ. Contamin. Toxicol.*, 69, 430- 435.

Chapter 58

Use of a Novel Super Absorbent Polymer for Colour Removal from Aqueous Solutions

R. Dhodapkar, N.N. Rao, S.P. Pande and S.N. Kaul

CSIR-NEERI, Nagpur

ABSTRACT

A super absorbent biodegradable polymer, Jalshakti, used for agricultural purposes is tested for colour removal from dye containing aqueous solutions. It is found to decolourize solutions containing basic dyes only, by adsorption process. The JS has a unique property to swell by 50 times its dry volume in water. The effects of pH, particle size, agitation, contact time, concentration of adsorbent and adsorbate have been examined. The data is in conformity with Langmuir, Freundlich and intraparticle diffusion models for adsorption. The mechanistic aspect has also been explored wherein, spectroscopic and potassium ion release studies reveal that the basic dyes are removed by adsorption-ion exchange mechanism involving the carboxylic acid groups of JS. The swelling of JS in water also facilitates adsorption of basic dyes due to opening of carboxylic acid functional groups. The use of Jalshakti has been illustrated for decolourization of solutions containing basic dyes and its performance is compared to the conventional adsorbent *i.e.* activated carbon.

Keywords: *Jalshakti polymer, Super absorbent, Basic dyes, Adsorption-ion exchange, Colour removal.*

Introduction

Several dyes make their presence strikingly visible as they impart colour to the water bodies. These compounds can be effectively removed by adsorption process in which dissolved organic compounds attach themselves to the surface of adsorbents. The surface charges on the adsorbent as well as the polarity of organic compounds predominantly control adsorption phenomenon. While a number of adsorbents, have been developed; only activated carbon (AC) found wide industrial applications as an adsorbent. The adsorption technique may be used as a polishing step or as a tertiary unit at the end of treatment stage to meet discharge standards. The industrial waste materials and by-products have been used as adsorbents for the removal of organic compounds, dyes, colour, surfactants, metals etc. from the wastewater (Gupta *et al.*, 1990 (a); 1990 (b); Matusda *et al.*, 1975; Bousher *et al.*, 1997). In addition, the agricultural waste material like natural polymeric adsorbents: chitin and chitosan (Juang *et al.*, 1996), wool carbonized waste (Malmary *et al.*, 1985), peat (Poots *et al.*, 1976 (a), Bousher *et al.*, 1997), bagasse pith (McKay *et al.*, 1987; Al-Duri *et al.*, 1990), maize cob (El-Guendi, 1991), biogas slurry, banana pith (Namasivayam *et al.*, 1993), hard wood saw dust (Poots, 1976 (b); Bousher, 1997) cotton waste, hair, bark, rice husk (McKay, 1987); tea waste ash (Balalsubramanium and Muralishankar, 1987); agricultural residues (Nigam *et al.*, 2000; Kumar *et al.*, 2000; Rao *et al.*, 2002) have been investigated. These adsorbents are biodegradable, and hence, their use is beneficial and poses no disposal problems.

Decolourization using adsorption is a result of adsorption and ion-exchange mechanism (Slokar and Marechal, 1997). This is influenced by many physico-chemical factors such as dye/sorbent interaction, sorbent surface area, particle size, temperature, pH, and contact time (Kumar *et al.*, 1998). The acid dyes are mostly adsorbed in a reversible process while basic dyes (cationic) can be removed via ion-exchange process (Bousher *et al.*, 1997). A particular adsorbent is applicable to a particular class of dyes for example negatively charged adsorbent has a high adsorption capacity for cationic dyes (Hao *et al.*, 2000).

This paper addresses the colour removal properties of Jalshakti (JS), a starch polymer, in aqueous medium. Detailed studies on characterization of JS, its efficiency for colour removal and adsorptive behaviour in the liquid phase were undertaken. The effects of particle size, pH, agitation speed and contact time on the adsorption process, verification of adsorption models *viz.*, Langmuir, Freundlich and intraparticle diffusion have been conducted. Finally, a tentative mechanism of adsorption of dyes over JS is discussed. The adsorption property of JS for removal of basic dyes has been elucidated in this work for the first time.

Methods

Materials

Jalshakti is an organic super absorbent polymer manufactured by India Organic Chemicals Ltd. Khopoli, Raigad, India. It is used for the increase of crop production due to its water retaining properties ('Jalshakti', 1988). JS was used as an adsorbent in this study. The acid dyes used in experiments were Patent Blue VS, Light Green and Brilliant Milling Violet, which were gifted by Colour Chem, Ahmedabad. The basic dyes Safranine T (ST), Methylene Blue (MB), Crystal Violet (CV), were purchased from E. Merck India Pvt. Ltd, Mumbai. Nuchar C-190 (Virginia, U.S.A.) activated carbon (AC) was used, which is known to remove colour from wastewater (Reife and Freeman, 1996).

Physico-chemical Determinations

The absorption spectrum of the dye was taken on Perkin Elmer (Model λ900) Spectrophotometer. Colour was measured as absorbance at the wavelength of maximum absorbance (λ_{max}). An Orion pH meter/specific ion meter (Model 720A) equipped with combination pH electrode was used to measure pH of all solutions as well as for potentiometric titrations. Sodium and Potassium ions were estimated using Mediflame Flame Photometer (Systronics, India). Infra Red spectra were recorded as KBr pellets after drying in an oven at 60°C using Perkin Elmer Model 983 IR Spectrometer. The transmittance mode IR spectra were recorded in the working range 300 to 4000 cm^{-1} at 3.0 cm^{-1} resolution.

Characterization of Jalshakti

A 5g sample of JS was sieved through ASTM standard sieves (US Standard Sieve Series, Endecotts (Filters) Limited, London) to obtain various particle size fractions.

The swelling characteristics of JS were determined for a 10 mg weight of JS of particle sizes >210 μm and 840 μm in a 100mL graduated measuring cylinder and the dry and wet volumes after addition of water were noted at 5 min interval.

The potentiometric titration was performed with the addition of 0.02 N H_2SO_4 and 0.02 N NaOH to a dispersion of 0.1 g of JS in 100 mL distilled water, which was continuously stirred by a magnetic stirrer. The potentiometric curves were obtained by plotting the observed changes in pH against the volume of acid or alkali. These curves were used to detect the presence of acidic and basic groups, if any, on the polymer.

Adsorption Experiments

For testing the preferential adsorption by JS, tests were carried out in Jar test apparatus (Phipps and Bird, USA). The stock solutions of 1 g/L of all the selected dyes and dye solutions of known concentrations were prepared by appropriate dilutions with distilled water. Adsorption tests were undertaken with acid and basic dyes of 10 mg/L concentration and equilibrated with 1 g/L of JS. The mixture was stirred at 100 rpm and existing pH 5.5± 0.2 for a contact period of 2h. Residual absorbance was determined spectrophotometrically at λ_{max}.of each dye.

Optimization of Particle Size, pH, Agitation and Dose of JS

JS (1g/L) of varying particle sizes was added to six jars containing fixed concentrations of MB and CV. The experiment was carried out for a time period of 2h, at existing pH *i.e.* 5.5 ± 0.2, and stirring speed of 100 rpm and residual colour in the filtrate was measured. The effect of pH on colour removal was examined in the range 2 to10 for fixed concentrations of two dyes MB and CV. The pH was adjusted to 2, 4, 6, 8 and 10 using sulphuric acid and sodium hydroxide, as the case may be, and 1g/L dose of JS of optimum particle size was added. The experiments were performed for a time period of 2h and residual colour was measured. The optimum pH was determined graphically by plotting per cent colour reduction vs. pH.

The effect of agitation speed on adsorption has been studied by fixing the concentration of MB at 50mg/L and that of JS at1g/L. The pH was set at the determined optimum and the agitation speed was varied from 20 to 100 rpm. The time period was 2 h and residual colour in the filtrate was determined.

The concentration of JS of was varied from 100 to 2000 mg/L for fixed concentrations of all the basic dyes at the determined optimum pH. The mixture was stirred for 2 h. The optimum dose was determined graphically for each dye by plotting per cent colour reduction against concentration of JS. At concentrations greater than 2000 mg/L, JS was swollen to absorb most of water.

Effect of Initial Concentration of Dye and Contact Time

The adsorption experiments were carried out at room temperature *i.e.* 25± 2°C at determined optimum conditions of particle size, pH, dose of JS and stirring speed. The initial concentration of dye was varied in each jar and equilibrated with fixed quantity of JS, for a contact period long enough to reach the adsorption equilibrium. Residual dye concentration was monitored periodically at regular time intervals. The JS was then removed by filtration and the filtrate was analyzed for residual colour. Prior to the concentration analysis, calibration curves were established for each dye at the λ_{max}. The kinetic study was used to determine contact time for maximum removal of colour and the results were then used to build the adsorption isotherms.

Comparative Colour Removal Studies with Activated Carbon (AC)

The efficiency of JS was compared with AC for varying initial concentrations of one representative dye (MB) by performing adsorption experiments at pH 6.0, 40 rpm, 0.8 g/L of adsorbent and a contact period of 2 h.

Data Analysis

The data obtained for dye removal at equilibrium by JS was fitted into Langmuir and Freundlich isotherms and intraparticle diffusion models. The rate constants were derived using linear regression of data in the plots and analysis of variation (ANOVA) of the data obtained using Analysis Tool Pack (Microsoft Excel).

Results and Discussion

Characteristics of JS

JS swells into a gel approximately 50 times of its dry volume when mixed with water. The swelling was higher in early stages of contact time for JS (840 μm) in comparison to JS (210 μm). However, the saturation swelling obtained at 20 min is the same for both the sizes. The amount of moisture present in JS is 16 to 18 per cent. About 3.82 mg/g sodium and 84.21 mg/g potassium is estimated in JS as determined by flame photometer. The potentiometric titration of the polymer (Figure 58.1) shows that the isoelectric point is at 6.5 pH wherein the surface charge is neutral. Each data point represents an average of three sets of experiments. There exist two plateaus,

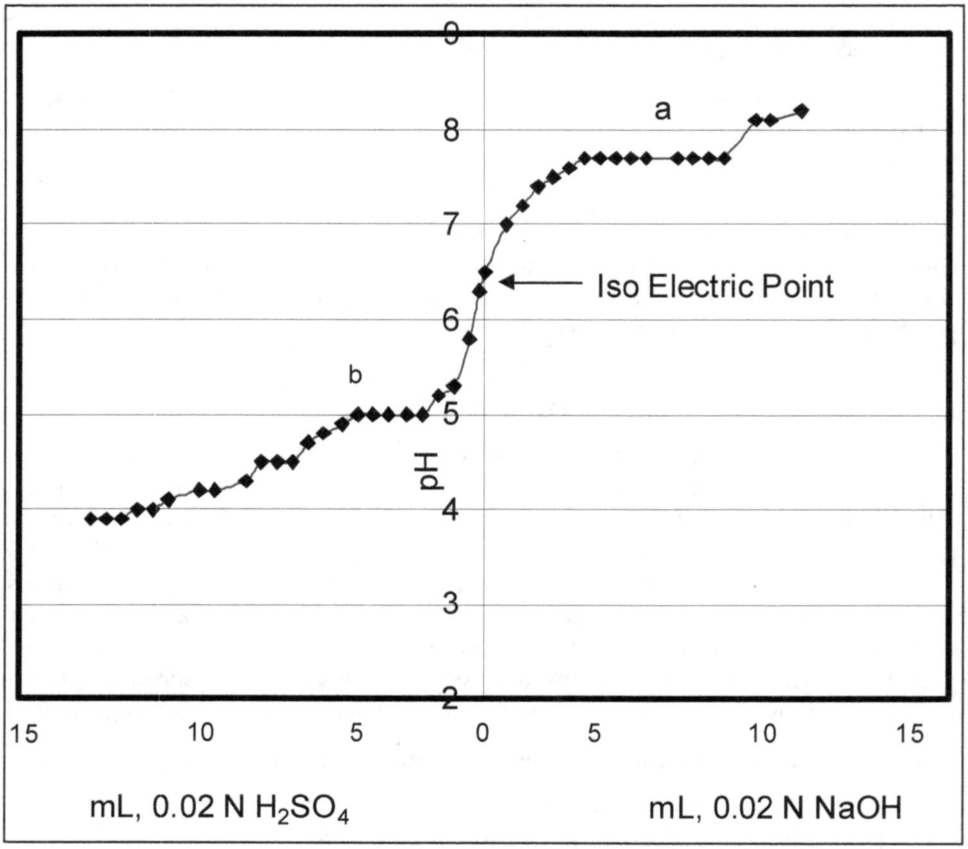

Figure 58.1: Potentiometric Titration Curve of Jalshakti.

indicated by 'a', and 'b' in the Figure 58.1. The invariance of pH shown at 7.7, and 5.0 indicates polyfunctionality of chemical structure of JS. The structure of JS is elucidated through IR spectra (Figure 58.2). The JS showed IR bands at 1714 cm^{-1} and 1081 to 1022 $cm^{-1.}$ This is attributed to the stretching of C=O and C-OH groups present in the chemical structure of JS. JS has a starch (cellulose) backbone on which –COOH pendant groups are present along with some – CH_2OH groups. The surface functionality appears to be dominated by weak acidic groups *viz.*; -COOH, followed by weak basic groups, *viz.* –CH_2OH. The former is represented by region 'a' and latter by 'b' in Figure 58.1.

Effect of Particle Size, pH, Agitation, Dose of Jalshakti

The λ_{max} for Safranine T (ST), Methylene Blue (MB), Crystal Violet (CV), Light Green, Brilliant Milling Violet and Patent Blue VS are 520,667, 590, 620, 545 and 640 nm respectively. Significant reduction in absorbance was observed in the case of all the basic dyes *viz.*: 93 per cent for Safranine T (ST), 98 per cent for Methylene Blue (MB) and 84 per cent for Crystal Violet (CV). At the same time the acid dyes *i.e.* Light

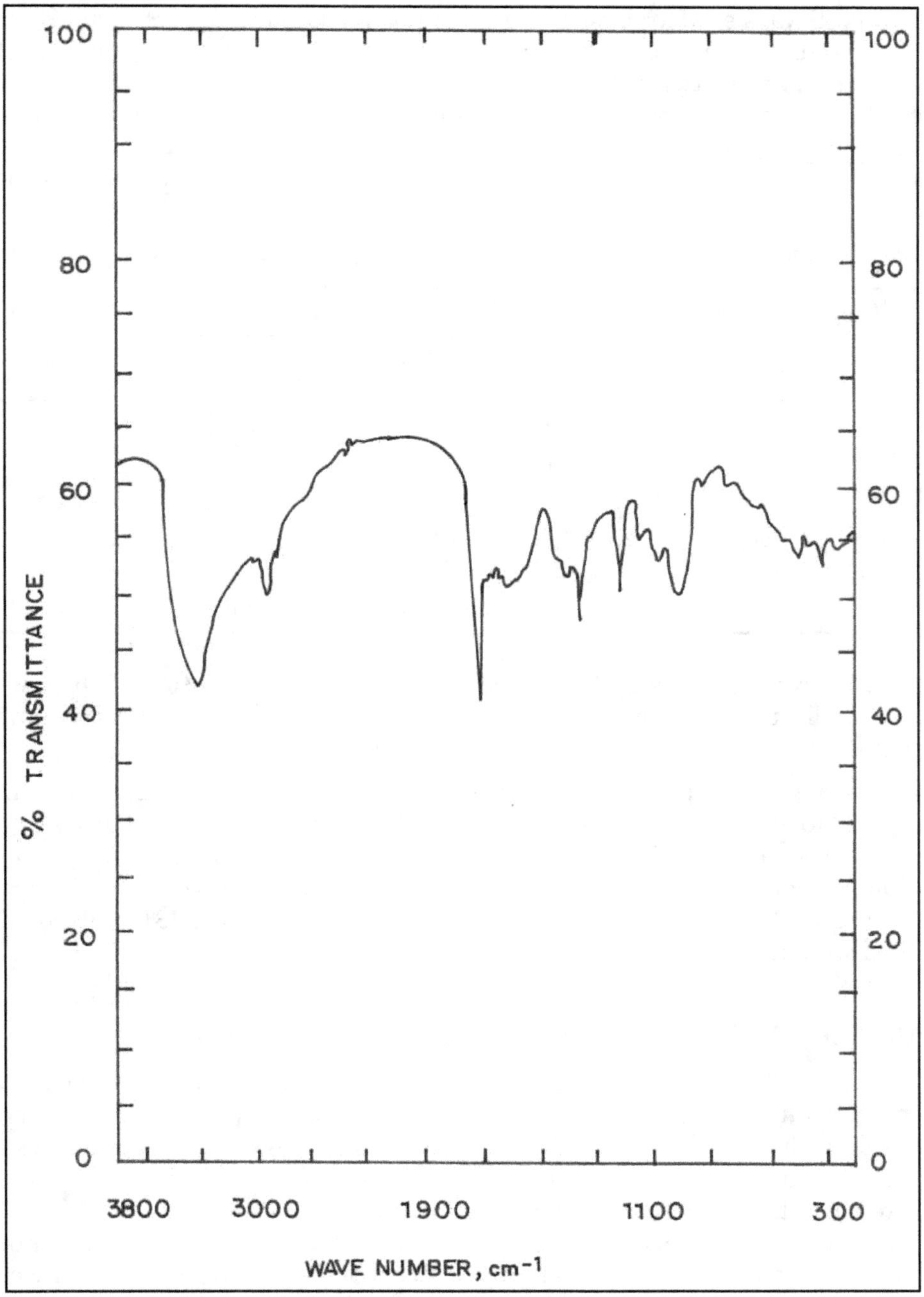

Figure 58.2: Infra Red Spectra of Jalshakti.

Green, Brilliant Milling Violet and Patent Blue VS, showed no removal of colour using JS. Hence, the detailed investigations were carried out with basic dyes only.

The JS was sieved to obtain five particle size fractions *viz.*, 840, 595, 420, 297 and 210μm were collected separately. The results of colour removal experiments using JS of different particle sizes for the dyes MB and CV are presented in Table 58.1. It can be seen that the per cent colour decreased with decreasing particle size. The smaller particle size of JS favoured the adsorption of dye onto it because of a larger surface area and a smaller diffusion distance expected over smaller particles. Further experiments were performed using JS of 210μm particle size only.

Table 58.1: Effect of Particle Size of JS on Colour Removal (JS: 1g/L)

Sl.No.	JS, Particle Size,μm	MB (50mg/L)		CV (5mg/L)	
		a.u.	Per cent Colour Reduction	a.u.	Per cent Colour Reduction
1	–	2.12± 0.14	–	0.54±0.04	
2	840	0.454±0.043	78±2	0.244±0.013	55±2.4
3	595	0.429±0.038	80±1.8	0.17±0.009	69±1.6
4	420	0.265±0.029	88±1.4	0.11±0.011	80±2.0
5	297	0.14±0.019	93±0.9	0.07±0.005	87±0.9
6	210	0.097±0.023	95±1.1	0.03±0.005	95±0.9

It is observed that the optimum pH range for adsorption of MB and CV by JS was 6 ± 0.5 as illustrated in Figure 58.3. It is found that adsorption of dyes is maximum at the isoelectric point of JS.

The adsorption process over JS was affected by the speed of agitation in the range 20 to 100 rpm. An agitation speed of d" 40 rpm at pH 6 ± 0.5 resulted in decolourization up to 97 per cent from aqueous solution of MB after 60 minutes. Below 40 rpm the efficiency of colour removal decreased to 91 per cent at the same contact period. The colour reduction increased up to 0.8g/L dose of JS for MB, ST and CV as seen in Figure 58.4. Further increasing the quantity of JS did not result in increased colour removal. This dose of JS was fixed as the optimum for further studies.

Effect of Initial Concentration of Dye and Contact Time

The results indicate that the extent of adsorption is dependent on the initial concentration of MB, ST, and CV. A representative figure showing the results obtained for various initial concentrations of MB are included (Figure 58.5). The values of absorbance plotted are an average of three sets of experiments. The colour removal decreases with the increasing initial concentrations. The removal of dyes is rapid in initial stages of contact period and thereafter it gradually decreases. At an equilibrium time of 60 minutes, 96 per cent of MB, 88 per cent of CV and 96 per cent of ST were removed as presented in Table 58.2. The uptake of adsorbate species is faster in initial stages *i.e.* 15 min of contact time and becomes slower thereafter. A large number of surface sites are available for the adsorption of pollutants in the initial stages; however later the remaining surface sites may be difficult to be occupied due to steric repulsion between the solute molecules on the surface. Therefore, adsorption of basic dyes becomes progressively slower.

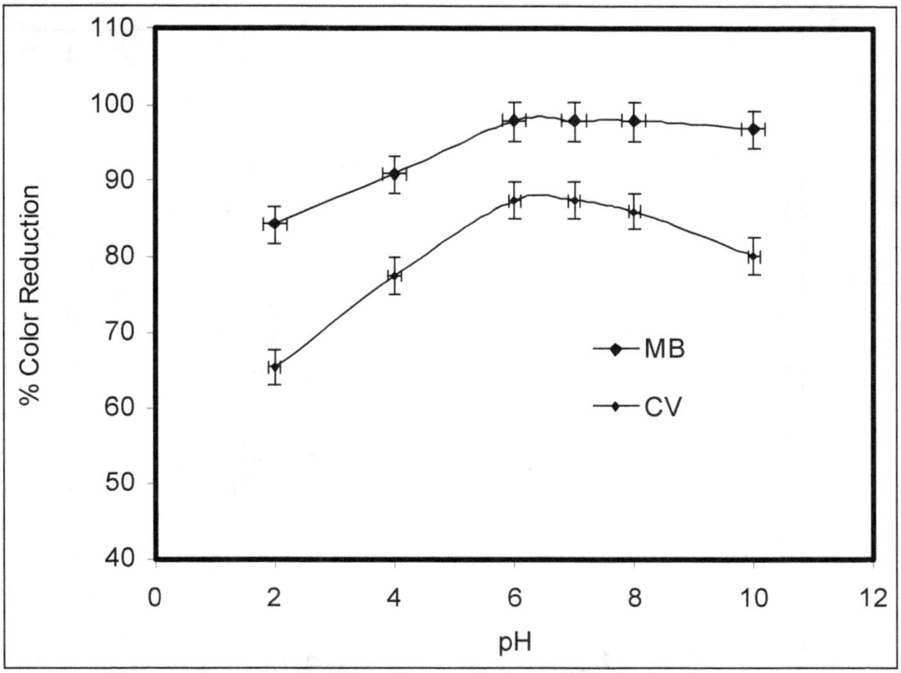

Figure 58.3: Effect of pH on Colour Removal of MB and CV by JS.

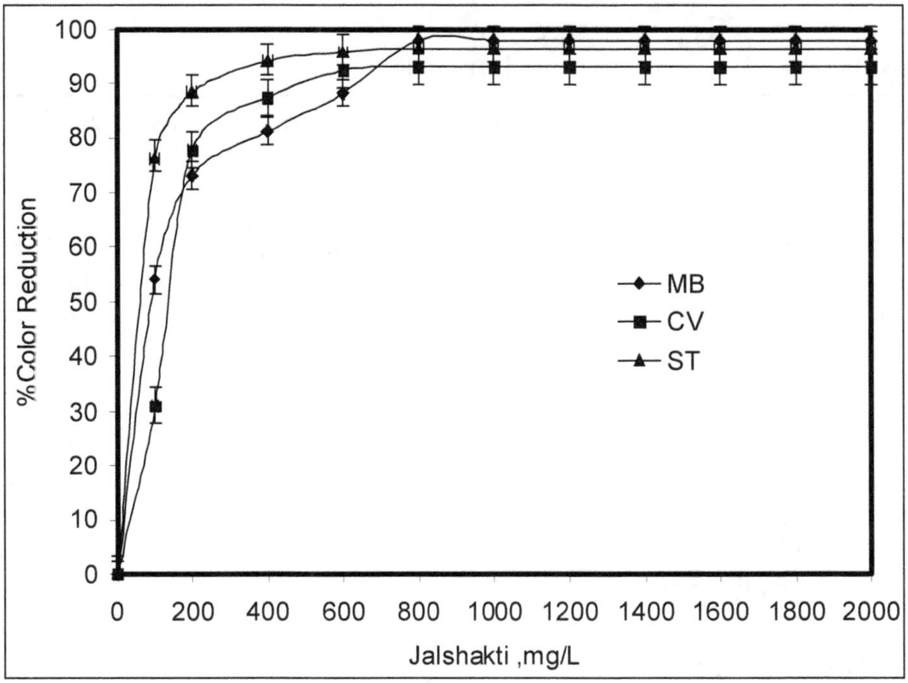

Figure 58.4: Optimization of Dose of JS for Removal of MB.

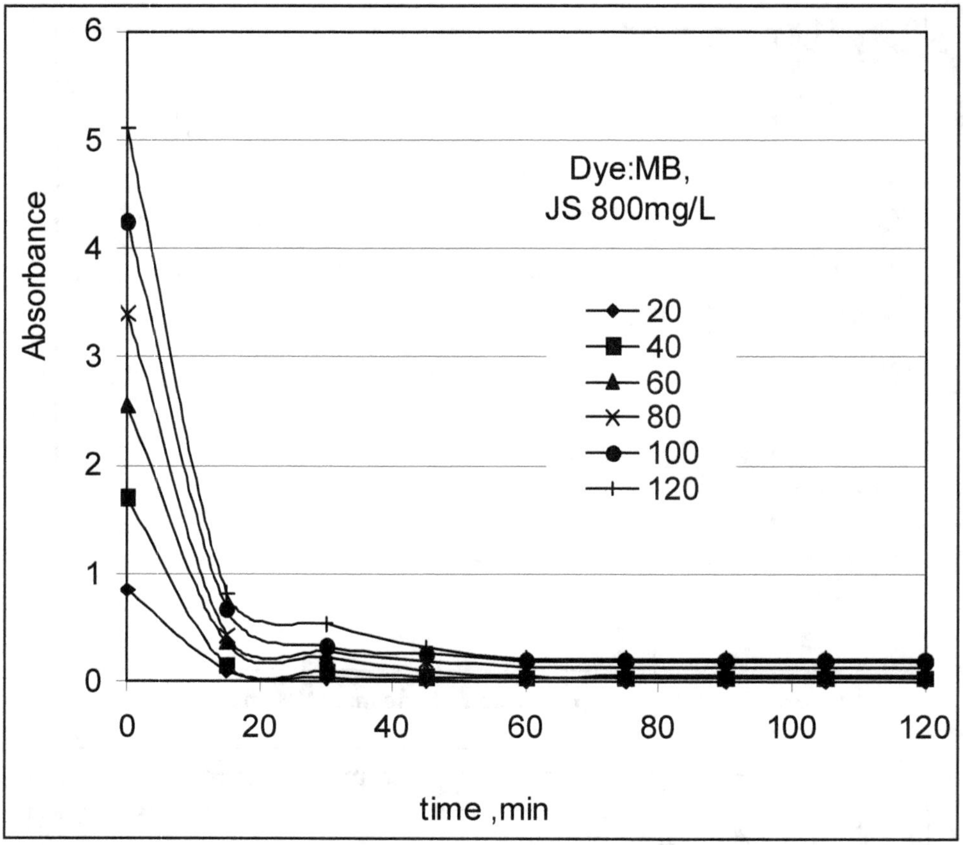

Figure 58.5: Effect of Initial Concentration of MB on the Colour Removal (JS 0.8g/L).

Adsorption Dynamics

The Langmuir adsorption isotherm plots for the basic dyes indicated that the adsorption does not reach a saturated monolayer of solute molecules on the surface of JS. The linearized Langmuir isotherm is represented by the following equation:

$$\frac{1}{q_e} = \frac{1}{Q^0} + \left[\frac{1}{Q^0 b}\right]\frac{1}{C_e}$$

where, Q^o= limiting adsorption capacity, b = adsorption constant, q_e = residual molar concentration of dye per gram of adsorbent at equilibrium, C_e = residual molar concentration of dye at equilibrium.

The Freundlich adsorption Isotherm is represented by the following equation:

$$\log q_e = \log k_f + \frac{1}{n}\log C_e$$

where k_f = adsorption rate constant, $1/n$ = constant related to adsorption intensity. The linearized Langmuir plot and Freundlich adsorption isotherm for one representative dye, MB is given in Figures 58.6 (a) and (b) respectively. Table 58.3 gives constants derived from Langmuir and Freundlich isotherms for basic dyes. The rate constants obtained from Freundlich adsorption isotherms indicate that the adsorption varies as MB> ST>CV. The triphenylmethane group of dye, CV is larger and bulkier than thiazine and azine groups of dyes of MB and CV (Chart 58.1) and the presence of phenyl group causes steric hindrance. This results in reduction in adsorption of these dyes on JS. MB exhibits highest adsorption rate constant having the smallest molecular size amongst the dyes investigated. The presence of dimethylamine $(CH_3)_2N$- and amine groups gives the dye enhanced basicity, which may assist in adsorption. The adsorption capacity *i.e.* k_f, of JS is 15.4 and 8.6 mgg^{-1} for ST and CV respectively. In the case of the dye adsorption, the relative affinity to JS is influenced by basicity of dye molecules and adversely by increase in molecular size.

Table 58.2: Effect of Contact Time on Colour Removal (JS: 0.8 g/L, 210 μm)

	MB (120 mg/L)		*CV (8 mg/L)*		*S T (100mg/L*	
JS (210μ)	*0.8g/L*		*0.8g/L*		*0.8g/L*	
Time, min	*a.u.*	*Per cent Reduction*	*a.u.*	*Per cent Reduction*	*a.u.*	*Per cent Reduction*
0	5.1±0.4	–	0.87±0.09	–	3.7±0.33	–
15	0.82±0.095	83.9±1.8	0.144±0.022	83.4±2.5	0.98±0.09	73.5±2.4
30	0.53±0.087	89.6±1.7	0.128±0.020	85.3±2.2	0.63±0.12	83.0±3.2
45	0.31±0.056	93.9±1.1	0.128±0.017	85.3±1.9	0.31±0.087	91.6±2.3
60	0.21±0.034	95.9±0.7	0.109±0.011	87.5±1.2	0.132±0.093	96.4±2.5
75	0.21±0.034	95.9±0.7	0.109± 0.011	87.5±1.2	0.132±0.078	96.4±2.1
90	0.2±0.017	95.9±0.7	0.109±]0.011	87.5±1.2	0.132±0.054	96.4±1.4
105	0.2±0.011	96.1±0.2	0.107±0.010	87.7±1.1	0.132±0.069	96.4±1.9
120	0.2±0.011	96.1±0.2	0.107±0.010	87.7±1.1	0.132±0.054	96.4±1.4

Intraparticle Diffusion

According to this model, a plots of q_e vs. $t^{0.5}$ was constructed as can be seen in Figure 58.6(c) for the representative dye MB. The intraparticle diffusion plot consists of an initial curved portion and a final linear portion. The multilinearity of these plots may be due to varying extent of adsorption in initial and final stages. The initial curved portions of the plots are due to the boundary layer diffusion effect (Lagergren, 1973); while the final linear portion is attributed to the intraparticle diffusion effects (McKay, *et al.*, 1980). The slope of the linear portion of these plots has been defined as a rate parameter k_{id}, a characteristic of the rate of adsorption in the region where intraparticle diffusion is rate-controlling step (Weber and Morris, 1962). The values of k_{id} for different systems were calculated from the slopes of these plots and are given

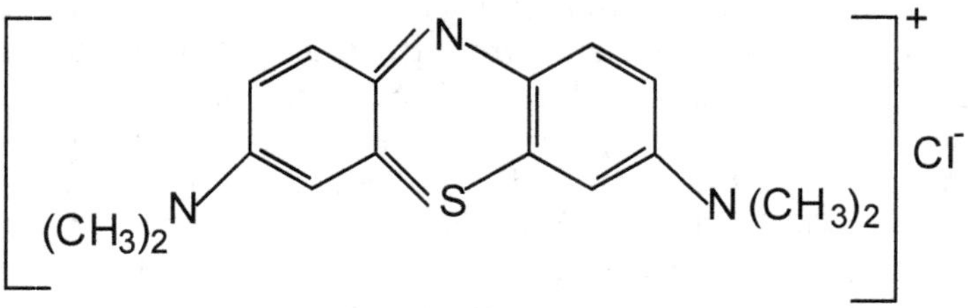

Methylene Blue

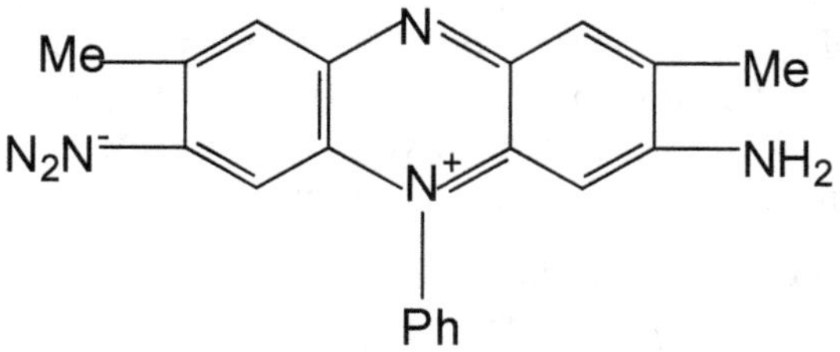

Safranine T

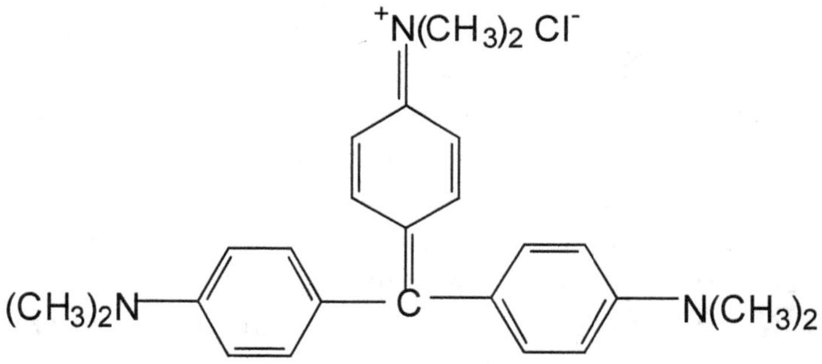

Crystal Violet

Chart 58.1: Chemical Structure of Basic Dyes.

in Table 58.3. The intraparticle diffusion is dominant in the adsorption of basic dyes over JS polymer.

The intraparticle diffusion coefficient 'D for the adsorption of dyes was calculated using the following equation (Bhattacharya and Venkobachar, 1984) and are given Table 58.3:

$$D=0.03\,(r_o)^{2}/t_{½}$$

where, r_o = radius of adsorbent particle *i.e.,* 0.105mm, $t_{1/2}$ = time of half adsorption

It is seen from the Table 58.3 that the values of pore diffusion coefficients are in the order of 10^{-8} and 10^{-9} $cm^2\ sec^{-1}$. D values in this order have been reported to indicate the intraparticle diffusion as the rate-controlling step (Khanna, 1977).

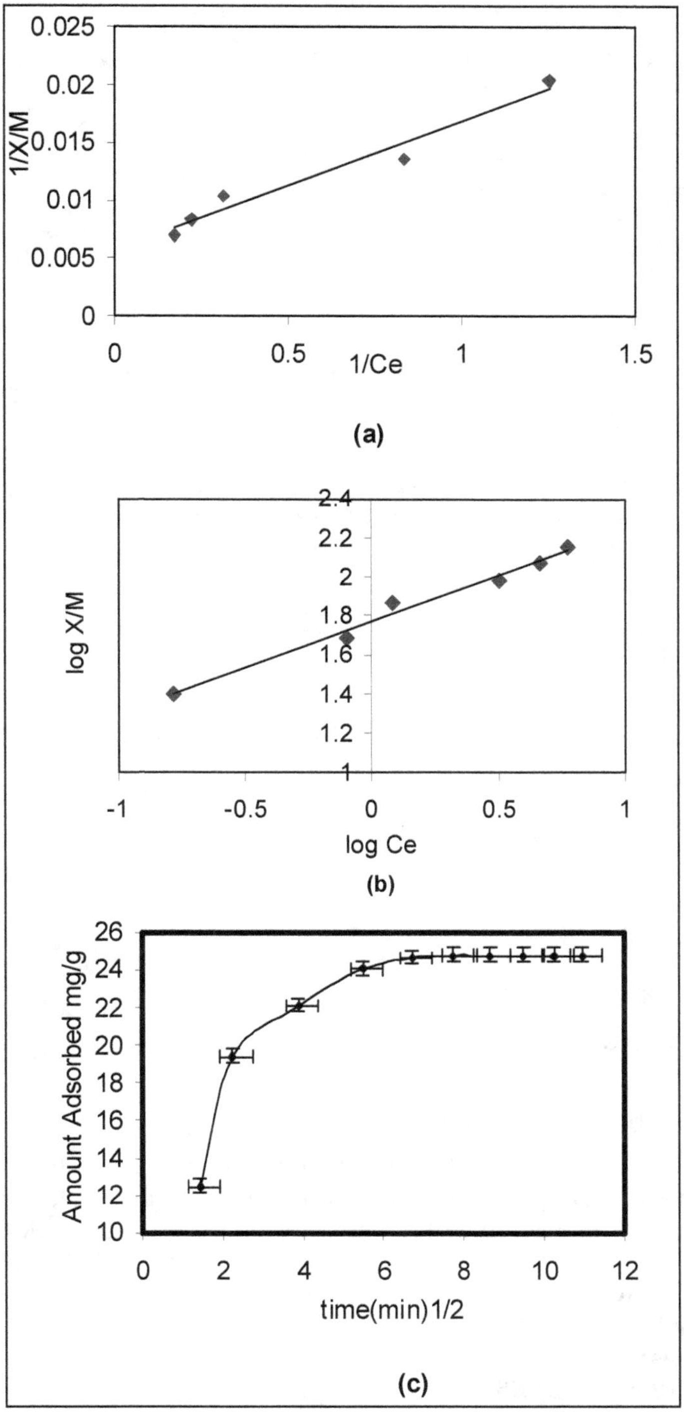

Figure 58.6: Adsorption Kinetics of Jalshakti with Methylene Blue (a): Linearized Langmuir isotherm (b) Freundlich isotherm (c) Intraparticle rate constant.

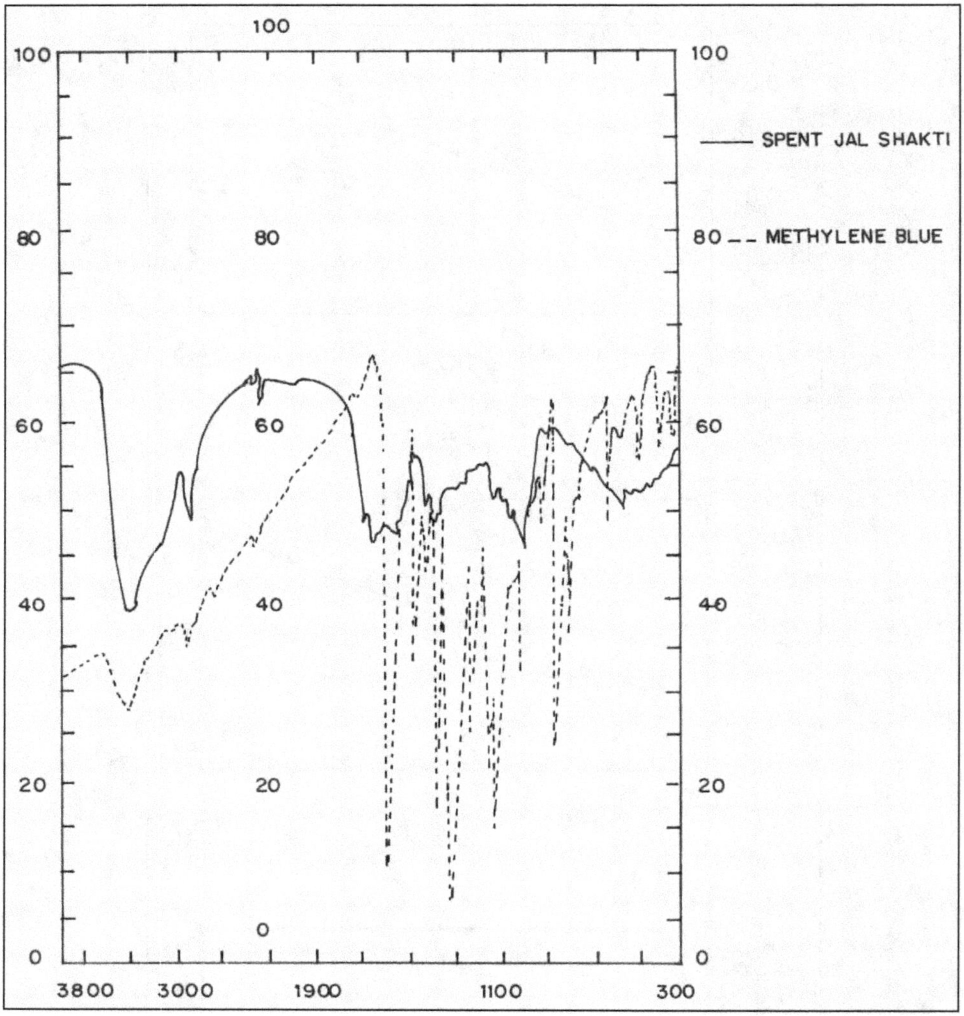

Figure 58.7: Infra Red Spectra of Methylene Blue and Spent JS.

Activated Carbon vs. Jalshakti

The colour removal properties of JS are found to be comparable to AC for MB concentrations d" 80mg/L. JS is more efficient in removal of colour from higher concentrations of Methylene Blue; for example, a 100mg/L concentration of dye with 0.8g/L of adsorbent gives 95 per cent reduction of colour with JS and 86 per cent reduction with AC. The Freundlich adsorption rate constants for adsorption of MB over activated carbon and Jalshakti are comparable *i.e.*, 63.8 and 58.9mg/g (Table 58.3) and JS has better affinity for the dye as indicated by 1/n values. The $t_{1/2}$ value for JS with MB is 2 min and with AC it is 17 min. The intraparticle rate constant is 5.17 $mgg^{-1}min^{-1/2}$ with AC and 0.92 $mgg^{-1}min^{-1/2}$ with JS.

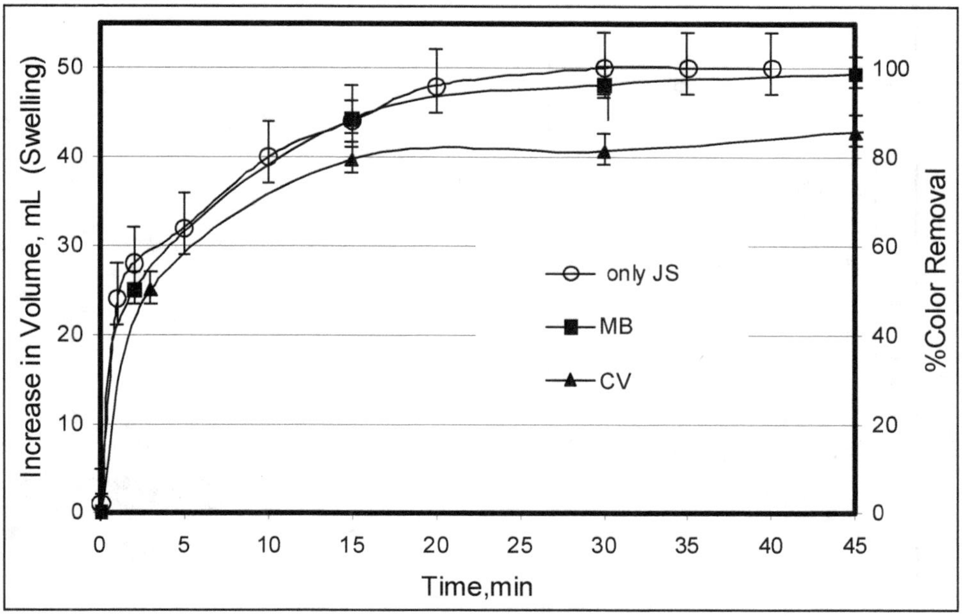

Figure 58.8: Swelling of JS as Compared with Per cent Colour Removal of Dyes as a Function of Time.

Tentative Mechanism of Adsorption of Basic Dyes over Jalshakti

JS has an isoelectric point of charge (IEP) at pH 6.5 and displays polyfunctionality. Some of the carboxylic acid appears to exist as its potassium salt *i.e.* $-COO^{-}K^{+}$. Interestingly, JS also shows 50 times increase in its volume due to swelling in aqueous medium.

In the light of the above characteristics of JS and in view of the fact that JS is effective in removing basic dyes only, the following tentative reaction step may account for removal of basic dyes using JS adsorbent.

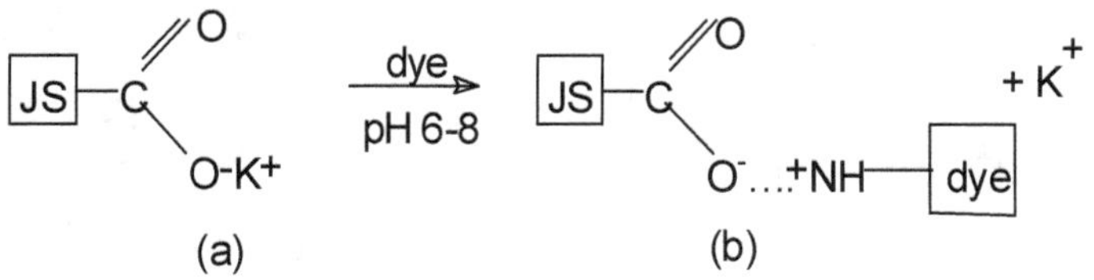

(1)

The above reaction step also explains the observed pH effect on the adsorption of basic dyes over JS. At lower pH, the pendant –COOK as shown in Equation1 (a) gets protonated to form undissociated –COOH group. This exhibits a weaker force compared to $-COO^{-}$ (Equation1 (b)) group, hence adsorption of basic dyes at lower pH is found to be poor.

Table 58.3: Langmuir and Freundlich Adsorption Constants and Intraparticle Diffusion Rate Constants

	Langmuir			Freundlich			Intraparticle Diffusion			
	Q^o mgg^{-1}	b Lmg^{-1}	R^2	k_f mgg^{-1}	$1/n$	R^2	$t_{1/2}$ (min)	Intraparticle Rate Constant k_{id} $mgg^{-1} min^{-1/2}$	Intraparticle Diffusion Coefficient D $cm^2 sec^{-1}$	R^2
Dye/JS										
ST	181.8	0.094	0.99	15.5	0.78	0.99	9	0.667	6.13×10^{-9}	0.99
MB	172.4	0.523	0.96	58.9	0.48	0.98	2	0.92	2.75×10^{-8}	0.95
CV	12.9	2.220	0.98	8.6	0.54	0.99	3	0.203	1.83×10^{-8}	1.00
Dye/AC										
MB	104.2	0.032	0.96	63.8	0.30	0.9	17	5.17	2.38×10^{-7}	0.98

When JS was equilibrated with distilled water, the concentration of K^+ was 10.7 mg/g of JS. To this, when MB was added, the concentration of K^+ increased to 16.03 mg/g of JS. The amount of MB adsorbed was 0.159 mM/g of JS and amount of K^+ released thereby was 0.15 mM/g of JS. Thus, for every mole of dye adsorbed, a mole of K^+ is released into medium. This confirms that $-COO^-K^+$ groups on the surface of JS are involved in the dye uptake. They were probably involved in a stochiometric reaction (Equation 1) and the adsorption mechanism can be likened to ion-exchange-adsorption or activated adsorption.

The above referred reaction step also explains why acidic dyes with anionic groups *viz.*, $-SO_3^-$ and -COOH groups were not removed by JS. The anionic groups on the acidic dyes tend to be repelled by $-COO^-$ groups on the surface of JS and hence no adsorption can be expected. The absence of adsorption of acidic dyes over JS further confirms the adsorption mechanism of basic dyes follows the above-referred step (involvement of substituted amino groups).

IR spectra for spent JS and of MB dye were taken after drying and making potassium bromide pellets as shown in Figure 58.7. The spectra clearly reveal decrease in intensity of band at 1710 cm^{-1} as compared to JS and attachment of dye at the C=0 group of JS. The peaks at 1350, 1326, 1246, and 1217cm^{-1} of MB, indicate the presence of -C-N stretching vibration and presence of tertiary amine in MB. These peaks are absent in spent JS indicating interaction through these groups. The $C=N^+$ band of MB at 2341 cm^{-1} is also present in spent Jalshakti with an additional 1661 cm^{-1} -C=N stretch vibration present in both spent JS and MB.

It has been found that per cent colour removals over AC and JS are comparable at low solute concentration, < 80 mg/L. However, the per cent colour removal using AC decreased sharply at higher solute concentration, > 100 mg/L. In contrast, the per cent colour removal using JS was quite steady in the entire concentration range *i.e.* 20 to120 mg/L of MB. The comparable adsorption efficiency of JS to AC as well as better performance of JS at higher solute concentration may be attributed to swelling of JS. In order to illustrate the relation between adsorption and swelling of JS, a plot of increase in volume due to swelling and per cent colour removals for two of the dyes was constructed as a function of contact time (Figure 58.8). From this plot it is evident that swelling saturates in about 20 to 30 min., while per cent colour removal also follows nearly the same pattern. Thus, at a period during which maximum swelling occurred, maximum colour removal also occurred. Approximately, 90 per cent and 80 per cent colour removal was found using MB and CV respectively, during 20 to 30 min of contact time. Thus, it may be inferred that swelling of JS and dye removal occur simultaneously and the two processes appear to be in a dynamic equilibrium at least till the saturation-swelling regime is attained. It may be concluded that during swelling, $-COO^-K^+$ groups open up increasingly for reaction with basic dye. As a result, dye removal progressively increases till saturation swelling. Supplementary evidence was gathered to support this conclusion. In this MB in acetonitrile was used for adsorption. In acetonitrile solvent, JS did not swell. True to this, only 8 per cent MB dye was adsorbed over JS from acetonitrile in contrast to 90 per cent removed from aqueous medium. Thus, swelling of JS is correlated to better adsorption properties of

JS. Jalshakti is a unique adsorbent with swelling property that aids in adsorption of basic dyes.

Conclusion

Jalshakti developed as a super absorbent polymer can be used to remove colour from basic dye containing aqueous solutions. Its unique swelling property aids in adsorption of these dyes. The affinity of JS for dyes is directly related to the basicity of dye and inversely to the molecular size. The results reveal that the basic dyes are removed through adsorption-exchange mechanism involving carboxylic acid groups and Potassium ions of JS. The adsorption of dyes does not reach a saturated monolayer on the surface of JS as evident from Langmuir adsorption isotherms. The mechanism of removal follows Langmuir, Freundlich adsorption and intraparticle diffusion models.

References

1. Al-Duri, B., McKay,G., El-Geundi, M.S., Abdul Wahab, M.Z., 1990. Three resistance transport model for adsorption onto bagasse pith.' *J. Environ. Engg*. 116, 487-502.

2. Balasubramnaian M.R., Muralisankar I., 1987. Utilization of fly ash and tea waste ash as decolorizing agents for dye effluents. *Ind. Journal of Technol*. 25, 471-474.

3. Bhattacharya, A. K., Venkobabchar, C., 1984. Removal of cadmium (II) by low cost adsorbents. *J. Environ. Engg*. 110, 110-122.

4. Bousher, A., Shen, X., and Edyvean, G.J., 1997. Removal of coloured organic matter by adsorption onto low-cost waste materials. *Water. Res*. 31, 2084-2092.

5. El Geundi, M.S., 1991. Colour removal from textile effluents by adsorption techniques. *Water Res*. 25, 271-273.

6. *Environmental Chemistry of Dyes and Pigments* (A. Reife and Harold S. Freeman, eds.), Published by John Wiley and Sons Inc., 1996.

7. Gupta, G.S., Prasad, G., and Singh, V.N., 1990 (a). Removal of chrome dye from aqueous solutions by mixed adsorbents: Fly ash and coal. *Water Res*. 24, 45-50.

8. Gupta, G.S., Prasad, G., Singh, V.N., 1990 (b). Mordant blue-13, removal from aqueous solutions by fly ash-II. *Asian Environ*. 40-48.

9. Hao, O. J., Kim, H., Chiang, Pen-chi, 2000. Decolourization of wastewater. *Critical Rev. Environ. Sci. Technol*. 30,449-505.

10. Jalshakti Polymer, 1988, *Technical Report*, National Chemical Laboratory, Pune, India,

11. Juang, R. S., Tseng, R. L., Wu, F. C., Lin, S. J., 1996. Use of chitin and chitosan in lobster shell wastes for colour removal from aqueous solutions. *J. Environ. Sci. Health* A 31, 325-338.

12. Khanna, P., Malhotra, S.K., 'Kinetics and mechanism of phenol adsorption on fly ash.' 1977, *Ind. J. Environ. Health* 19, 224-237.

13. Kumar, A., Rao, N. N., Kaul, S. N., 2000. Alkali treated straw and insoluble straw xanthate as low cost adsorbents for heavy metal removal- preparation, characterization and application. *Bioresource Technol.* 71,133-142.

14. Kumar, M. N. V. R., Sridhari, T.R., Bhavani, K. D., Dutta, P.K., 1998. Trends in colour removal from textile mill effluents. *Colourage*, 40, 25-34.

15. Malmary, G., Perineau, F., Molinier, J., Gaset, A., 1985. A continuous process for dye removal from liquid effluents using carbonized wool waste. *J. Chem. Technol. Biotechnol.*, 35A, 431-437.

16. Matsuda, Hisayuki, Shinela Katsuro, Yasuhara, Minoru, Sumiyoshi, Tsukasa, May 1975. *Treatment of coloured wastewater*. Japan, Kokai 75, 60 468.

17. McKay, G., El Guendi, M., Nassar, M.M., 1987. Equilibrium studies during the removal of dyestuffs from aqueous solutions using bagasse pith. *Water Res.* 12, 1527-1533.

18. Namasivayam, C., Kanchanna, N., Yamuna, R.T., 1993. Waste banana pith as an adsorbent for the removal of rhodamine-B from aqueous solutions. *Waste Managmt.* 13, 89-95.

19. Nigam, P., Armour, G., Banat, I.M., Singh, D., Marchant, R., 2000. Physical removal of textile dyes from effluents and solid-state fermentation of dye adsorbed agricultural residues. *Bioresource Technol.* 72, 219-226.

20. Poots, V.J.P., McKay, G., Healy, J. J., 1976a. The removal of acid dye from effluent using natural adsorbents I: Peat. *Water Res.* 10, 1061-1066.

21. Poots, V.J.P., McKay, G., Healy, J. J., 1976b.The removal of acid dye from effluent using natural adsorbents. II: Wood. *Water Res.* 10, 1067-1070.

22. Rao, N.N., Kumar, A., Kaul, S.N., 2002. 'Chromium exchanged insoluble straw xanthate (ISX-Cr+3) for removal of free cyanide: Combined effect of ligand displacement reaction and sorption. Sep. *Sci. Technol.* 37, 2167-2182.

23. Slokar, Y.M., Le Marechal, A.M., 1997. Methods of decolouration of textile wastewaters. *Dyes Pigments*, 37, 335-356.

24. Thacker, N.P., Vaidya, M.V., Dhodapkar, R.S., Sipani, M., Tankhiwale, S., Nov. 1993. GAC treatment for the removal of organochlorine pesticide. *Proceedings on Alternate Methods of Water Treatment*, Nagpur, 199-211.

Chapter 59

Fenton and Electro-Fenton Methods for Oxidation of H-acid and Reactive Black 5

N.N. Rao, Gaurav Bose, Pradnya Khare and S.N. Kaul

CSIR-NEERI, Nagpur

ABSTRACT

Oxidative treatment of H-acid (HA) and Reactive Black 5 (RB5) using Fenton reagent (Fe^{2+}/H_2O_2) and electro-Fenton (EF) method was reported. Optimization of doses of ferrous iron and hydrogen peroxide was carried out in each case using HA; and the oxidation of RB5 was also carried out under the optimized conditions. Approximately 71 per cent COD was removed in 2 h using conventional Fenton method at optimized doses; Fe^{2+}= 0.3 g/L (5.37 mM), H_2O_2 = 6 mL/L (53.0 mM); H_2O_2/Fe^{2+} = 10. In contrast, more than 92 per cent COD was removed in 15 min. using EF method with optimized Fe^{2+} dose of 0.130 g/L (2.34 mM) and 8 ml/L (70.6 mM) of H_2O_2. The pseudo-first order rate constants (k) for Fenton reagent and EF method were 0.054 and 0.38 min^{-1}. The COD removal through EF method was 7 times faster. The calculated energy requirement of the EF method was 0.82 kgCOD/kWh at the minimum applied current (0.25 A) when approximately 92.5 per cent COD was removed. In the case of RB5, about 67 and 87 per cent COD was removed under optimized Fenton and electro-Fenton conditions, respectively. The higher efficiency of the EF method was attributed to incremental addition of Fe^{2+} and accompanying higher H_2O_2/Fe^{2+} molar ratio. The results are discussed in the light of the mechanism for Fenton's oxidation.

Keywords: *Oxidation, Fenton reagent, Electro-fenton method, H-acid, Reactive black-5.*

Introduction

Approximately 10,000 different synthetic dyes weighing more than 10^6 tons are produced annually worldwide. Roughly 28-30 per cent of annual production of dyes is discharged into the environment (Willets and Ashbolt, 2000). Release of the dye containing effluents into water bodies is unacceptable at least on two counts: i) their intense color and ii) poor decolorization and biodegradation in conventional wastewater treatment plants. Various commercially available dyes carry sulphonic acid groups as substituents, apart from others, in their molecules mainly to increase water solubility. But this also increases recalcitrance to environmental breakdown as the C¾S bond is thermodynamically stable (Kertesz *et al.*, 1994). Consequently, aromatic sulphonates have spurred great environmental concern (Schwitzguebel *et al.*, 2001), since they require treatment at extreme physico-chemical conditions, such as high pressure and temperature, and special reactor configurations. H-acid, an aromatic di-sulphonate compound, is an important dye intermediate. Some commercial dyes, for example, Reactive Black 5, are the derivatives of H-acid. Conventional biological processes can not effectively degrade these substances (Chen and Pignatello, 1997). The H-acid wastewater is currently treated through evaporation-incineration technology that is expensive as well as energy intensive (Kaul *et al.*, 1998).

There is substantial research published with the objective of finding alternative treatment technologies for such hard to treat organic compounds (Vogelpohl *et al.*, 2001). Most of it deals with the application of hydroxyl radicals mediated reaction chemistry to oxidize such compounds exhaustively. The hydroxyl radical is a powerful oxidant and provides a relatively inexpensive mechanism for the oxidation of a variety of aqueous phase organic compounds. One such approach is based on the iron (II)-catalyzed peroxide-based oxidation, popularly known as Fenton reagent (Merli, 2002). A range of phenol compounds, for example, chloro-, nitro-, amino-phenols, cresols and substituted cresols can be oxidized using Fenton's oxidation catalyst (Wang, 1992, Casero *et al.*, 1997). Further, Fenton's treatment of printing and dyeing wastewater (Smiss, 1983), and synthetic textile wastewater (Kuo, 1992; Gregor, 1994; Solozhenko, 1995) has been reported. The application of ferrous-hydrogen peroxide for the treatment of H-acid manufacturing wastewater is also reported (Wanpeng, 1996).

The conventional Fenton reagent has undergone several modifications, for example, substitution of ferrous catalyst with other transition metal oxides/salts as pseudo-Fenton catalysts (Dongre *et al.*, 2002; Teel *et al.*, 2001; Huang *et al.*, 2001; Weiss, 1935); use of chelated iron complexes for increasing the operational pH range (Watts and Stanton, 1994; Aplin, 2001); incremental addition of peroxide to maximize oxidation efficiency (Merli, 2002); *in-situ* generation of Fe^{2+} catalysts by application of direct current between suitable electrodes thus avoiding the need for iron salt addition (Huang *et al.*, 1999; Chiarenzelli *et al.*, 1997); *in-situ* generation of hydrogen peroxide at suitable anodes and through cathodic reduction of molecular oxygen (Alvarez-Gallegos and Pletcher, 1998a,b; Qiang *et al.*, 2002). The latter two modifications are achieved through application of electrochemical principles and are generally referred to as electro-Fenton processes. They provide an alternative to the production of hydroxyl radicals via Fenton reagent. Both, conventional Fenton and electro-Fenton oxidation processes have attracted great deal of attention for the treatment of groundwater contaminated with chlorinated solvents (USEPA, 1998); subsurface storage tank water and slurry contaminated with PCBs and VOCs (Chiarenzelli, 1997); contaminated soils (Watts and Stanton, 1994) and wastewater (Szpyrkowicz, 2001; Kang and Hwang, 2000; Schaefer *et al.*, 2000) is concerned.

In the present study, we report effectiveness of electrochemically generated ferrous iron catalyst in the presence of hydrogen peroxide, for the treatment of aqueous solutions containing H-acid and Reactive Black 5. Reactive Black 5 is a derivative of H-acid. It comprises H-acid molecule at the center, and two azo- (p-vinyl sulphonic acid) pendant groups are attached to it. It would be interesting to know the difference in reactivity of parent molecule and derivative because the derivatives are commonly used in textile dyeing. The study revealed that electro-Fenton method is more efficient and requires much less contact time for the removal of significant amounts of the compounds when compared to the conventional Fenton reagent.

Methods and Materials

1-amino-8-hydroxynaphthalene-3,6-disulphonic acid, disodium salt (H-acid, $C_{10}H_7O_7NS_2Na_2$) was procured from M/s Rohan Dyes and Intermediates Ltd., Vadodara, India. 1-amino-8-hydroxy-2,7-bisazo[(p-vinylsulphonic acid)]-naphthalene- (3,6-disulphonic acid), tetra sodium salt, $C_{26}H_{21}O_{19}N_5S_6Na_4$, C.I. Reactive Black 5) was purchased from M/s Atul Dyes, Ahmedabad, India. Sodium tetrafluroborate (97 per cent +) was obtained from Loba Chemie, Mumbai, India. Other chemicals *i.e.*, ferrous sulfate hepta hydrate (98 per cent +), hydrogen peroxide (30 per cent w/v, 99.75 per cent), sulfuric acid (98 per cent, sp.gr. 1.84, 36N) were analytical grade chemicals and procured from E. Merck India Pvt. Ltd, Mumbai, India. A stock solution of 0.90 M $FeSO_4$ (50 mgFe^{2+}/mL) was used as the source for Fe^{2+} ions. Hydrogen peroxide was used as received from E. Merck India Pvt. Ltd.

Fenton Oxidation Procedure

Conventional Fenton oxidation was performed in a 6-beaker (1L) assembly of Jar Testing Apparatus (Phipps and Bird, PB-700 TM, USA). A 500 mL solution of 250 mg/L H-acid or RB5 was placed into these beakers. Initially the pH was adjusted to 4.0 using H_2SO_4. Then, different doses of ferrous iron were introduced by adding requisite amounts of $FeSO_4$ stock solution. Subsequently, a fixed dose of hydrogen peroxide (4 mL) was added to each of the beakers. Thus, the peroxide to Fe^{2+} ratio was varied from 5.0 to 20. The solutions were stirred for 2 h and the final pH was recorded. The test samples were collected and analyzed for concentration of H-acid or RB5 and chemical oxygen demand. The reaction was optimized with respect to doses of Fe^{2+} and H_2O_2.

Electro-Fenton Procedure

In this case 2 g/L dose of sodium tetrafluroborate was added in the stock solutions of 250 mg/L H-acid or RB5, as supporting electrolyte. Since, the H-acid and RB5 solutions were made in de-mineralized water, addition of a supporting electrolyte was necessary in order to make solutions sufficiently conductive. Electro-Fenton oxidation of 500 mL of H-acid or RB5 solutions was carried out in a 650 mL capacity Perspex reactor using square shaped iron (8 cm ´ 8 cm) and titanium (8 cm ´ 8 cm) plate electrodes. These electrodes were cut out from iron and titanium sheets supplied by M/s Vidharbha Metals, Nagpur. The electrodes were placed vertically parallel in the reactor at an inter-electrode distance of 4 cm. Then, a requisite volume of H-acid or RB5 solutions was introduced and DC current was applied using Testronics 34 D DC Power Supply (M/s Testronics, Mumbai). The power supply was equipped with in-built Voltmeter and Amperemeter that facilitated application of constant current as well as measurement of operational voltage. The iron electrode was connected to positive pole (anode) while Ti electrode was joined to negative pole (cathode). The reactor was equipped with an overhead stirrer for mixing the reactants at a controlled speed. Appropriate adjustments were made in the reactor for the employment of continuous addition of acid or hydrogen peroxide. The pH was monitored continuously using a pH probe. Since the solution pH was increasing during EF treatment, concentrated sulfuric acid was added and the pH was readjusted to 3.0 ± 0.1. Initially, the dose of hydrogen peroxide was fixed (4 mL) and the dose of ferrous iron was varied and optimized. The desired doses of ferrous iron were introduced by applying specific and different magnitudes of DC current for a finite period. Typically, the cell voltage remained at 5 V. All the electro-generated iron was assumed to be in the ferrous form since Fe^{2+} is stable at pH less than 4.5-5.0. The amount of ferrous iron generated was calculated from Faraday's equation:

$$w = z \times I \times t \qquad (1)$$

where,

w= amount of ferrous iron in grams, z = electrochemical equivalent weight of iron (2.8939 ´ 10^{-4} g $Coulomb^{-1}$), I = current (A), and t is time (sec).

The COD of test samples was estimated as a function of ferrous iron dose.

Optimization of Hydrogen Peroxide dose

This was performed by varying the dose of peroxide from 1 to 8 mL (17.6 – 141 mM) at the optimized dose of Fe^{2+}. A blank run was also performed without adding H_2O_2 and keeping the rest of the conditions identical. Generally, hydrogen peroxide was applied as a single dose *i.e.*, batch mode. However, some experiments also were conducted by continuous addition of hydrogen peroxide with the objective of comparing the efficiencies of the two different modes of peroxide addition.

Utilization ratio of hydrogen peroxide

The per cent utilization ratio of H_2O_2 (E_H), can be calculated using Equation (1) (Huang *et al.*, 2001).

$$E_H \text{ (per cent)} = 100 \times [(COD)_0 - (COD)_t]/[H_2O_2] \times 0.47 \qquad (2)$$

where,

COD_0 = initial COD

COD_t = COD at time t

The denominator corresponds to the theoretically available oxygen expressed as a fraction of initial concentration of hydrogen peroxide. Considering that each mole of hydrogen peroxide produces 1 mole of OH radicals (half a mole of O_2), (16/34 = 0.47) moles of oxygen would be available from 1 mole of hydrogen peroxide.

Analysis

The initial concentration of H-acid and RB5 was measured by monitoring their wavelength of absorption (λ_{max}) at 358 nm (e = 5.742 ´ 10^3 $M^{-1}cm^{-1}$) and 590 nm (e = 4.0543 ´ 10^4 $M^{-1}cm^{-1}$), respectively. Absorbance was determined using a Systronics 119 UV-Vis Spectrophotometer. However, due to rapid decrease in the intensity of the above absorption maxima under Fenton and EF conditions, the spectrophotometric method of estimating time course variation of HA or RB5 could not be done. Necessarily, tests for chemical oxygen demand (COD) were performed according to Standard Methods (APHA-AWWA-WEF, 1998). The pH of the treated solution was increased to the pH range of 7 to 8 by adding 1 N NaOH. The Fe^{2+} was removed as hydroxide sludge. Decomposition of the unreacted H_2O_2 in water and O_2 gas was also favored. Both ferrous iron and unreacted H_2O_2 interfere with COD determination. The residual iron (total) concentration in the treated solutions was determined using Model 24 Inductively Coupled Plasma Atomic Emission Spectrometer (JOBIN YVON EMISSION Instruments S. A., France). The concentrations of total iron and Fe^{2+} were also measured as per Ferrozine Method (Hach, 1998) and 1,10-Phenanthroline method (APHA-AWWA-WEF, 1998), respectively using Hach DR/2010 Spectrophotometer. Cyclic voltammograms (CV) of H-acid were recorded using PGSTAT 20 (Autolab, Netherlands) and General Purpose Electrochemical Software (GPES) in a three-electrode assembly (Pt anode; Pt mesh cathode and Ag/AgCl reference electrode). A 25 mL solution of 100mg H-acid in pH 4.0 buffer was employed for recording CV. The solution was stirred well using magnetic stirrer and was purged with N_2 gas (10 mL/min). The potential was scanned between –1.0 V and +1.25 V, starting from 0 V at a rate of 10 mV/sec.

Results and Discussion

Conventional Fenton Oxidation

The following control experiments were conducted. Firstly, addition of 6 mL (105.8 mM) of H_2O_2 to H-acid solution (without Fe^{2+}); and secondly, addition of 0.150 g Fe^{2+} in the absence of H_2O_2 to H-acid solutions. In both the cases, neither concentration of H-acid nor COD was found to decrease with increasing contact time.

Optimization of Fe^{2+} and H_2O_2

In this case, the dose of Fe^{2+} was varied from 0.1 to 0.4 g (3.58-14.32 mM) at a fixed dose of H_2O_2 (4 mL, 70.6 mM) and Fenton oxidation was performed at pH 4 for a period of 2 h as described previously. It conventional to control pH at ≤ 4 and peroxide: Fe^{2+} molar ratio is generally kept > 5-10 in order to simulate Fenton oxidation conditions (Walling 1975; Sims 1983). The decrease in the COD, per cent COD removals and hydrogen peroxide utilization ratio (E_H) at different doses of Fe^{2+} for the Fenton treatment of H-acid solutions are presented in Table 59.1. Approximately 60-68 per cent COD was removed at all applied doses of Fe^{2+}. A maximum of 68.3 per cent COD was removed at 0.150 g Fe^{2+} dose, where the peroxide utilization ratio was 33.7 per cent. The average E_H was 32.3 per cent. The utilization ratio of H_2O_2 is a measure of peroxide actually utilized in COD removal experimentally when compared to the theoretically removable COD by the total hydrogen peroxide dose. The final pH of the Fenton reagent treated solutions was in the range 3.5-3.9. The residual iron in treated solutions at optimum Fe^{2+} dose was 9 ± 1 mg/L.

Table 59.1: Conventional Fenton Oxidation-Optimization of Dose of Iron

Sl.No. (1)	*[$FeSO_4$], mL* (2)	*[Fe^{2+}], mM* (3)	*COD*, mg/L* (4)	*Per cent R#* (5)	*E_H* (6)
1.	0	0	278 ± 12	0	0
2.	2	3.580	96 ± 8	65.5	32.3
3.	3	5.370	88 ± 6	68.3	33.7
4.	4	7.162	92 ± 8	66.9	33.0
5.	5	8.950	92 ± 8	66.9	33.0
6.	6	10.74	96 ± 7	65.5	32.3
7.	8	14.32	112 ± 7	59.7	29.4

Concentration of H-acid	:	250 mg/L
[H_2O_2]	:	4 mL (70.6 mM)
Treatment time	:	2h
Solution volume	:	500 mL
Stock Solution of $FeSO_4$	:	50 mg Fe^{2+}/mL

* Average of two sets of measurements.

Average percentage calculated using lowest and highest percentage values.

Optimization of H_2O_2 dose was performed by varying H_2O_2 dose from 1 to 6 mL (17.6-105.8 mM) at a fixed dose of Fe^{2+}; *i.e.*, 0.150 g (5.37 mM). The decrease in the COD, per cent COD removals and hydrogen peroxide utilization ratio (E_H) at different doses of H_2O_2 for the Fenton treatment of H-acid solutions are presented in Table 59.2. Approximately 47-74 per cent COD was removed at all applied doses of H_2O_2; however a maximum of 74 per cent COD was removed at 53 mM H_2O_2 dose. The E_H varied from 92.2 per cent at 17.6 mM H_2O_2 dose to 17.7 per cent at the highest peroxide dose (105.8 mM). The higher the inlet peroxide concentration, the lower the E_H. At the optimum H_2O_2 dose, the peroxide utilization ratio was 48.7 per cent (Table 59.2).

Electro-Fenton Oxidation of H-Acid

Initially, H-acid solution having an initial pH of 3.0 was subjected to electro-Fenton oxidation at an arbitrary current of 0.5 A and a dose of 70.6 mM H_2O_2. The reaction was carried out for 30 min. However, some brown flocs were formed approximately at treatment time higher than18 minutes. This was due to formation of ferric hydroxide [$(Fe(OH)_3]$ caused by the reaction of ferric ions and the OH^- released at the cathode during hydrogen evolution. There was vigorous evolution of hydrogen gas at the cathode. The pH of the test solution quickly increased to above 6 and caused floc formation. The hydrogen gas that evolved at cathode caused flotation of some Fe $(OH)_3$ flocs. Additionally, some sludge also was found at the bottom of the cell signifying that a part of Fe $(OH)_3$ settled to the bottom. Some H-acid may be removed by Fe $(OH)_3$ flocs. Under the conditions of EF method employed in the present study, no direct oxidation of H-acid at Fe anode could be observed. The applied current was utilized for generation of Fe^{2+} catalyst at the iron anode. This was checked by measuring total iron and Fe^{2+} separately according to published procedures (Hach, 1998; APHA-AWWA-WEF, 1998). The difference between concentration of total iron and Fe^{2+} was less than 5 per cent, which was within the experimental error.

Addition of hydrogen peroxide to H-acid solution containing Fe^{2+} resulted in intense reddish brown coloration. However, this color disappeared gradually with the treatment time and solutions were colorless at the end. These color changes represent formation of some intermediates that also undergo oxidation as the contact time increased. The disappearance of colored intermediates from the reaction mixture leads to the colorless solution at the end. The colour changes found were discussed in detail elsewhere in the text.

Optimization of Contact Time

The rapid variation of pH referred in the EF experiments was avoided by controlling and maintaining pH at 3.0 ± 0.1 by continuous addition of concentrated sulfuric acid. Further, it was desirable to conduct the reaction for a short period to minimize acid addition. Hence, electro-Fenton oxidation was carried out for 15 min.

In a preliminary experiment of electro-Fenton oxidation of 250 mg/L HA solution at pH 3.0 ± 0.1 and at an arbitrary current of 0.5A and a dose of 70.6 mM H_2O_2, the concentration of COD was reduced from 278 to 36 mg/L. Due to rapid decrease in the intensity of Approximately, 30-50 mL/L sludge was also formed upon raising pH to 7 by addition of alkali to treated H-acid solutions. The COD was gradually reduced by the use of Fenton reagent for the first 80-90 minutes of treatment. Although, the COD removal was decreased after this time period, the reaction continued until 2 hours were completed. Approximately 64 per cent of initial COD were removed during first 80 minutes of Fenton treatment. In contrast, the COD decreased rapidly, and more than 92 per cent COD was removed at EF reaction time within 15 minutes. Thus, the optimum reaction time for Fenton reagent was considered as 2 h, while that for EF method was 15 minutes.

Control Experiments

Considering that application of current results in dissolution of iron anode during EF treatment, the different control experiments performed were a) No H_2O_2; Current (I) = 0.5 A; pH = 3; reaction time =15 min. and b) 4 mL H_2O_2 (70.6 mM): I = 0; pH = 3; reaction time =15 min. The concentration of Fe^{2+} generated was estimated as 0.130 g (4.68 mM) in the former case, In both the cases, neither concentration of H-acid nor COD was found to decrease. These studies revealed that the presence of both Fe^{2+} and H_2O_2 was necessary for achieving oxidation of the chosen substrates, and it was necessary to control pH at 3.0 ± 0.1 in order to avoid electroflotation and electrocoagulation processes that otherwise may contribute to substrate removal.

Table 59.2: Conventional Fenton Oxidation-Optimization of Dose of H_2O_2

Sl.No. (1)	*$[H_2O_2]$, mL (2)*	*$[H_2O_2]$, mM (3)*	*COD*, mg/L (4)*	*Per cent R# (5)*	*E_H (6)*
1.	0	0	278 ± 12	0	0
2.	1	17.60	148 ± 10	46.8	92.2
3.	2	35.30	96 ± 8	65.5	64.54
4..	3	53.00	72 ± 7	74.1	48.70
5.	4	70.60	84 ± 7	69.8	34.40
6.	5	88.20	108 ± 8	61.2	24.10
7.	6	105.80	128 ± 9	54.0	17.70

Concentration of H-acid	:	250 mg/L
Treatment time	:	2h
Solution volume	:	500 mL
Dose of Fe^{2+}	:	0.150 g (5.37 mM)

* Average of two sets of measurements.

Average percentage calculated using lowest and highest percentage values.

Optimization of Electro-Generated Fe^{2+} Dose and H_2O_2 Dose

Optimization of dose of Fe^{2+} was performed by varying current from 0.25 to 1.75 A, which yielded different Fe^{2+} concentrations in the range 0.0653 to 0.588 g (1.17-9.36 mmol). The current was applied for 15 minutes. A fixed dose of 4 mL (70.6 mM) of H_2O_2 was added at each current. The quantity of sulfuric acid that was required for maintaining pH 3.0 ± 0.1 was in the range 0.6-1.3 mL at different applied currents. The optimum dose of Fe^{2+} was the dose at which maximum COD removal was found. In the electro-Fenton process this may also be considered as the minimum current at which maximum COD removal occurred. The decrease in COD, per cent COD removals and hydrogen peroxide utilization ratio (E_H) at different doses of Fe^{2+} for the electro-Fenton treatment of H-acid solutions are given in Table 59.3. More than 90 per cent COD was removed at all applied doses of Fe^{2+}. However, a maximum COD reduction of 96.8 per cent was found at 0.196g Fe^{2+} dose (7.02 mM Fe^{2+}; current, 0.75A). At the minimum current, 0.25A, 92.5 per cent COD was removed. Thus, a Fe^{2+} dose of 0.196g was considered optimum. Nevertheless, 0.25A at which 0.0653 g Fe^{2+} (2.34 mM Fe^{2+}) was generated may also be considered as optimum from the practical point of view. Because the energy requirement by EF method at minimum current would be lower, this dose was also considered for estimating the energy requirement for EF method. The average E_H was 46 per cent. The residual iron in treated solutions at the optimum Fe^{2+} dose was found 8.5 ± 1 mg/L.

Optimization of H_2O_2 dose was performed by varying H_2O_2 dose from 1 to 8 mL (17.6-141 mM) at a fixed dose of Fe^{2+}; *i.e.*, 0.196 g (7.02 mM). The decrease in COD, per cent COD removals and hydrogen peroxide utilization ratio (E_H) at different doses of H_2O_2 for the electro-Fenton treatment of H-acid solutions are given in Table 59.4. A maximum of 92.8 per cent COD removal occurred at a dose of 4 mL of H_2O_2 (70.6 mM). The E_H varied from 97.8 per cent at 17.6 mM H_2O_2 dose to 18.6 per cent at the highest peroxide dose (141 mM). The higher the inlet peroxide dose, the lower the E_H was. At the optimum H_2O_2 dose, the peroxide utilization ratio was 45.7 per cent.

Table 59.3: Electro-Fenton Method–Optimization of Dose of Fe^{2+}

Sl.No. (1)	*Current, A* (2)	*Fe (g)* (3)	*Fe (mM)* (4)	*COD*, mg/L* (5)	*Per cent R#* (6)	E_H (7)
1.	0.00	0	0	278 ± 12	0	0
2.	0.25	0.0653	2.340	20.6 ± 3	92.6	45.6
3.	0.50	0.1306	4.680	14.5 ± 3	94.8	46.7
4.	0.75	0.1960	7.020	8.9 ± 2	96.8	47.7
5.	1.00	0.2612	9.360	13.5 ± 3	95.1	46.9
6.	1.50	0.3920	14.04	21 ± 3	92.4	45.6
7.	1.75	0.5880	18.72	23.5 ± 3	91.5	45.1

Concentration of H-acid	:	250 mg/L
Treatment time	:	15 min.
Solution volume	:	500 mL
Volume of H_2O_2	:	4 mL (70.6 mM)

*: Average of two sets of measurements.

#: Average percentage calculated using lowest and highest percentage values.

Addition of H_2O_2: Batch versus Continuous

In this case, the optimized dose of H_2O_2 (4 mL, 70.6 mM) was added either in a single lot or continuously drop wise throughout the reaction period chosen for electro-Fenton method and the final COD in each case was determined. The COD was reduced to 10 and 33 mg/L from the initial 278 mg/L, when 4 mL H_2O_2 was added in a single lot and continuously, respectively. This accounts for 96.4 per cent COD reduction for single lot addition and 88 per cent COD reduction for continuous addition of peroxide. In the single lot addition of peroxide, higher H_2O_2/Fe^{2+} ratio is expected to prevail when compared to continuous addition. This may have resulted in higher per cent COD removal in the former case. On the contrary, under continuous addition of H_2O_2, low H_2O_2/Fe^{2+} ratio would prevail and it may not vary significantly as concentrations of both H_2O_2 and Fe^{2+} increase gradually with time of EF treatment. The low H_2O_2/Fe^{2+} ratio may favour formation of some intermediate compounds. This may have resulted in lower per cent COD reduction in this case (Table 59.4).

Table 59.4: Electro-Fenton Method–Optimization of Dose of H_2O_2

Sl.No. (1)	$[H_2O_2]$, mL (2)	$[H_2O_2]$, mM (3)	COD*, mg/L (4)	Per cent R# (5)	E_H (6)
1	0	0	278 ± 12	0	0
2	1	17.60	140 ± 9	49.6	97.8
3	2	35.30	28 ± 4	89.9	88.6
4	4	70.60	20 ± 3	92.8	45.7
5	5	88.20	52 ± 5	81.3	32.0
6	6	105.80	60 ± 5	78.4	25.8
7	8	141.00	68 ± 5	75.5	18.6

Concentration of H-acid	:	250 mg/L
Treatment time	:	15 min.
Solution volume	:	500 mL
Current	:	0.75 A
Dose of Iron	:	0.196 g (7.02mM)

Relative Efficiency for the Treatment of H-Acid Solutions: Electro-Fenton versus Conventional Fenton Oxidation Methods

The comparison of the results of both electro-Fenton and Fenton oxidation methods for the treatment of H-acid solutions is presented in Table 59.5. For obtaining the efficiency (h) of COD removals by electro-Fenton method ($\eta_{Fe\ (EF)}$) and that by conventional Fenton method ($\eta_{Fe\ (F)}$), the COD removals were normalized with respect to applied Fe^{2+} dose *i.e.*, grams of COD removed per gram Fe^{2+} dose was estimated in each case. The $\eta_{Fe\ (EF)}$ was in general higher than $h_{Fe\ (F)}$, especially so at the lower Fe^{2+} doses. The $\eta_{Fe\ (EF)}$ was 3.94 gCOD/gFe^{2+}, while $\eta_{Fe\ (F)}$ was 1.82 gCOD/gFe^{2+} at the lowest applied doses. However, the $\eta_{Fe\ (EF)}$ was 1.37 gCOD/gFe^{2+} at the optimized dose of Fe^{2+} (0.196 g), where as $\eta_{Fe\ (F)}$ was 1.26 gCOD/gFe^{2-} at the optimized dose of Fe^{2+} (0.150 g). The efficiency of COD removal in each case decreased with increasing Fe^{2+} dose; it being comparable with each other at higher doses of Fe^{2+}.

The comparison of the efficiencies of Fenton reagent and EF method was made by deriving pseudo-first order rate constants in each case using the COD removal profiles. Accordingly, the plots of–log[$(COD)_t/(COD)_0$] versus reaction time (using the data between 0 and 30 minutes only was considered) were prepared and pseudo-first order rate constants (k) constants for Fenton reagent ($k = 0.054 \pm 0.002$ min^{-1}) and EF method ($k = 0.38 \pm 0.002$ min^{-1}) were deduced. The COD removal by the use of EF method was 7 times faster.

Some other investigators (Huang *et al.*, 1999, 2001) have also found the EF method more efficient. When applied to the treatment of bioeffluent of petrochemical manufacturing wastewater, the EF method showed 80 per cent COD removal efficiency while Fenton oxidation resulted in less than 50 per cent COD reduction. These authors have also reported somewhat modified EF method where Fe^{2+} was electro-generated via the electrochemical reduction of ferric sulfate or ferric hydroxide sludge. When applied to treatment of PVC stabilizer processing wastewater having high COD (11000 mg/L) and high heavy metal concentration (Pb=7500 mg/L), this method showed 89 per cent COD removal at 4.0 gFe^{2+}/L and 28.0 gH_2O_2/L. In contrast, only 36 per cent COD was removed using Fenton reagent (at 4.0 gFe^{2+}/L and 28.0 gH_2O_2/L). Similarly, hydroxylation of chlorobenzene and phenol has been reported using in-situ electro-generated Fenton reagent (Hsiao and Nobe 1993a,b).

Table 59.5: Comparison of COD Removal Efficiency as a Function of Iron Dose Electro-Fenton and Fenton Methods

Sl.No.	*Electro-Fenton Method*			*Fenton Oxidation*		
	Fe^{2+} (g)	*COD (g/L)*	*$gCOD/gFe^{2+}$*	*Fe^{2+} (g)*	*COD (g/L)*	*$gCOD/gFe^{2+}$*
(1)	*(2)*	*(3)*	*(4)*	*(5)*	*(6)*	*(7)*
1.	0	0.2780	–	0.00	0.278	–
2.	0.0653	0.0206	3.9400	0.10	0.096	1.82
3.	0.1306	0.0145	2.0180	0.15	0.088	1.26
4.	0.1960	0.0089	1.3730	0.20	0.092	0.93
5.	0.2612	0.0135	1.0130	0.25	0.092	0.74
6.	0.3920	0.0210	0.6556	0.30	0.096	0.61
7.	0.5880	0.0235	0.4328	0.40	0.112	0.41

Energy Requirement of Electro-Fenton Method

The energy requirement of electro-Fenton method was estimated using the following Equation (3).

$$\text{Energy Requirement } (E_R) = [\text{CODr}]/\text{Energy consumed} \quad (3)$$

The electrical energy consumed was calculated using the Equation (4)

$$\text{kWh} = V * I * t/1000 * \text{Volume} \quad (4)$$

Where, V = voltage (5 V); I = 0.75 A; t = treatment time (0.25 h) and volume of sample = 0.5 L. The electrical energy consumed was 1.875 ´ 10^{-3} $kWhL^{-1}$ or 18.75 kWh m^{-3}. As the COD removed was 0.27 g per 0.5 L (or 5.4 $kgCOD/m^3$), the energy requirement can be estimated from Equation (3) as 0.288 kgCOD/kWh. On the other hand, at the minimum applied current (0.25 A), approximately 92.5 per cent COD was removed. This corresponds to an energy requirement of 0.82 kgCOD/kWh. The energy requirement by EF method at minimum current is lower.

Comparison of Electro-Fenton and Conventional Fenton Oxidation Methods under Optimized Conditions – H-acid versus Reactive Black 5

The removal of Reactive Black 5 from aqueous solution was examined under optimized conditions for electro-Fenton (2.34 mM Fe^{2+}, 70.6 mM H_2O_2) and conventional Fenton oxidation (5.37 mM Fe^{2+}, 53 mM H_2O_2) methods. The details of reaction conditions and efficiencies for the removal of H-acid and Reactive Black 5 are presented in Table 59.6. Approximately, 92.5 and 82.2 per cent COD removals were obtained for H-acid and Reactive Black 5, respectively using electro-Fenton method. The rate constant (k) for HA and RB5 degradation was 0.38 and 0.31 min^{-1}. respectively. Similarly, 71.2 and 67.2 per cent COD removals were obtained using conventional Fenton method for H-acid and Reactive Black 5, respectively. The per cent COD removals with EF as well as Fenton reagent are lower for RB5 compared to H-acid. This may be attributed to complex molecular structure of RB5.

Table 49.6: Comparison of Electro-Fenton and Conventional Fenton Oxidation of H-acid and Reactive Black 5 Under Optimized Conditions

Compound (Conc.) (1)	*Parameters* (2)	*Fenton Oxidation* (4)	*Electro-Fenton* (3)
H-acid (250 mg/L)	Dose of Fe^{2+} (mM)	5.37	2.34
	Dose of H_2O_2 (mM)	53.0	70.6
	E_H (per cent)	33.7	47.7
	Treatment time, (min.)	120	15
	$COD_{initial}$ (mg/L)	278	278
	COD_{final} (mg/L)	72	21
	per cent COD removal	71.2	92.5
	$Fe_{residual,}$ (mg/L)	9.0	8.5
Reactive Black 5 (250 mg/L)	Dose of Fe^{2+} (mM)	5.37	2.34
	Dose of H_2O_2 (mM)	53.0	70.6
	E_H (per cent)	39.72	38.65
	Treatment time, (min.)	120	15
	$(COD)_{initial}$, mg/L	292	292
	$(COD)_{final}$ (mg/L)	82.0	52.0
	per cent COD removal	67.2	82.2
	$Fe_{residual,}$ (mg/L)	8.9	8.0

Higher Efficiency of Electro-Fenton Method: Progressively Changing H_2O_2/Fe^{2+} (Mole Ratio)

The reactions that occur at anode and cathode (5-8), Fenton reaction in solution (9-11) and other reactions (12-15) that are mainly responsible for catalyst dissipation/ OH radical consumption in side reactions may be written as

At anode:

$$Fe \rightarrow Fe^{2+} + 2e^- \quad (5)$$

$$H_2O \rightarrow \tfrac{1}{2}O_2 + 2H^+ + 2e^- \quad (6)$$

At cathode:

$$H_2O + e- \rightarrow \tfrac{1}{2}H_2 + OH^- \quad (7)$$

$$Fe^{3+} + e^- \rightarrow Fe^{2+} \quad (8)$$

Fenton Process (In solution):

$Fe^{2+} + H_2O_2 \rightarrow Fe^{3+} + {}^{\bullet}OH + {}^{-}OH$ (OH formation) (9)

$RH + {}^{\bullet}OH \rightarrow H_2O + R^{\bullet}$ (10)

$R^{\bullet} + Fe^{3+} \rightarrow R^{+} + Fe^{2+}$ (11)

(RH = a generic organic compound)

Other reactions:

$Fe^{2+} + {}^{\bullet}OH \rightarrow Fe^{3+} + {}^{-}OH$ (k = 3.0 ´ 10^8 $M^{-1}s^{-1}$) (12)

${}^{\bullet}OH + H_2O_2 \rightarrow H_2O + HO_2^{\bullet}$ (k = 3.3 ´ 10^7 $M^{-1}s^{-1}$) (13)

$HO_2^{\bullet} + Fe^{3+} \rightarrow Fe^{2+} + O_2 + H^{+}$ (14)

$Fe^{3+} + OH^{-} \rightarrow Fe(OH)_3\downarrow$ (15)

At a dissolving anode such as iron, reaction (5) may be assumed to be predominant. On the other hand, the major complimentary reaction at cathode was hydrogen gas generation (7). Here, the Fe^{3+} formed during OH radical formation in reaction (9) would be reduced to Fe^{2+}. This may be regarded as catalyst regeneration step. The OH radicals formed in reaction (9) oxidize a generic compound (RH) via formation of several intermediates that degrade gradually and eventually into carbon dioxide. The reaction steps (12)-(15) mostly represent Fe^{2+} dissipation and oxidant consumption in undesirable side reactions.

During electro-Fenton process the dose of Fe^{2+} increases progressively with the time of treatment. This implies that higher H_2O_2/Fe^{2+} molar ratio prevails initially and this decreases gradually with treatment time. However, in the experiments using electro-Fenton reactor, it may be predicted that the concentration of peroxide decreases, while that of Fe^{2+} increases, as the reaction progresses. Hence, the actual trend of change in H_2O_2/Fe^{2+} ratio as a function of reaction time would be unlike that represented by inverse function of Fe^{2+}. The H_2O_2/Fe^{2+} ratio would indeed decline steadily. Nevertheless, higher H_2O_2/Fe^{2+} ratio may be expected to prevail initially in this case too. The higher efficiency of electro-Fenton method at lower doses of Fe^{2+} (Table 59.6) may be attributed to this incremental addition of Fe^{2+} and accompanying higher H_2O_2/Fe^{2+} mole ratio. This would avoid competition of Fe^{2+} for OH radicals (reaction (11)) which otherwise would become available for substrate. Further, the excess H_2O_2 would also remove OH radicals in a reaction as shown in (13). However, based on the reported rate constant data (Merli 2002) for reaction (12) and (13), the former can be expected to be 9 times faster. The electro-Fenton process is highly complex and its effectiveness greatly depends on the specific reaction conditions; for example, the application of optimized doses Fe^{2+} and H_2O_2 and continuous adjustment of pH at 3.0.

In contrast, the Fe^{2+} catalyst is dosed in single lot in conventional Fenton method leading to side reaction (12) and consequently the efficiency was lower.

Reaction of Fenton reagent with H-acid

The reaction of H-acid with Fenton reagent was monitored by collecting UV-vis spectra for the solutions at $[H_2O_2]/Fe^{2+}$ mole ratios ranging from 1-10. In each case, the test samples were withdrawn at different time intervals up to 2 h and UV-vis spectra were recorded. The changes in spectral features were examined at fixed concentration of Fe^{2+} and H-acid but varying the dose of H_2O_2. H-acid displayed multiple absorption at 330-400 nm. Its intensity decreased and a new broad absorption feature centered at 474 nm (400-600 nm) appeared at $[H_2O_2]/Fe^{2+}$ = 1,2 and 5. The solutions were red-brown colored; and this color faded only very slowly at $[H_2O_2]/Fe^{2+}$ = 1,2 and 5. It was seen that the multiple absorption at 330-400 nm due to H-acid was completely removed; as well as, the new broad absorption feature centered at 474 nm also disappeared at $[H_2O_2]/Fe^{2+}$ = 10. The red-brown color completely disappeared at this high $[H_2O_2]/Fe^{2+}$ mole ratio. The above spectral changes may be ascribed to oxidation of $-NH_2$ chromophore in H-acid molecule into $-NO_2$. This transformation is responsible for the visible absorption centered at 474 nm. The spectral changes similar to the above were also reported previously (Wanpeng *et al.*, 1996). These authors reported that amino group in H-acid is first oxidized to nitro group as indicated by decline in intensity of peak at 348 nm and increase in absorption in the range 400 to 700 nm. However, with the increasing oxidant concentration, hydroxyl groups substitute the nitro group. These transformations accompany deepening of colour. In the present study also, the persistence of red-brown color at $[H_2O_2]/Fe^{2+}$=1,2 and 5 suggests the presence of nitro- intermediate. At higher $[H_2O_2]/Fe^{2+}$ ratio, *i.e.*, 10, this intermediate was quickly oxidized through a series of OH radical substitutions/ring-opening events, eventually leading to complete oxidation of H-acid. The mechanism of Fenton oxidation was discussed elsewhere (Wanpeng *et al.*, 1996; Merli, 2002; Solozhenko *et al.*, 1995; Vogelpohl, 2001). Thus, the higher $[H_2O_2]/Fe^{2+}$ ratio favors exhaustive oxidation of H-acid. It is inferred the degradation of the target compounds can be affected by a variety of competing reactions and poor Fe: H_2O_2 stoichiometry. Concerning Fenton oxidation of H-acid wastewater having 60000 mg/L initial COD, Wanpeng *et al.* (1996) reported 80 per cent COD removal efficiency at very high H_2O_2/Fe^{2+} ratio (3.6 mmol Fe^{2+}/L and 1.76 mol H_2O_2/L).

The oxidation of $-NH_2$ into $-NO_2$ was further verified by recording the CV of H-acid without and with Fenton reagent addition. The irreversible oxidation peak labeled as '(a)' found at +1.08 V was attributed to oxidation of $-NH_2$ group into $-NO_2$ group in H-acid. The intensity of this peak slightly increased upon addition of 1-mL H_2O_2 (30 per cent w/v) and after allowing for more than 1 h contact time indicating increase in the concentration of nitro-intermediate. This solution turned dark brown when 2 mL of $FeSO_4$ (50 mg/mL) was added. But the feature at + 1.0 V remained unaffected. This peak disappeared only upon adding excess H_2O_2, say 4 mL (see CV (b)). This indicated that the strongly colored nitro intermediate undergoes further degradation when Fe^{2+} and H_2O_2 were present in appropriate ratio. In the corresponding CV, the Fe^{3+}/Fe^{2+} redox couple was also detected at –0.61 V. The CV data supplements the interpretation provided on the basis of UV-vis data.

The conventional formulation for Fenton reagent (reaction (9)) may be rewritten as in Step 16, considering that $[Fe^{II}(H_2O)_6]^{2+}$ is the most likely species in aqueous acidic solutions.

$$[Fe^{II}(H_2O)_6]^{2+} + H_2O_2 \rightarrow [Fe^{III}(H_2O)_5(OH)]^{2+} + {}^{\bullet}OH + H_2O \qquad (16)$$

When peroxide was added to Fe^{2+} solutions, particularly at higher Fe^{2+} concentration, light yellowish color developed. These solutions displayed an absorption peak at dikka 300 nm (e = 2200 $M^{-1}cm^{-1}$. The solutions of H-acid plus Fe^{2+} or H-acid plus H_2O_2 did not possess color. This absorption peak may be attributed to iron-peroxo complex (Equation 17).

$$[Fe^{II}(H_2O)_6]^{2+} + H_2O_2 \rightarrow [Fe^{III}(H_2O)_5(O_2^{2-})]^+ + H_2O + 2H^+ \quad (17)$$

Several investigations dedicated previously to characterization of peroxo complexes reported that $O_2^{2-} \rightarrow M^{n+}$ ligand-to-metal charge transfer transitions (LMCT) in 300-360 nm region (Biswas *et al.*, 1994, Kieninger 1999). While the involvement of hydroxyl radicals is generally accepted (Brillas *et al.*, 1998a,b; Walling 1975; Zepp *et al.*, 1992); mechanisms implicating, peroxo (O_2^{2-}), hydroperoxo ($^{\bullet}OH_2$) and ferryl ions (Fe^{IV}=O) (Watts *et al.*, 1999; Wink 1994; Ensing 2003) are also reported. In the present case also it is believed that not only hydroxyl radicals are the primary oxidants, but peroxo (O_2^{2-}), hydroperoxo ($^{\bullet}OH_2$) species may have also participated in the mechanism. The formation of latter species is favored at high H_2O_2 concentration, as hydroxyl radicals react with hydrogen peroxide resulting in a series of propagation reactions *(e.g.*, reaction (13)) that yield peroxo (O_2^{2-}) and hydroperoxo ($^{\bullet}OH_2$) species. In the light of the above, further investigations to understand the complete mechanism of Fenton's oxidation are important.

Conclusion

Both Fenton reagent and electro-Fenton methods are effective towards oxidation of H-acid and Reactive Black 5. The optimized doses are: Fe^{2+} = 0.300 g/L (5.37 mM); H_2O_2 = 6 mL/L (53 mM) and $[H_2O_2]/Fe^{2+}$ = 10 for conventional Fenton reagent and Fe^{2+} = 0.130 g/L (2.34 mM); H_2O_2 = 8 mL/L (70.6 mM) and $[H_2O_2]/Fe^{2+}$ = 30 for electro-Fenton method. The pseudo-first order rate constants (k) for Fenton reagent and EF method are 0.054 and 0.38 min^{-1}, respectively. The removal of COD by the EF method is faster. The energy requirement using EF method is 0.82 kgCOD/kWh at the minimum applied current (0.25 A) when approximately 92.5 per cent COD was removed. The higher efficiency of electro-Fenton method is attributed to incremental addition of Fe^{2+} and accompanying higher H_2O_2/Fe^{2+} ratio as this would avoid competition of Fe^{2+} for OH radicals which otherwise would become available for the substrates. The advantage with electro-Fenton method is, high treatment efficiency can be achieved within a shorter treatment period i.e, 15 minutes, in contrast to 2 h needed for conventional Fenton oxidation method. This implies small unit sizes and low capital costs for the electro-Fenton method.

References

1. Alvarez-Gallegos, A., and Pletcher, D. (1998a), " The removal of low level organics via hydrogen peroxide formed in a reticulated vitreous carbon cathode cell. Part 1: the electrosynthesis of hydrogen peroxide in aqueous acidic solutions." *Electrochimica Acta*, 44, 853-861.

2. Alvarez-Gallegos, A., and Pletcher, D. (1998b), " The removal of low level organics via hydrogen peroxide formed in a reticulated vitreous carbon cathode cell. Part

2: the removal of phenols and related compounds from aqueous solutions, *Electrochimica Acta*, 44, 2483- 2492.

3. APHA-AWWA-WEF, (1998). "*Standard Methods for Examination of Water and Wastewaters*". 20th Edition, Washington D.C, American Public Health association, Method No. 5220B

4. Aplin, R. Feitz, A. J. and Waite, T. D. (2001), " Effect of Fe (III)–ligand properties on effectiveness of modified photo-Fenton processes." *Water. Sci. Technol.* 44 (5), 23-30.

5. Biswas, S, Bhattacharjee, S., and Bhattacharyya, R. (1994). "Synthesis and reactivity of peroxomolybdenum (VI) complexes of the types [MoO_2 (O_2) $(QO)_2$] (QO = 8-quinolate anion) and [MoO_2 (O_2) $(L\text{-}L)_2$] (L-L = neutral N-N donors) and crystal and molecular structure of (PPN{Mo_2O_7]". *Ind. J. Chem.* 33A, 10-19.

6. Brillas, E., Mur, E., Sauleda, R.,Sanchez,L., Peral, J., Domenech, X., and Casado, J. (1998a). "Aniline Mineralization by AOP's: anodic oxidation, photocatalysis, electrofenton and photoelectrofenton processes." *Applied Catalysis. B.* 16, 31-42.

7. Brillas, E., Sauleda, R., and Casado, J. (1998b). "Degradation of 4-chlorophenol by aniodic oxidation, electro-fenton, photoelectro-Fenton, and peroxi-coagulation processes." *J. Electrochem. Soc.* 145, 759-765.

8. Casero, I., Sicilia, D., Rubio, S., and Perez-Bendito, D. (1997). "Chemical Degradation of Aromatic Amines by Fenton,'s Reagent." *Wat. Res.* 31 (8), 1985-1995.

9. Chen, R., and Pignatello, J. J. (1997). "Role of quinone intermediates as electron shuttles in Fenton and photoassisted Fenton oxidations of aromatic compounds". *Environ Sci. Technol.* 31, 2399-2406.

10. Chiarenzelli, J., Scrudato, R., Fabrizio, M., and Wunderlich, M. (1997) "Electrochemical peroxidation of PCBS and VOCs in subsurface storage tank water and slurry." *Environmental Applications of Advanced Oxidation Technologies, Electric Power Research Institute*, September, 1997,4- (77-89).

11. Chou, S., Huang Yao-Hui, Lee Shen-Nan, Huang Gaw-Hao, and Huang Chihpin, (1999) "Treatment of high strength hexamine containing wastewater by electro-fenton method". *Wat. Res.* 33 (3), 751 – 759.

12. Dongre, R. S., Rao, N. N., and Kaul, S. N. (2002). "Fenton's oxidation systems for chemical treatability of H-acid Wastewater", *Proc. Natl. Conf. on Pollution Prevention and Control in India*, 2-3 March, 2002, held at VRCE, Nagpur, India, 341-344.

13. Gregor, K. H. (1994). "Oxidative decolorization of textile wastewater with advanced oxidation processes". *In Chemical Oxidation Technologies for the Nineties*, Vol. 2, Eckenfelder, Jr W. W., Bowers A. R., and Roth J. A. (eds), Technical Publishing Co. Inc.,.161-193.

14. Hsiao, Y.L., and Nobe, K. (1993a). " Hydroxylation of chlorobenzene and phenol in a packed bed flow reactor with electrogenerated Fenton reagent." *J. Appl. Electrochem.*, 23, 943 – 945.

15. Hsiao, Y.L., and Nobe, K. (1993b). " Oxidative reactions of chlorobenzene and phenol with *in situ* electrogenerated Fenton reagent." *Chem. Eng. Commun.*, 126, 97- 110.

16. Ensing, B. (2003). http: //www.cmm.upenn.edu/~ensing/thesis/node42.html

17. Hach DR/2010 Spectrophotometer Handbook, Method 8147, P.315, Hach Company, U.S.A., 1998

18. Huang, H. H., Lu, M.C., and Chen, J.N. (2001). "Catalytic decomposition of hydrogen peroxide and 2-Chlorophenol with iron oxides". *Wat. Res*, 35 (9), 2291-2299.

19. Huang, Yao-Hui, Cheng, C.C., Huang, G. H., and Chou, S. (2001). "Comparison of a novel Electro-Fenton method with Fenton reagent in treating highly contaminated wastewater." *Wat. Sci. Technol.* 43 (2), 17-24.

20. Huang, Yao-Hui, Chou, S., Perng Ming-Ging, Huang Gaw-Hao, and Cheng Sheng-Shung. (1999). "Case Study on the bioeffluent of petrochemical wastewater by electro-Fenton method." *Wat. Sci. Technol.* 39 (10-11), 145-149.

21. Kang, Y. W., and Hwang, K. Y. (2000). "Effects of reaction conditions on the oxidation efficiency in the Fenton process". *Wat. Res.* 34 (10), 2786-2790.

22. Kaul, S. N., Nandy, T, Deshpande, C. V., Srivastava, A., Shastry, S and Szpyrkowicz, L. (1998). "Application of full scale evaporation-incineration technology for hazardous wastewater". *Wat. Sci. Technol.* 38, 363-372.

23. Kertesz M. A., Cook A. M., Leisinger T (1994). Microbial metabolism of sulfur- and phosphorus-containing xenobiotics. FEMS Microbiological Reviews 15, 195-215.

24. Kieninger, M (1999). http: //bilbo.edu.uy/dequifim/proyectos/kieninger/m_kieninger_dfgproposal.html.

25. Kuo, W. G. (1992). "Decolourizing dye wastewater with Fenton reagent". *Wat. Res.* 26 (7), 881-886.

26. Merli, C., Petrucci, E., Da Pozzo, A, and Pernetti, M. (2002). "Fenton-Type Treatment: State of the Art". *Proc. of Indo-Italian Workshop on Emerging Technologies for Industrial Wastewater and Environment*, Sept. 2-4, 2002, held at National Environmental Engineering Research Institute, Nagpur, India, 21-32.

27. Qiang, Z., Chang Jih-Hsing, and Huang Chin-Pao (2002). "Electrochemical generation of hydrogen peroxide from dissolved oxygen in acidic solutions."*Wat. Res.* 36, 85 – 94.

28. Schaefer, G.D., Jaynes, M.O. and Farrell, J.A. (2000). "Remedial action using Fenton's chemistry under state dry cleaning program" *ARG, May 2000. OXY-CAT™ Pilot Test Summary Report*, Swift cleaners, Jacksonville, Florida

29. Schwitzguebel Jean-Paul, Aubert S., Grosse W., and Laturnus F. (2001). Sulphonated aromatic pollutants-Limits of microbial degradability and potential of phytoremediation. *Environ Sci. and Pollut. Res.* 1-11 (Online First).

30. Sims, A. F. E. (1983). "Industrial Effluent Treatment with Hydrogen Peroxide". *Chemistry and Industry*, 18 (7), 555-558.

31. Solozhenko, E. G., Soboleva, N. M., and Goncharuk, V. V. (1995). "Decolorization of azodye solutions by Fenton's oxidation", *Wat. Res.*, 29 (9), 2206-2210.

32. Szpyrkowicz, L., Juzzolino, and Kaul, S. N. (2001). "A Comparative study on Oxidation of Disperse Dyes by Electrochemical Process, Ozone, Hypochlorite and Fenton Reagent" *Wat. Res.* 35 (9), 2129-2136.

33. Teel, A. L., Warberg, C. R., Atkinson, D. A., and Watts, R. J. (2001). "Comparison of mineral and soluble iron catalysts for the treatment of trichloroethylene", *Wat. Res.*, 35 (4), 977-984.

34. USEPA, (1998). "Field application of in-situ remediation technologies: Chemical Oxidation." *EPA* 542-R-98-008, 31.

35. Vogelpohl, A., Geissen, S. O., Kregert, B., and Sievers, M. (2001). "Oxidation Technologies for Water and Wastewater Treatment II". (Selected Proceedings of the 2nd International Conference on Oxidation Technologies for Water and Wastewater Treatment, held in Clausthal-Zellerfeld, Germany, 28-31 May 2000) *Water Science and Technology*, 44 (5), 1-345 (http: //www.iwaponline.com/wst/ 04455/05/default.htm.

36. Walling, C. (1975). "Fenton reagent revisited" *Acc. Chem. Res.* 8, 125-131

37. Wanpeng, Z, Zhihua, Y., and Li, W. (1996). "Application of Ferrous-Hydrogen Peroxide for the Treatment of H-acid Manufacturing Process Wastewater". *Wat. Res.* 30 (12), 2949-2954.

38. Watts, R.J. and Stanton, P.C. (1994). "Process conditions for the total oxidation of hydrocarbons in the catalyzed hydrogen peroxide treatment of contaminated soils" *WARD 337.1*, Washington State Department of Transportation, Olympia, Washington.

39. Watts, R. J., Bottenberg, B. C., Hess, T. F., Jensen, M. D., and Teel, A. L. (1999). "Role of reductants in the enhanced desorption and transportation of chloroaliphatic compounds by modified Fenton's reactions". *Environ. Sci. Technol.* 33, 3432-3437.

40. Weiss, J. (1935). "The Catalytic Decomposition of Hydrogen Peroxide on different metals." *Trans. Farad. Soc.* 3L, 1547-1557.

41. Willets J, Ashbolt N. (2000). Understanding anaerobic decolorization of textile dye wastewater: mechanism and kinetics. *Water. Sci. and Technol.* 42 (102), 409-415.

42. Wink, D. A., Nims, R. W., Saavedra, J. E., Untermahlen, W. E. Jr., and Forde P. C. (1994). "The Fenton oxidation mechanism: Reactivities of biologically relevant substrates with two oxidizing intermediates differ from those predicted for the hydroxyl radical". *Proc. Natl. Acad. Sci. USA.* 91 (14), 6604-6608.

43. Zepp, R. G., Faust, B. C., and Hoigne, J. (1992). "Hydroxyl Radical Formation in Aqueous Reactions (pH 3-8) of Iron (II) with Hydrogen Peroxide: the Photo-Fenton Reaction." *Environ. Sci. Technol.* 26, 313-319.

Chapter 60

Need to Protect Steep Slopes as Ecological Sensitive Areas in India

T.K. Ghosh, S.B. Merchant and S.N. Kaul

CSIR-NEERI, Nagpur

ABSTRACT

Slopes failure and erosion of soil due to extensive developmental activities in hilly areas are major environmental hazards during last few decades in mountains. Fragility of a mountainous ecosystem is dependent on interaction of geology, drainage, soil and degree of slope with natural processes, and man's developmental and economic activities. Amongst seven categories of slopes; starting from 'flat to almost flat' (0-3°) to 'extremely steep slope' (above 40°), steep slopes (17-31°) which are prone to failures, are considered ecological sensitive and need protection. Certain illustrative examples, towards landuse and sensitivity of slopes in different parts of India have been cited. Suggestive measures with reference to buffer zone, sustainable land use, plantation, restricted anthropogenic activities etc., to protect steep slopes in ecologically sensitive areas, have been delineated.

Keywords: *Steep slope, Ecological sensitivity, Buffer zone.*

Introduction

A major concern in India is the environmental degradation, and threat to nature due to various anthropogenic activities. Most of the present activities, threatening the natural habitats and ecosystems, can be brought under five categories, *viz.* harvesting of renewable resources, pollutants, changes in land use pattern, introduction of exotic species and extreme natural events occurring with great irregularities. During the last five decades, extensive developmental activities have taken place in hilly areas, and as a result, slope failure and erosion of soil due to such activities are major environmental hazards. The Himalayas, which undergo fast and multiple developmental activities, focus everyone's attention towards the repeated and growing landslide hazards.

Slopes, aspects and altitude are important terrain parameters from land utilization point of view and also in assessing disaster prone zones. Among these three parameters of the terrain, slope is very important for assessing land capability, erodibility, stability and irrigability. Mass movement can occur anywhere steep slopes or slide prone areas and at locations encountered with weak materials.

The presence of water in the slope also plays critical role, both in increasing driving forces and in reducing resistance to sliding. Slope failures affect and disrupt transportation routes, mining, urban development and timber harvesting operation. Individual slope failure may not be like certain other natural hazards such as earthquakes, major floods and tornadoes. Slope failures are collectively more widespread and total financial loss from slope movements exceeds that of most other geological hazards. In view of this, steep slopes, which are prone to failures, are considered ecological sensitive areas and need special attention.

Concept of Slope

The "Slope" of a land area could generally be defined as its upward or downward inclination to horizontal plane. Therefore, the slope is usually measured as an angle in relation to the horizontal plane. In addition, the slope could also be measured as the proportion or percentage inclination to the horizontal plane. Thus, slope of a land area could be denoted as either "x" degree slope, or as l: y slope or as "z percent" slope. The slopes of the land varies from place to place and thus, in the process of mapping slopes of a large land mass, classification of slopes into suitable ranges or categories become a practical necessity.

The slope categories that can be derived from Survey of India toposheets on the guidelines of All India Soil and Land Use Survey (AIS and LUS) are mentioned below:

Sl.No.	*Slope Category*	*Slope (per cent)*
1.	Nearly level	0-1
2.	Very gently sloping	1-3
3.	Gently sloping	3-5
4.	Moderate sloping	5-10
5.	Strong sloping	10-15
6.	Moderate steep to steep sloping	15-35
7.	Very steep sloping	> 35

However, slope ranges, classified by United States Geological Surveys (USGS) are flat to almost flat (0-3° or 0-5 per cent), gently sloping (3-6° or >5-10 per cent), sloping (6-11° or >10-20 per cent), moderately steep (11-17° or >20-30 per cent), steep slope (17-31° or >30-60 per cent), very steep (31-40° or >60-84 per cent) and extremely steep (above 40° or above 84 per cent).

Ecological Importance and Threats

Slopes are integral parts of any mountainous or hilly terrain, which has evolved through various geological processes. Deteriorating mountainous environment, in the forms of natural hazards and anthropogenic activities, has become a matter of great concern. In terms of ecological sensitivity the features of hilly terrain are outlined hereunder.

Unique Ecosystem

The ecology at different mountainous zones is complicated in view of diversity of physical conditions. Increase in altitude results in the decrease in temperature, rainfall, development of drainage and the rates of stream dissection. Major biotic communities generally appear as irregular bands, often with very narrow ecotones. In a mountain, usually four or five major biomes with many zonal sub-divisions may be present. Consequently, there is closer contact between biomes, and more interchange of biota between different biomes than occurs in non-mountainous regions. On the other hand, similar communities are more isolated in the mountains, since mountain ranges are rarely continuous. As a result of isolation and topographic differences, many species are unique to the mountain communities. Growing human activities (faulty landuse and deforestation) may increase the frequency of instability of the slopes resulting in loss of species and biodiversity.

Anthropogenic Activities in Slope Regions of Mountainous Ecosystems

One of the major components of highland degradation with respect to anthropogenic utility or activity is the steep slopes, which are also under continuous modification by geomorphic processes, thus encumbering the formation and retention of soils. It is always important to distinguish the anthropogenic impacts from the natural hazards while addressing major activities of the slope regions of the mountainous ecosystems. Anthropogenic ventures with reference to channel manipulation, watershed modification, urbanization, water pollution, alteration of vegetation etc. lead the hilly slopes ecologically sensitive. The detrimental effects in the form of flood hazards, damage in river region, poor water quality and reduced availability of groundwater are conspicuous in affected zones. Some of the major activities include:

Extensive Agricultural Activities

The ever increasing pressure of the population has driven the people to greater heights, above 2000 m, to bring under plough the steep slope of once-forested tracts. Most of the inhabited villagers and their farms are located on the fan-shaped accumulations of debris of landslides of the past, especially in the Himalayan region of the country[1]. These fans are made up of loose, fragmentary material and frequently form gentler slopes, supporting rich vegetation on rich soil carpets. The inhabitation on these not-very stable geo-morphic features at higher altitudes has led to extensive damage to the ecological sensitive regions. However, agricultural activity through terrace development (irrigated terraces) on the slopes can reduce the inherent instability of slopes, which range from gentle to extremely steep (>35 per cent). The lower the slope angle, the deeper is the soil cover with enhanced possibility of being covered by vegetation, and vice versa.

Decrease in the Vegetative Cover

The role of vegetation cover in stabilizing the slopes has been discussed widely by various authors[2–4]. In general, landslides are more in clear areas and by roadsides rather than forested lands. Increasing demands of the local population and the industrialists for the fire, fuel-wood and fodder; intensive livestock grazing, timber extraction and felling of the trees for the road constructions have visualized reckless deforestation in the Himalayan zone causing extensive loss of vegetative cover leading to massive soil erosion and sedimentation in the catchment areas of the Himalayan rivers[5]. Once the vegetal cover is stripped off, the soils loose the capacity of absorbing rainwater. Thus, any decrement in vegetative cover of the slope results in its amplified steepness and a shallower soil cover, adding to the ecologically sensitive areas.

Road Construction Activities

The most important factor causing destructive landslides is road construction. The network of roads in the Himalayan region alone is approximately 44,000 km long. Unscientific planning of the road construction often leads to destabilization of the hill slides and production of colossal amount of debris which is normally allowed to slide down the slopes into the valleys. The debris is ultimately carried away by the water to the plains, where it covers the fertile fields during floods. It is estimated that the construction of just one km long road requires removal of about 40,000 to 80,000 m^3 of debris. Operational costs and land acquisition problems often result in construction of road on the steep slopes especially at a degree of 25-30°, thus forming one of the major cause of destabilization and landslides catastrophe in such sensitive zones[6]. Most of the landslides are concentrated between 1200-2500m altitude, *i.e.* the altitudinal belt of 1200-1600m (41.77 per cent), 1600-2000m (8.23 per cent) and in 2000-2500m (30.38 per cent). The elevations comprising of weak geological structures (Shales, Limestone, Dolomite, Slates, Phyllits and crushed Quartzites) are the stages of vigorous landslides.

Overgrazing of the Hillslopes

The number of the cattle population, especially in the Himalayan regions, far exceeds the supporting capacity of forests and pastures. Extensive grazing leads to destruction or reduction of the regeneration capacity of the flora of the region. Defoliation by lopping for the young fodders is also a continuous process in the hill slopes. According to Singh and Saxena[7], the plant cover of the land under intensive grazing pressure in Himalayan zones has dwindled to merely 16.8, 24.0, 25.8 and 38.0 per cent at altitudes 763, 1080, 1650 and 1800 respectively.

Organized Industry

The industrial capacity has been permitted to grow well beyond the capacity of the resource base, often against sound technical advice. Consequently, there has been a serious depletion of the forest stock in mountain. The real intention of bringing industries to backward areas of Himalayas and Western Ghats is to bring benefits of enhanced economic opportunities and modernization to the locality.

Mining

The hilly areas of Himalayas and Western Ghats bear slate, bauxite, iron and manganese in substantial quantities and there has been a great deal of open cast mining for these ores. When the ore is extracted, it often leads to landslides and leaves behind substantial quantities of tailings, which get in waterways causing serious problems of siltation.

Tourism

The hilly areas of Himalayas and Western Ghats offer scenic resource as well as places of religious and historical interest. The tourism industry has brought some employment and economic gains to the local people, although outsiders largely control it. The lodging and transportation of tourists need careful planning to avoid adverse environmental consequences on slopes.

Drainage/Hydrology

Any zone with a water surplus and downstream movement, known as draining. Erosion and transfer of sediment occur through the action of water moving through the terrestrial phases of the hydrological cycles, and as a result the drainage from catchment of mountainous ecosystem has been considered a fundamental geomorphological unit. As drainage basins are catchment areas for the stream water, any variation in these areas affects the stream characteristics. Drainage density has a close relation to precipitation, run off and the topography of the drainage catchment. Vegetation and rainfall tend to be closely related, so that the area of high rainfall are often characterized by a dense vegetation cover, as the vegetation enhances the infiltration and reduces run off.

Slope Categorization for Landslide Hazard Zonation

Landslides are considered to be natural phenomena. However, evidences reveal that occurrence of landslides has increased due to unplanned development of sloping lands. The United States Geological Surveys (USGS) has established an approximate relationship between slope and soil slip failures on lands. This co-relation classifies the sloping lands into six categories and specifies susceptibility of failure of soils on each category of slopes.

Studies carried out in Sri Lanka[8–9] reveal that great majority of land slides occur in slope ranges 15-40°, with a mode between 25 and 30°. No landslides were observed in the slope range 0-5°. Fewer slides on slopes above 40° could be explained by general lack of soil cover in such area. Normally, only rocks can stand in above 40°, and rockfalls are the major hazards.

Indicators of Landslide Prone Areas in Steep Slopes

Several visible indicators can be applied to identify slides, fall, and flow of rocks, earth and debris in steep slopes. Formation of cracks and fractures, especially as transverse cracks and fractures, encourages seepage of rain-water down to the weathered bedrock layer. If there are iron-rich clay materials between bedrock and the top soil layer, lubrication of weathered material and blocks of cracked rock move downwards by force of gravity.

Muddy water springs are another indicator of landslides. Sometimes, water flows through bedrock and the top soil, outward from valley bottoms, intersects the roads and building foundations. The iron-rich clay materials in these muddy springs indicate that the bond between bedrock and the top soil is broken. This feature appears on steep slopes, sometimes even on 15° slopes subject to downward movements.

When the soils of steep slopes are saturated, the bond between bedrock and weathered top soil is broken. Consequently, the vegetation cover would wither due to the petrifaction of the root-system. This is also an indication of slope failure during the heavy rains.

Illustrative Examples

Fragility of a mountainous ecosystem is dependent on interaction of geology, drainage, soil and degree of slope with natural processes and man's developmental and economic activities. A few examples from Indian sub-continent are cited hereunder.

Kashmir Himalayas

Gulmarg and Pahalgam tahsils are the centres of developing urban regions and have become a service centre for growing tourism industry in the Kashmir valley. The slope and land use pattern of 2 tahsils, *viz.* Gulmarg and Pahalgam of Kashmir Himalayas are as follows:

- ☆ The region having slopes 5°-10° follows the valley floor land use types and except some land under settlements and roads, the area is used for cultivation of crops such as rice and vegetables.
- ☆ Region of 10°-20° slopes is a transition between valley floor and hills and has both land-use types. A good portion of land is under cultivation and equally sizeable amount is under forest cover. At higher altitudes, forest is the only land-use.
- ☆ Region of 20°-30° slopes is totally hilly, with forest and grazing land use types prevailing, and as the angle of slope increases the preference is given to grazing where forest cover becomes sparse. This region is an attraction to trans–humans and their herds.
- ☆ Region of 30°-40° slopes is not practiced for any specific land use due to very steep nature. However, occasional grazing is found on slopes facing the rivers. The region, otherwise, is devoid of vegetation.

☆ Region of slopes above 40° is usually covered with perpetual snow. The region is the ultimate source of water in these tahsils.

Further, on the basis of geomophologically related parameters, such as slope, soil, geology, drainage, altitude etc. to ascertain the suitability of Gulmarg and Pahalgam tahsils of Kashmir Himalayas, the following four suitability groups were designated by Singh and Pirazizy[10].

☆ Highly suitable (HS), with no significant limitations to sustained application of tourism.

☆ Moderately suitable (MoS), having limitations which are moderately severe

☆ Marginally suitable (MaS), in aggregate severe limitations for sustained application of tourism.

☆ Currently not sustainable (CNS), limitations are very severe and can not be restored with existing knowledge.

Table 60.1: Suitability for Development in Gulmarg and Pahalgam Tahsils of Kashmir Himalayas on the Basis of Geomrophic Items and Ecological Fragility (after Singh and Pirazizy[10])

Feature	*Suitability Groups*			
	HS	*MoS*	*MaS*	*CNS*
Geomorphology				
Slope (Degree)	3-11	11-17	17-22	>22
Area under each slope type (km²)	149.25	180.42	199.72	72.6
Topography	Flood Plain	Dry Farm-Land	Alpine	Rocky Slopes
Zone	Paddy belt	Forested slopes	Pastures	Glacial heights
Geological formation	Older alluvium, Karewa	Shale, limestone, clay	Quartzite, sand-stone	Metamorphosed rocks
Altitude (m)	1500-2000	2000-2500	2500-3000	> 3500
Natural drainage	Poor	Moderately good	Good	Very good
Soil erosion	Practically negligible	Moderate to heavy	Severe	Bed-rock exposed
Per cent of top soil	25	25-75	dikka50	Nil
Use type				
Settlement	Good	Not recommendable	Dangerous	Impossible
Construction	Good	Not suitable	Dangerous	Impossible
Preservation	Poor	Moderate	Good	Very good
Productions	Good	Moderate	Not feasible	Impossible

HS, MoS, MaS and CNS: Please refer text.

In these two tahsils, four types of use of tourism, *viz.* settlement (hotel, housing etc), construction (roads, high ways etc.), preservation (wild life, natural landscape etc.) and production (agriculture, horticulture etc.) were identified with regard to the fragility of the area (Table 60.1). The rapid growth of the industry with little regards to suitability and capability of the tahsils to support the activities has imposed negative environmental effects.

Mussoorie Region

Based on detailed environmental studies in Mussoorie region, Prasad[11] divided the region into following five zones with varying degrees of fragility at different slope ranges.

Severe Hazardous

This zone occupies the Nagdhar and Gorphadhar ranges with upper catchment of Paligad and Gochugad. This area possesses very steep slope, high altitude, fine drainage with dense reserved forest. The zone also enjoys high mountain peaks and ranges, *i.e.* Nagtibba (3022 m), Singhudhar, Gorphadhar and Koridhar with rugged topography and highly dissected terrain. This zone has practically no settlement with an areal coverage of 68 km^2.

Moderate Hazardous

This zone is visualized all over the region covering 213.6 km^2 area. The zone is well marked with Yamuna valley, lower Bin, upper parts of Kotinadi and Saurna rivers.

Less Hazardous

This zone possesses settlement density 2-3 houses per km^2 with an areal coverage of 159.5 km^2. This zone covers Bin and Koti valley in west, lower catchments of Arnigad, Bandal and Baladi nadi in south, Patalgad in centre, middle and lower parts of Paligad and surrounding areas of Kyari village in north.

Moderately Favourable

This zone contains 3-4 houses per km^2 and only 36.44 km^2 of area lies under this region. It is found mostly in the central part of valley regions of Arnigad, Bandal nadi, Koarkuli nadi and lower part of Paligad.

Favourable Zone

This zone is most suitable for settlement. This category includes Mussoorie – Landaur urban complex. Only 45.3 km^2 of area falls in this category. Central ridge with top of the hills, which are more or less flat and terraced with gentle slopes in some localities, are in favorable zone. Mussoorie and Landaur towns are mainly responsible for raising the settlement density.

Dehradun

Dehradun is one of the border districts of Uttaranchal lying on the S-W region of the state. The slope map of the district exhibits basically four broad slope ranges. The steep slopes (17° and above) appear in practically whole of Chakrata, the hills of Sahaspur and Raipur blocks. The eastern bulge of Chakrata (along its border with

Uttar Kashi) has a slope of more than 31°. Most of the other portions of the district possess 'not so steep slope' (6-17°), except the Siwaliks, which forms the southern border of the district Saharanpur[12].

Nainital

The much-deformed Nainital area is in an unstable condition. The acceleration of erosion and increased instability of hill slopes are attributable primarily to human activities – indiscriminate construction of heavy buildings in a very unplanned way. Auden[13] had counseled against construction of houses on slopes steeper than 26.5° of slates and shales. The sher-ka-danda slope made up of lower krol slates and marlites, and characterized by bedding plane cleavage steeper than 26.5°, is now densely dotted with multistoried extra-heavy houses with flat concrete roofs. The natural protection provided by oak trees has diminished, as trees have been felled. The crown cover of the trees being less than 20 per cent, the forest is unable to protect the soil from erosion.

The drainage system developed by Britishers to ensure safety of the slopes is choked with garbage and debris. The result is that storm water infiltrates the ground through multitudes of joints, fissures, fractures and cavities, thus reducing the shear strength and frictional resistance in rocks. The dumping of debris on slopes, in front of the foundation, modifies the slope substantially, making them steeper and unstable.

Joshimath and Almora

Seven types of slope instabilities, *viz.* surface erosion (15.41 per cent), rock fall (21.64 per cent), collapsed terraces (25.26 per cent), slide (32.25 per cent), flow (2.21 per cent), slump (2.13 per cent) and complex (1.10 per cent) have been recorded by Pande[14] in this area. As none of the seven types of instabilities have more than 35 per cent of area affected, it can be claimed that degradation is still within well approachable reclamation limits.

Structurally fragile zones are more prone to instability. In Berinag Fm the sedimentary window rocks of calc-zone have been subjected to various phases of deformation, which has resulted in the generation of NW-SE, N/NNE-S/SSW and EW trends of floods. This is a very sensitive zone and almost all types of instabilities are noticed here. The major joint sets developed in rocks (50/75 SE, 130/62 SW, 360/50W, 300/V, 205/40 NE, 300/40) have also developed slope instabilities.

The barren area is characterized by 58.83 per cent of slope instability, while agricultural land and forest areas have registered 30.4 and 10.73 per cent respectively of instability affected areas. Among the forested areas, pine forest registers maximum cases of instability (62.42 per cent) while oak and mixed forests have 9.32 and 28.26 per cent of affected areas respectively. Out of the total instability incidence, 42.88 per cent cases were related to old instability while rest of the 57.12 per cent was reported to be newly developed. The rectilinear slope forms exhibit 56.82 per cent of instability incidences followed by concave (22.06 per cent), steep (9.11 per cent) and convex (6.41 per cent) slope forms.

Shimla and Kullu

The steep slopes, in northwest of Shimla and Kullu districts in Himachal Pradesh, cover about 24.84 per cent of total geographical area. This belt at higher elevations comprises the richest meadows utilized by Trans-human's herds for grazing and forests cover. The slopes in Shimla and Kullu districts are under continuous modification by geomorphic processes such as slides, avalanches, rockfalls and creep. The steep slopes prevent the formation and retention of soils, due to inherent instability. The area is comprised of a wide range of gradient categories varying from gentle in low- lying areas, to very steep on the high inclinations[15]. The slopes are categorized in followed six groups.

- ☆ ***Flat or almost flat (0-3°):*** These slopes hardly cover one per cent of the zone's total geographical areas.
- ☆ ***Gently sloping (3-6°):*** This category covers about 5.2 per cent of the zones. The slopes correspond to the densely populated areas of foothills and some tracts of Ani Tahsil in Kullu district.
- ☆ ***Moderate slope (6-8.5°):*** The category covers about 18.70 per cent of the total geographical area. This belt also hosts major anthropogenic ventures with regards to industry. North of Shimla district and Mandi valley of Kullu district are placed in this category.
- ☆ ***Sloping (8.5 – 11.3°):*** This category is confined to the entire Kullu valley with exception of some patches near Manali, Bhuntar and Larji in Kullu district, and Shimla, Narkanda, Jubbal, Rohru, Kufri and Theog in Shimla district. The category covers about 41.1 per cent of the total geographical area. The major land uses in this belt are horticulture, terrace cultivation and tourism industry. This belt is the most vulnerable zone which experiences majority of the land slides.
- ☆ ***Moderately steep (11.3-17°):*** This category prevails in NW of Shimla and Kullu districts, covering about 25 per cent of total area. This belt at higher elevations comprises the richest meadows utilized by Transhumant's herds for grazing, and forest cover in this belt changes to Alpine vegetation mixed with cool temperate species, with domination of *Betula* and *Rhododendron* genera.
- ☆ ***Steep and very steep (Above 19.3°):*** This category covers 9.1 per cent of the total geographical area and confined to glacial and mountain heights all along the boundary of Kullu with Pangi, Lahul and Spiti.

Suggested Steps to Protect Steep Slopes

The relevant area for protection would need to take into account of certain destructive features which are commonly present including various combinations of steep slopes, seismicity, residual soil, high pore water pressure, thick and deeply weathered soil cover, undercutting of the base of the slope, and weak material outcropping below stronger material. Since the horizontal planes near the top and base of a slope are prone to landslides and receive boulders/debris of a slide respectively, suitable buffer zones should be provided. In general, a minimum horizontal distance of 500m at both the ends of a slope can be considered as buffer zones. In mountainous ecosystems, buffer zones may be extended further in landside-prone slopes. The areas of immediate interest are as follows:

- ☆ Identification of landslide prone areas, and preparation of comprehensive micro-landuse map and terrain analysis.
- ☆ Gradation of different kinds of slopes and detailed investigation of geological features.
- ☆ Avoidance of steep slopes for human settlement, agriculture, pasture etc., and identification of suitable lands for natural forests, water sheds, human habitation and other anthropogenic activities in mountainous ecosystems.
- ☆ Encouragement of plantation using indigenous species and strict compliance of forest laws including maintenance of forest cover as applicable in hilly areas.
- ☆ Strict vigil towards unauthorized quarrying of rocks and other mining activities, and also restriction of use of explosives in steep slopes.
- ☆ Providing contour drains and terracing against soil erosion and earthslips.

References

1. Messerli, B (1985). Stability and instability of Mountain Ecosystems: An interdisciplinary approach. *in* Singh, T.V. and Kaur, J. ed. *Integrated Mountain Development*. Himalayan Books, New Delhi, p.72-97.
2. Bandarathillake, H. M. (1980). Water shed afforestation for controlling soil erosion and landslides. *The Sri Lanka Forester* 18 (3/4): 131-136
3. Dimantha, S. (1986). Influence of land use on landslides. Proc. Multidisciplinary Seminar on Landslides, June 1986, Inst. of Fundamental Studies, Sri Lanka; p. 15-19
4. Bahuguna, S. (1986). Deforestation and its impact. *in* Radhakrishna, B. P. and Ramachandran K.K. ed. *India's Environment: Problems and Perspectives*, Geological Society of India, Bangalore, p. 167-173
5. Ghosh, T. K. (2002). From forest to fish- a catchment modification approach for the environmental management in wetland ecosystem, p. 378-393, *in* G. Tripathi *ed*. Modern Trends in Environmental Biology, CBS Publishers and Distributors, New Delhi.
6. Valdiya, K.S. (1985). Accelerated Erosion and Landslide-Prone Zones in the Central Himalayan Region, *in* J.S. Singh ed. *Environmental Regeneration in Himalaya: Concepts and Strategies*. Central Himalayan Environment Association and Gyanodaya Prakashan, Nainital, India, p. 12-38
7. Singh, J.S. and Saxena, A.K. (1980). *The Grass Cover in the Himalayan Region,* In Proc. National Seminar on Resources, Development and Environment in Himalayan Region, DST Govt. of India, New Delhi, *p*. 164-203.

8. Bandara, R. M. S., Haridharan, S. and Cruickshank, R. D. (1994). Landslides in Badula district of Sri Lanka, *Proc. National Symp. on Landslides* in Sri Lanka, March 17-19, 1994. National Building Res. Organisation, Colombo; p. 127-132

9. Wijewickrama, M. I. D. H., Kotuwegoda, W. P. P. K., Wickramathanthri, N. C. and Karunaratne, R. A. P. De A. M. (1994). Landslides in Nuwara Eliya District. *Proc. National Symp. on landslides* in Sri Lanka, March 17-19, 1994. National Building Res. Organisation, Colombo; p. 119-126

10. Singh, R. B., and Pirazizy, A. A. (1990), Anthropogenic impact on landscape synthesis in Kashmir Himalayas: a study of land suitability and capability classification *in* R. B. Singh ed. *Contribution to Indian Geography XI Environmental Geography,* Heritage Publishers, New Delhi, p. 112-120

11. Prasad, H. (1986). Physical environment and settlements in Mussoorie region *in* H. H. Singh *et al., ed. Geography and environment: issues and challenges.* Concept Publishing Company, New Delhi, p. 211-230

12. Mukherjee, A. and Bhoosnurmath, K. (1993). *A perspective plan for a hill district,* vol. 4. Heritage Publishers, New Delhi.

13. Auden, H. (1942). Geological report on the hillside of Nainital, Report of GSI, Calcutta.

14. Pande, R. K. (1991). Slope instability problems in Karmi area of Central Himalayas, India. *in* H. S. Sharma ed. *Indian Geomorphology.* Concept Publishing Co., New Delhi, p. 165-176.

15. Pirazizy, A. A. (1992). *Environmental Geography and Natural Hazards.: Exigencies of Appraisal in Highland–Lowland Interactive Systems*. Concept Publishing Company, New Delhi. 216 p.

Chapter 61

Unaccounted For Water (UFW) Management of Water Distribution System in Selected Cities of India

S.N. Dhone

CSIR-NEERI, Nagpur

Background

Providing conditions for adequate quantity of water flow through the pipelines and preservation of hygienic quality of water in the distribution system are the twin objectives of water supply system management. Unaccounted For Water (UFW) which comprises of physical and non physical losses, therefore, assumes significances in view of diminishing water resources.

Over a period of three decades, NEERI has developed expertise in conducting studies on PMWDS with special emphasis on prevention of water losses. Appropriate procedures for conducting the studies have been developed and tested in the field.

NEERI Scientist, in association with BMC Engineers, Conducted World Bank aided project on waste assessment, detection and control in water distribution system in 1971. This was the first project on conservation of water in this Institute. The Institute has undertaken various sponsored projects on the subject for several cities in India since 1972.

As a part of these projects, the Institute has also organized annual All India Training Programmes on the subject under the sponsorship of CPHEEO, Ministry of Urban Development, Govt. of India, New Delhi. The first programme was at Bombay in the year 1971.

Place :
WWM District:
Date :

S. No.	Particular	Details	Remarks
1.	Populations (Souls)		
2.	Total number of connection		
3.	Total number of properties		
4.	Total Length of Mains (m)		
5.	No. of meters		
	Working		
	Not working		
6.	No of psp's		
7.	Water supply to the WWM District (m^3)		
8.	Supply hours (hrs)		
9.	Normal pressure (ave), (kg/cm^2)		
10.	Water consumption (lpcd)		

Figure 61.1: Details of WWM District.

Objective

Estimation of Unaccounted For Water (UFW) (which Comprises of physical and non physical losses) (Figures 61.1 to 61.3).

Waste (Leakage) Assessment, Detection and Control

Impart field oriented training to the engineers on the various facets of water distribution system including instrumentation.

Unaccounted For Water (UFW)

The term Unaccounted For Water (UFW) represents the difference between net production and consumption.

Place :

WWM District:

Date :

S. No.	Valve No.	From to	Valve No	Dia (mm)	Length (m)	Material	Year of laying	No. of inter conne-ction	Street Name	Alignm-ent offset	Ave dept h (m)	Soil type	Remar -ks

Figure 61.2: Inventory of Pipes.

Place :

WWM district :

Date :

S. No.	Valve No.	Type of Main	Size of main (mm)	Size of valve (m)	Location	Offset	Total no. of turns	Existing operating condition	Direction of closing

F.O. – Full open, P. C. – Permanently closed, X – Junction, B. V. – Boundary valve, S.V. – Step valve, A. C. W. – Anticlockwise, C. W. – Clockwise

Figure 61.3: Inventory of Valves.

The Two main Categories of UFW are physical losses (due to leakage) and non physical loss.

In the absence of Installation of reliable production meters on the supply lines and correct domestic water consumption, it is extremely difficult to estimate the non physical or administration losses (incorrect billing, faulty meters, illegal connections, theft).

Waste Assessment, Leak Detection and its Control (Figures 61.4 to 61.9)

Instruments/Equipments

i) Underground pipe locator ii) Valve Box cover locator iii) Measuring Wheel (Road measurer) iv) Ultrasonic Flow meter (clump on type) v) Pressure gauges/ recorders vi) Sounding rod/Digital Sound detector/Electronic Leak Detector

Waste Assessment Techniques

Stop Cock Technique

Water supply in the majority of the cities is intermittent, thus, Stop Cock Technique was adopted for assessing waste levels (leakage upto corporation limit only).

In this method the total leakage is assessed in the waste meter (WWM) district upto consumers premises is carried out by closing boundary valves and all the stop cocks or taps of the service connections. The supply of water is allowed through a single feeder pipe where the flow meter installed. The rate of flow into the district is measured with the help of revenue meter, waste detecting meter (kent type) or Ultrasonic flow meter. This rate of flow indicates the initial waste level in the wwm district.

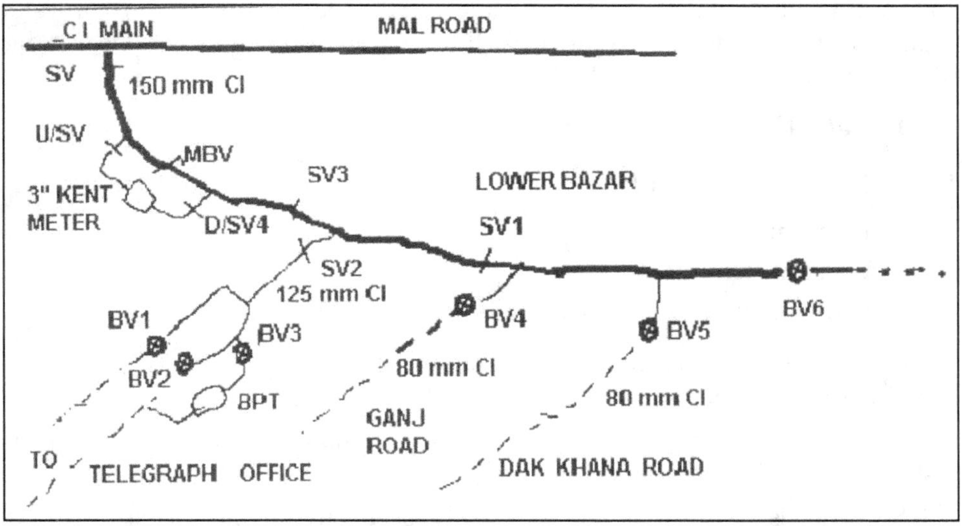

Figure 61.4: Skelton Plant of WWM District at Lower Bazar, Shimla.

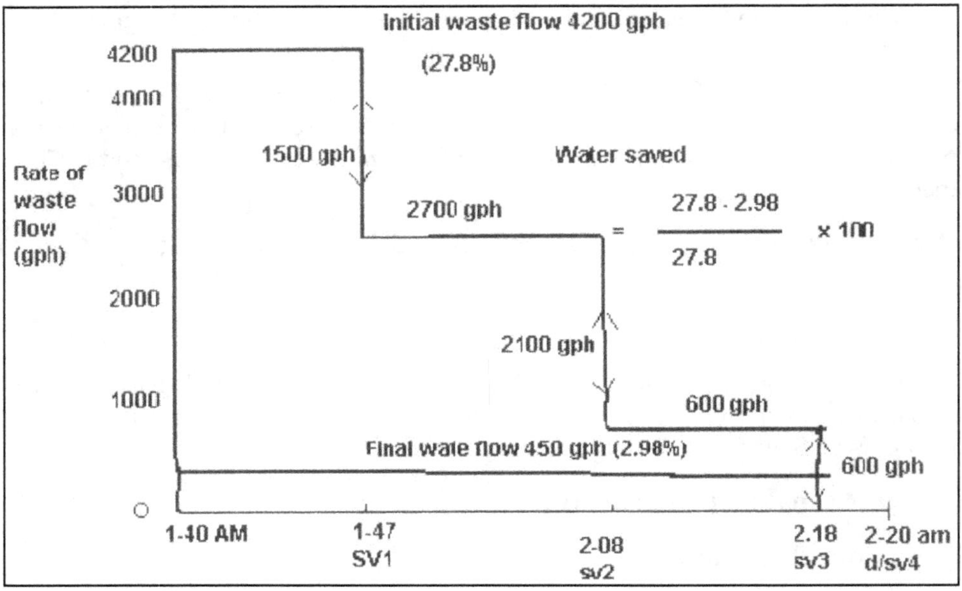

Figure 61.5: Waste Assessment and Step Test at Lower Bazar, Shimla (Stop Cock Method).

Minimum Night Flow (MNF) Technique

In continuous water supply systems, substantial quantity of water goes as waste in the absence of appropriate maintenance and corrective measures.

The flow of the water registered by a flow meter especially during late night hours could be termed as MNF, if and only if consumers demand is satisfied. The water is allowed to enter a single feeder through the flow meter. All the boundary valves and ball cocks of the overhead tank over the roof top are checked for water tightness. The recorded Minimum Night flow in the flow meter depicts the wastage due to leakage in the WWM district.

Leak Detection and its Control

Leak Detection Methods

- ☆ Visual observation
- ☆ Use of instruments
- ☆ Use of nitrous oxide gas, or radioactive isotopes, etc.

Visual detection is easily possible when the leaking pipe is at a shallow depth and when marshy land is formed around the leaking pipe.

Sounding rod, Electronic Leak detector and Leak noise correlator are some of the instruments used in leak detection surveys.

Sounding is the most common and most cost effective method of locating leaks. There are two ways in which sounding can be undertaken. (i) Direct Sounding (ii) Indirect sounding (surface sounding).

Direct sounding consists of making soundings directly on valves, hydrants, stopcocks and other convenient fittings listening for the characteristic noise of leaking water.

Indirect sounding is on the surface sounding consists of making soundings on the ground surface above the line of the pipeline to determine the point of maximum sound intensity.

Sounding rod alone or with the electronic leak detector can be used to pinpoint the leaks in the buried pipes. The experience has shown that approximately 80 per cent of the leaks could be located by using sounding rod.

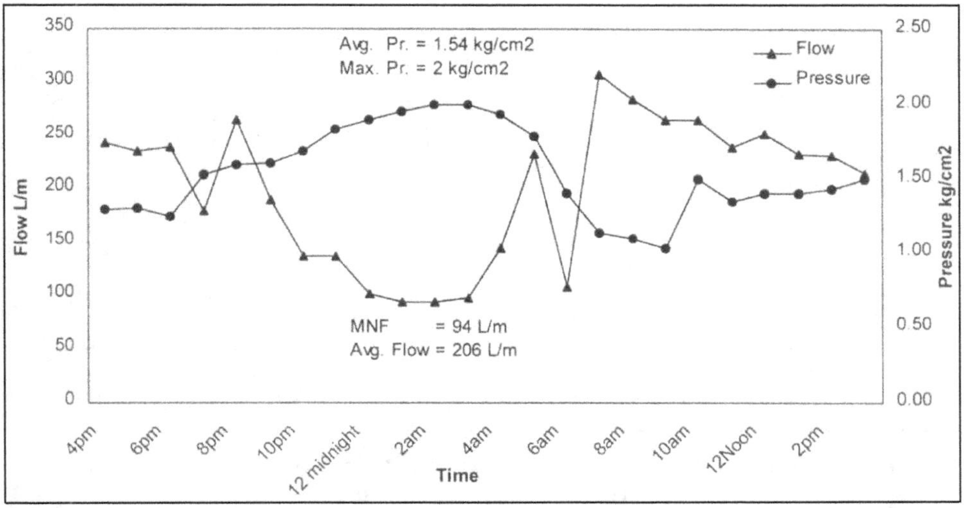

Figure 61.6: Hourly Flow and Pressure Pattern for WWM District South Fort, Trivandrum (Initial waste assessment).

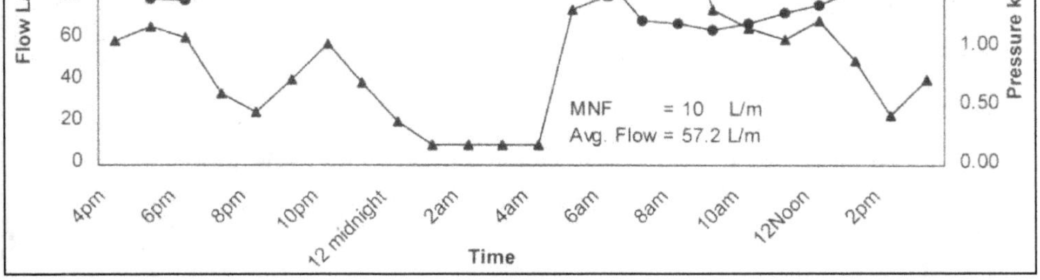

Figure 61.7: Hourly Flow and Pressure Pattern for WWM District South Fort, Trivandrum (Final waste assessment).

Leak Noise correlator (LNC) is leak locating instrument which can be used in water mains, transmission mains. In this instrument many peaks of leakages observed on the screen with distance from left or right transducers.

If leakage in water mains cannot be traced by traditional techniques then nitrous oxide gas could be used.

After leak detection survey, the leaks could be repaired and listed the types of leaks. Then again carried out the waste assessment test till the waste level would be brought down to permissible level.

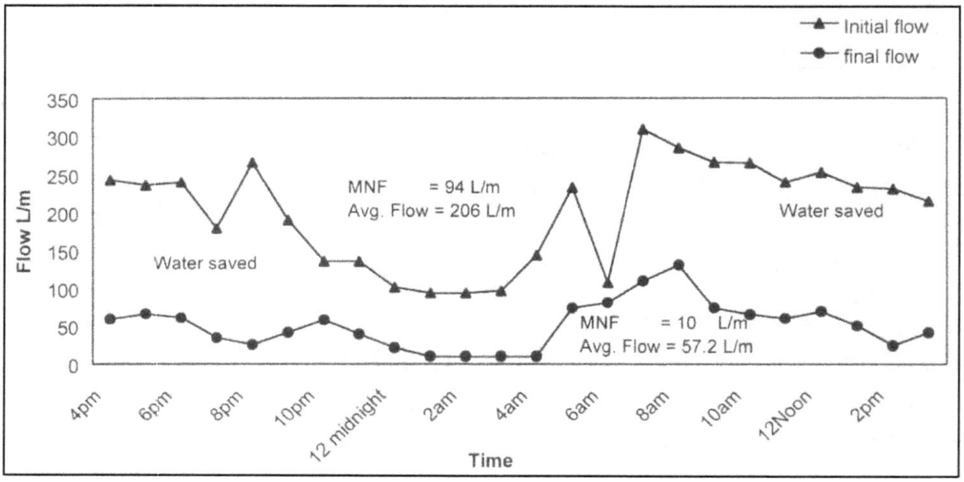

Figure 61.8: Hourly Flow Pattern for WWM District South Fort, Trivandrum (Before and after control).

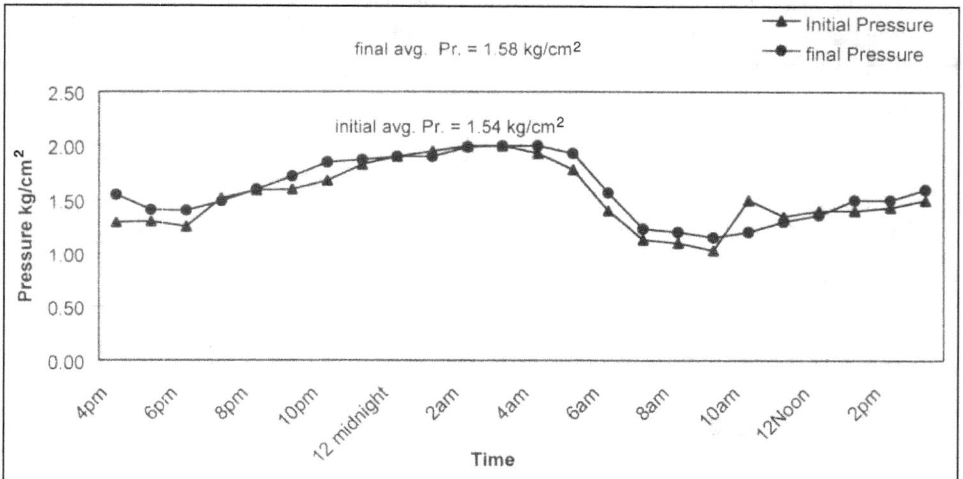

Hourly Pressure pattern for WWM district South fort, Trivandrum
(before and after control)

Figure 61.9: Hourly Pressure Pattern for WWM District South Fort, Trivandrum (Before and after control).

Indian Scenario

In the selected Indian Cities during the study period, representative WWM Districts (2 or 3) were identified and the investigations on waste assessment, detection and its control were carried out (Figure 61.10). Each district would generally cover 150-300 house service connections. The cities covered in the studies so far are:

Table 61.1: Indian Scenario

S. No.	City	Population of Pilot study district	No. of service connection	Hours of Supply	Per-capita Consumption (lpcd)	Waste Flow as Percentage of total water supply	
						Initial	Final
1.	Bombay	8000 7067	241 281	15.0 5.0	216.0 200.0	28.5 33.0	8.5 7.4
2.	Aurangabad	6177	377	3.5	56.0	28.5	12.0
3.	Delhi	1562	260	13.0	155.0	17.7	10.0
4.	Madras	2000	170	6.0	135.0	36.0	7.0
5.	Lucknow	1517	149	24.0	135.0	19.0	6.0
6.	Calcutta	15552 7619	177 216	24.0 24.0	139.0 176.0	24.6 25.0	6.0 3.1
7.	Ahmedabad	1555 1703	184 150	6.0 5.5	192.0 108.0	20.1 22.5	2.0 2.5
8.	Surat	3747 2625 2580	311 301 318	5.5 5.5 7.0	196.2 135.2 199.3	24.0 25.0 26.0	1.0 1.8 3.7
9.	Nagpur	2502	393	24.0	202.4	44.0*	11.0
10.	Cochin	3639 2207	302 296	15.0 15.0	115.0 115.0	21.0 41.0	3.0 8.0
11.	Jaipur	1440	179	8.0	210.0	26.5	3.5
12.	Shimla	1239 129	198 82	3.0 8.0	120.0 120.0	27.8 45	2.98 7.36
13.	Ludhiana	1100	149	17.0	208.0	33.0	15.00
14.	Trivandrum	1105 322	288 88	24.0 24.0	144 159	29.3* 52.6	11.30 19.00*
15.	Coimbatore	5859	365	4.0	123.57	38.63	16.32
16.	Cuttuk	2739	248	10.0	135	21.0	10

* Includes wastage in consumer premises

Waste Due to Leaks: Indian Scenario (1971-90) (Based on Pilot Studies).

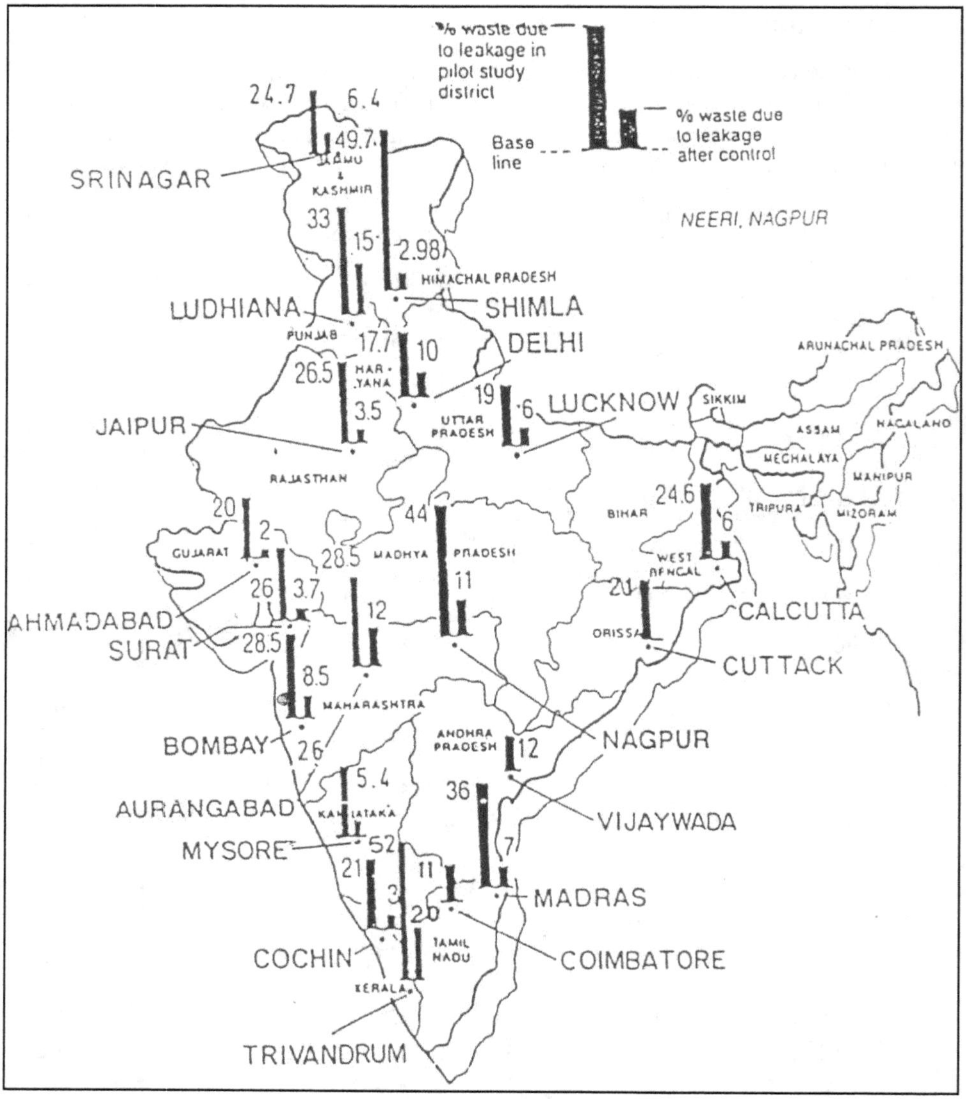

Figure 61.10: Waste Due to Leakages: Indian Scenario (1971-90) (Based on Pilot Studies in Water Distribution System).

Ahmedabad, Aurangabad, Bombay, Bhubaneshwar, Bangalore Cochin, Coimbatore, Calcutta, Cuttack, Delhi, Gandhinagar, Hyderabad, jaipur, Ludhiana, Lucknow, Madurai, Mysore, Madras, Nagpur, Srinagar, Shimla, Surat, Salem, and Trivandrum (Table 61.1)00000000.

Water Supply in the majority of the above cities is intermittent, thus Stop Cock technique was adopted for assessing waste levels. Minimum Night Flow (MNF) technique was also used successfully in Trivandrum and Nagpur as 24 hours of water supply was available during the study period. It was also demonstrated by

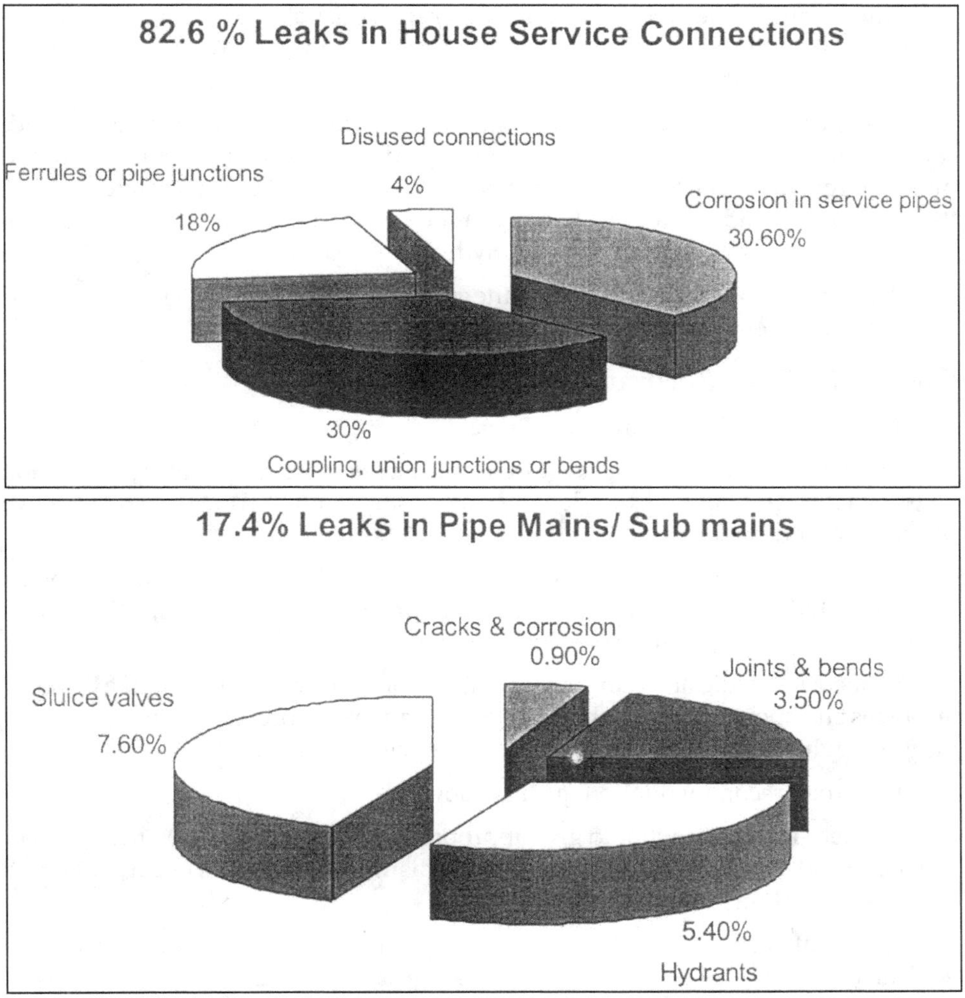

Figure 61.11: Types of Leaks–Percentage Distribution

NEERI through case studies that MNF technique could be utilised (even if the supply is not 24 hours) as for extended hours of supply, on fulfilling minimum specified conditions such satisfaction of water needs of the consumers, etc.

Salient Finding (Figure 61.11

Extent of water loss is estimated between 11 and 52 per cent of the total water supply. Expressed in terms of per capita basis, this works out to be between 16 and 102 lit/day.

Leaky pipes were detected success fully by traversing sounding rod and electronic leak detector over the alignment of the pipelines. LNC has limited use due inadequate pressures in the distribution network. However LNC could be used to pinpoint the leaks from transmission mains.

After remidial measures (Leak detection and repairs), the extent of waste was reduced to 4-12 per cent of the total water supply. It improved water flow rate and water pressures in the system.

83 per cent of the leakage in the distribution system occur in the house service connections from ferrule to the consumer taps due to corrosion of under ground pipelines (galvanized iron), at the junction of the ferrule and submains and through the pipe joints. And remaining 17 per cent occur in pipe mains/submains due to cracks and corrosion, joints and bends, hydrants, sluice valves.

Benefit cost analysis confirms that the expenditure incurred on UFW and R and C could be recovered in 6 to 24 months period.

Conclusion and Recommendations

The broad conclusions are as follows:

The need and efficacy of leakage detection and control in the water distribution system has been established. Conducting routine studies on PMWDS is a cost effective proposition and needs concerted efforts.

The preventive maintenance programme results in indirect augmentation of the existing water supply in terms of quantity and pressure. It also prevents the ingress of extraneous pollution into the system.

Trained and dedicated staff, non availability of instruments, inadequate financial resources and above all the emphasis on conservation of processed water are some of the issues which need urgent attention of the decision makers.

The broad recommendations are as follows:

The decision makers should give appropriate importance to UFW management in general and conservation of water from the distribution systems in particular and ensure sustainability of these programmes.

A separate cell (unit) in the water supply undertakings is recommended the dedicated staff associated should be given a special allowance.

Town/city should be divided into wastewater meter districts and the appropriate method of waste assessment detection and control should be used on a routine basis.

Water supply undertaking embarking upon this programme should procure appropriate equipments/instruments.

Consumer participation in such programmes should be ensured which possibly could be achieved by organizing health and environmental education programmes.

Chapter 62

Estimation of Effluent Treatment Cost: Financial Apportionment

S.N. Kaul[1], D.R. Saini[2], J.K. Pandey[2], Vinay Prakash[3] and P. Kaul[4]

[1]*CSIR-NEERI, Nagpur*
[2]*CSIR-NCL, Pune*
[3]*HEMRL, Pune*
[4]*Symantec (USA), Pune*

There is general support throughout the world for the "Polluter Pays Principle", *i.e.* the industry must pay the cost of disposing of waste in an environmentally acceptable manner.

It, therefore, follows that the cost of operating a CETP must be borne by the industry, which the CETP serves. This is also generally accepted with the word of caution from especially the industry itself that increased production costs can adversely affect competitiveness, especially when a large proportion of the waste producing industry relies heavily on the export market.

Even through the "Pollution Pays Principle" is accepted as the desired model, there is a strong argument for outside support due to the competition from other industries not required to comply with these new demands, and the social and environmental benefits obtained by operating combined treatment systems.

A CETP does not produce products of saleable value. Even if sludges, and possibly the final effluent could be utilized, it would only result in the reduction of total expenses. Therefore, for a CETP to operate with a profit, or just to obtain a 'no-profit – no-loss' situation, it must obtain revenue from either levying of fees upon the industries delivering waste to the plant, outside assistance in the form of grants, low interest loans, tax benefits, etc. or a combination of these.

Even though it is accepted that standards of environmental pollution control also improve in countries competing with Indian exports, there is a strong opinion that some form of subsidy to the industry or to the owner of the CETP should be considered in order to ensure viability of the CETP and the competitiveness of the industries utilizing its capacity. At least this should be the case for the pioneering units now being discussed in India.

It is obvious that no single yardstick can be fixed for controlling pollution from the industrial estates under consideration and a thought should be given to the various alternatives available in terms of many interdependent factors, such as drainage system nature and composition of effluents, degree of treatment to which the effluent must be subjected to, mode of disposal of treated effluent and also cost of waste treatment. Based on these factors, the promoting agency may have to develop a suitable financial apportionment method with a view to realize the capital and recurring expenditure incurred in constructing waste collection, treatment and disposal systems [1].

Equitable Sharing of the Financial Burden

A Prime Requirement

Equitable sharing should be the main objective while developing a financial apportionment method. The direct as well indirect benefits derived by CETP should be proportionately distributed to the member industrial units. Though this is practically very difficult, attempts have to be made towards achieving this condition. No industry should be benefited at the cost of other industry as far as financial apportionment is considered.

Equitable sharing can successfully be carried out if there is sufficient competition in the market to prevent the charge from being posed on to the consumer and the level of charge is equivalent to the level of damage or at least to the cost of treatment. In implementing these charges, it is necessary to ensure that they are not merely taken as a license to pollute. An important aspect of effluent charges is that these encourage the industry to look at pollution as basically an economic problem. Insignificant as it may seem, there is a vast psychological difference between introducing a charge on effluents and applying a standard [2].

Number of formulations are now available to device sharing financial burden but probably none is the ideal method ensuring 100 per cent equitable or proportionate sharing. The methods available for cost-sharing are described in subsequent sections.

It is to be kept in mind that generally the method of arriving at a figure of cost-share range from the very simple to the highly complex. The simplest involves some inequities. The more complex methods eliminate some of these inequities at the sacrifice of simplicity and ease of administration. Therefore, there is no ideal method. In practice, a compromise is reached where little sacrifice is made in both.

Quantity Method

This is the simplest method and applicable for joint treatment of sewage and low strength industrial wastes. The total cost for debt service and costs of operation of the sewage treatment plant is divided by the total cubic meter of wastes handled to obtain a service charge in terms of cost per 1000 m^3 based upon volume alone. In areas having no industries, the wastes are generally uniform in composition so this approach is acceptable. In highly industrialized areas, however, the user with dilute wastes ends up subsidizing the user with wastes of high BOD and suspended solids contents and this is certainly not a fair or equitable method of obtaining of revenues[(3)].

Quantity-Quality Method

In this method, both the quality or concentration of the wastes in terms of BOD and suspended solids as well as the volume is considered in computing the sewer tax. An attempt is made in this method to levy charges proportioned to the use of and benefits received from the sewage treatment plant. This is in contrast to an ad valorem tax on property to finance sewage treatment works wherein the tax is merely based upon the ability to pay.

Table 62.1 lists each components of the sewage treatment plant with the annual operating costs of the component. These total annual costs are then apportioned to the three design parameters of a sewage treatment plant. For example, note that the operating costs of the interceptor sewers and pumping stations are Rs. 5.0 lacs and 1.6 lacs per year respectively. Since sewers and pumping stations are designed on the basis of volume alone and essentially their maintenance is a function of volume, these two costs are apportioned to volume of flow handled. Therefore, these two figures of Rs. 5.0 lacs and 1.6 lacs are charged to volume alone. On the other hand, a primary settling tank not only handles the total volume of flow but also collects the settleable solids and floating solids which are pumped to the digester. Therefore, Rs. 1.2 lacs per year operating cost is equally charged to suspended solids and volume. The same reasoning is carried through for remaining components of the treatment plant as indicated by Table 62.1.

Table 62.1: Apportioning of Operation Costs of a Sewage Treatment Plant to Volume, Suspended Solids and BOD

Components	*Annual Operating Costs Rs. in Lacs*	*Volume*		*Suspend Solids*		*BOD*	
		Per cent	*Rs. in Lacs*	*Per cent*	*Rs. in Lacs*	*Per cent*	*Rs. in Lacs*
Interceptor Sewers	5.00	100	5.00	—	—	—	—
Pumping Stations	1.60	100	1.60	—	—	—	—
Sedimentation	1.20	50	0.60	50	0.60	—	—
Trickling Filters	2.00	10	0.20	—	—	90	1.80
Sludge Digestion	4.00	—	—	100	4.00	—	—
TOTAL Rs.in lacs	13.80	53.6 per cent	7.40	33.4 per cent	4.60	13 per cent	1.80

The fixed annual costs are shown in Table 62.2 for each component of the sewage treatment plant. Again the same reasoning is used to assign the costs to the three pollution parameters.

Table 62.2: Apportioning Fixed Costs to Volume, Suspended Solids and BOD

Components	First Costs Rs. in Lacs	Annual Fixed (5 per cent) Debt Retirement	Volume		Suspended Solids		BOD	
			Per cent	Rs. in Lacs	Per cent	Rs. in Lacs	Per cent	Rs. in Lacs
Interceptor Sewers	200.00	10.00	100	10.00	—	—	—	—
Pumping Stations	10.00	0.50	100	0.50	—	—	—	—
Sedimentation	20.00	1.00	85	0.85	15	0.15	—	—
Trickling Filters	100.00	5.00	10	0.50	—	—	90	4.50
Sludge Digestion	40.00	2.00	—	—	100	2.00	—	—
Total Fixed Costs	370.00	18.50	64.1	11.85	11.6	2.15	24.3	4.50

Table 62.3: Total Annual Costs–Operating Plus Fixed Costs for Volume, Suspended Solids and BOD

	Volume	Suspended Solids	BOD	Total
Operating Costs, Rs. in lacs (from Table 62.1)	7.40	4.60	1.80	13.80
Fixed Costs, Rs. in lacs (from Table 62.2)	11.85	2.15	4.50	18.50
Total Cost (Rs. in lacs)	19.25	6.75	6.30	32.30

Table 62.3 summarizes and totals Tables 62.1 and 62..2 of the total annual costs of Rs. 32.3 lacs to operate the sewage treatment plant under consideration, Rs. 19.25 lacs can be charged to volume, Rs. 6.75 lacs to suspended solids, and Rs. 6.30 lacs to BOD. This particular plant handles, say, three MGD of wastes containing 300 mg/L suspended solids, and 350 mg/L BOD. Table 62.4 shows the final step in circulating the actual unit costs in terms of Rs. per 1000 m^3, per kg suspended solids and per kg of BOD.

Normally, the average domestic and commercial user is assumed to have so-called "normal" sewage and will be charged a flat rate based upon volume which in turn is merely derived from the water bill (Table 62.5 for example). Table 62.6 indicates the derivation of flat rate volume method (quantity method) or quantity metered wherein the charges are based upon volume alone.

Basically, this "Quality-Quantity" (Q-Q) method of financing is the ideal way of calculating the charges. However, there are certain inequities involving perhaps a philosophical discussion rather than a technical or financial approach.

The sewage treatment plant in this sample problem is handling three MGD of wastes, and all the costs of the plant, both fixed and operating, are charged to treating this three MGD of wastes. However, the plant may be designed with excess capacity for future growth. Let's assume for a moment that the municipal sewage treatment plant is designed for six MGD of wastes at 300 aand 350 mg/L of suspended solids and BOD respectively for future growth of the community. All the fixed costs of the sewage plant are dependent on the size of the plant alone and not upon the quantity of wastes, which the sewage treatment plant is receiving. Therefore, the donors of three MGD of wastes are paying at least in part for six MGD of wastes, a future sum. This excess capacity benefits the entire community, particularly undeveloped property. In one sense, then, the present users are paying for the amortization of capacity designed for future users and undeveloped properties.

Table 62.4: Unit Costs for Sewage Treatment Based on Volume, Suspended Solids and BOD

Design Parameter	*Design Basis of Plant*	*Design Basis in Terms of Annual Quantities*	*Unit Costs Total Costs/ Total Quantities*
Volume	3 MGD	1095 MG/yr	Rs.1925000/1095 MG = Rs.1758/MG = Rs. 465/1000 m^3
Suspended Solids	300 mg/L	1240000 kg/yr	Rs.675000/yr 1240000 kg/yr = Rs.0.544/kg SS
BOD	350 mg/L	1453000 kg/yr	Urs.630000/yr 1453000 kg/yr = Rs.0.434/kg BOD

Table 62.5: Unit Charges for 'Normal' Sewage Based on Water Bill

For billing small users, charges can be made on the basis of cost per 1000 m^3 assuming "normal" sewage as discharged by the small user contains 2000 mg/L BOD and 250 mg/L Suspended Solids

1000 m^3 x 0.2 kg/m^3 BOD x Rs.0.434/kg BOD	=	Rs. 87
1000 m^3 x 0.25 kg/m^3 SS x Rs.0.544/kg SS	=	Rs.136
Volume Cost from Table 62.4, Rs./1000 m^3	=	Rs.465
Volume Cost from Table 62.4, Rs./1000 m^3	=	Rs.465
Total Cost per 1000 m^3 "normal" sewage	=	Rs.688

Table 62.6: Flat Rate Method – Charges Based upon Volume Alone

Total Cost to Operate Plant (Table 62.3)	= Rs. 32.3 lacs/yr
Total Volume Influent to Plant (Table 62.4)	= 1,095 MG/yr
Rs. 32,30,000/yr 1,095 MG/yr = Rs. 2950/MG	= Rs. 780/1000 m^3

This fact has led many authorities to favour the method levying surcharges (discussed in this Chapter under "Sanitation District Surcharge Formula") in which users and other property owners alike pay charges proportional to the benefits received.

This method is in reality, an extension of the Quantity-Quality Method. Part of the revenues is obtained by charges based upon BOD, Suspended Solids, and Volume of wastes discharged to the sewer. The remaining money can be obtained by *ad valorem* tax on users, non-users, and undeveloped properties alike. All benefits from the existence of a sewage treatment plant. Access to sewers enhance the value of undeveloped property.

This method is certainly the most equitable method developed to date. It is also the most complex of the systems of sewer charges to administer [3].

Maize Formulation

'Polluter Pays' principle provides an acceptable measure for pollution control if a rational and simple method could be established to estimate the cost of fulfilling the legal obligations of each member polluter industry in the CETP scheme. A procedure based on the volume and degree of noxiousness of wastewater contributed to the system for an equitable distribution of annualized financial burden has been developed by Malz [4] and is used in USA.

Since the combined treatment is mostly biological in nature, wastewater dilution factor, V to ensure survival of fish (48 hours exposure) is expressed as,

$$V = \frac{A}{A_O} + \frac{B}{B_O} + \frac{TDS}{TDS_O} + F - 1$$

where,

A = Sedimentable matter, mg/L

B = COD of settled sample, mg/L

TDS = Total dissolved solids, mg/L

F = Fish toxicity

A_o, B_o and TDS_o = Compared with effluent discharge standards set by local governing body

Noxiousness degree, S, is then estimated from dilution factor, V, expressed in appropriate slabs rather than in discrete numbers to account for deviation in sampling and analysis of wastewater.

Dilution Value (V) and Noxiousness Degree (S)

V	*S*
1–2	1
3–4	2
4–6	3
6–7	4

Annualised cost apportioned to each polluter is calculated as follows:

$$\text{Cost Share} = \frac{S.Q.Z}{\Sigma(S.Q)}$$

where,

Q = Annual flow of an individual industry

Z = Total financial burden on collection and treatment

The formula for calculating dilution factor, V and slabs for determining degree of noxiousness, S can be changed appropriately to suit the specific case. Cheda *et al* (5) have modified the formula to use it for a cluster of tanneries in U.P. as follows:

$$V = \left[\frac{A}{A_o} + \frac{1}{2}\left[\frac{B}{B_o} + \frac{C}{C_o}\right] + F\right] - 1$$

where,

A = Settleable matter in wastewater, mg/L

B = BOD of settled wastewater, mg/L

C = COD of settled wastewater, mg/L

F = Fish toxicity factor

A_o, B_o, C_o = Effluent discharge standards set by U.P. Pollution Control Board for Settleable matter, BOD and COD respectively, mg/L

Noxious degree, S was determined from the following slabs of dilution factor, V values:

V	1-4	5-8	9-12	13-16	17-20	21-28	29-36
S	1	2	3	4	5	6	7

Annualised cost apportioned was calculated as:

Annual apportionment = S x annual flow x total annualized cost of joint wastewater collection and treatment system per unit flow.

Roman Formulation

Participation in costs of CETP should be conductive to the application of larger treatment plants in place of many small ones, which in total, are always more expensive. Roman (74) proposed the following formula, which is widely used in Poland, to calculate the share of each industrial plant participating in the costs of a CETP:

$$I_u = \frac{Q_o}{Q}\left[I_o\left(1+\left(\frac{S_u - S}{S}\right) - e\right) + I_d\right]$$

where,

I_u = Amount expressing share of the participating industrial plant

Q_o = Average daily effluent flow from the given industrial plant discharged to the CETP, in m^3/day

Q = Average daily effluent flow from all industrial plants discharged to the CETP, in m^3/day

I_o = Construction costs of the CETP

I_d = Construction costs of all additional facilities used for transport of effluents to the CETP, expressed as BOD_5 in kg O_2/m_3

S_u = Average strength of an effluent of the given industrial plant, expressed as BOD_5, in kg O_2/m^3

S = Average strength of the mixture of all effluent discharged to the CETP. expressed as BOD_5, in kg O_2/m^3

e = Costs of treatment plant units depending on the pollution load to construction costs of the whole treatment plant

This formula is based on the following assumptions:

- ✰ The share in costs of building of a CETP should correspond to the amount and strength of discharged effluent.
- ✰ The costs of facilities for transport of effluent to the CETP should be covered by all users of the plant proportionately to the amount of their effluents, but irrespective of distance of transport.

The second assumption creates a kind of reduced rate for participants situated at a greater distance from the common treatment plant and promotes enlargement of its territorial reach.

The amount of penalties for transgression of requirements for effluent discharge depends–according to the regulation being obligatory in Poland – on the pollution load contained in the discharged effluent above the admissible limit and on the type of polluting matter.

Some examples of the amount of fines to be paid by the 'guilty' industry in Poland are given below:

for 1 kg BOD_5	3 zloty
for 1 kg dissolved substances	3 zloty
for 1 kg suspended solids	3 zloty
for 1 kg fat	3 zloty
for 1 kg COD	2 zloty
for 1 kg heavy metals	50 zloty
for 1 kg phosphate (PO_4)	2 zloty
for 1 kg volatile phenols	10 zloty

The levels of the penalties were established in such a way that they are higher than the costs of removal of polluting matter in an effluent treatment plant (*e.g.* the cost of removal of 1 kg BOD_5 in a treatment plant ranges from 1.5 to 2.5 zloty whereas the penalty would be 3 zloty). Beside the fines for the amount of polluting matter paid by the industrial plant, Polish legislation provides also sentences of persons responsible for pollution of water. Penalties and imprisonment are applied depending on the size of damage caused by pollution.

Roman concluded that the economic aspects of water protection against pollution presented above should never be considered as the main factor influencing economic decisions in this field. They should always be regarded as one of many factors to be taken into account when making the decisions. Above all, the economic characteristics of solutions should only serve for determining that solution is close to the optimum in relation to costs, because the necessity of water protection results mainly from conditions beyond economics, such as environmental protection and a healthy water supply for people.

While developing the abovementioned formulation, Roman deduced certain important conclusions as described below:

In general, it is convenient to apply unit investment costs defined as the quotient of the sum of investment costs of a treatment plant and of its capacity:

$$i = \frac{J}{Q} \ (\text{cost/m}^3\text{/day})$$

where,

J = Investment costs of a given plant

Q = Average 24 hours amount of purified effluent, m^3/day

The sum of the unit investment costs is affected by various factors, the most important of which are:

- ✰ the size of the treatment plant expressed, for instance, as amount of effluent treated,
- ✰ the concentration of polluting matter in effluent and other properties of effluent affecting the technological scheme of a treatment plant and admissible loading of its devices, and
- ✰ the required effect of the treatment plant.

The relationship between unit investment costs and capacity of effluent treatment plants reveals that large treatment plants are more economical than small ones, which is in conformity with the general rule of building the economical effectiveness of investment projects increases with their size. From the fact, the following practical conclusions may be drawn:

- ✰ The use of a CETP working for several factories situated in the neighbourhood seems to be advantageous.

- ☆ Discharge of effluents from individual producers to a main municipal treatment plant seems to be reasonable.
- ☆ It is necessary to search for more economical solutions for small treatment plants, which cannot be, however, miniatures of the traditional systems in big treatment plants, but should be based on special solutions adjusted to the specific technical and economic character of purification of small amounts of effluent.

The data of the influence of wastewater strength on the amount of investment costs concern municipal effluent treatment plants. As may be seen, in such a case, the investment costs grow linearly with the increase of effluent strength. The growth of costs is marked, but relatively small, as a 5-fold increase of investment costs. On this basis, it can be concluded that from the economic point of view, it may be reasonable to strive at lowering the amount of effluent even in such a case, when it is accompanied by an increase of strength in reverse proportion to the amount of effluent. This statement, however, in case of specific industrial wastewater, may sometimes not be true.

A characteristic feature is the rapid change in growth of costs with the increase of purification effect. A particularly rapid increase of investment costs appears with the rise of the purification effect in the range about 90 per cent [6].

Mogden Formulation

This formulation is derived on the principle of payment on the basis of cost incurred by the equivalent sized sewage treatment plant in comparison with the CETP. This formula is extensively used in Germany, U.K. and other countries. The formula is as follows [7]:

$$C = V + \left(\frac{O_t}{O_s}\right) B + \left(\frac{S_t}{S_e}\right) S + \left(\frac{TDS_t}{TDS_s}\right) t$$

where,

C = total cost/m^3 of effluent

V = cost/m^3 which relate solely to volume

B = cost/m^3 which relate solely to bio-oxidation of settled sewage

S = cost/m^3 which relate to suspended solids removal

t = cost/m^3 which relate to total dissolved solids removal

O_t = oxygen demand of settled trade effluent

O_s = oxygen demand of settled sewage

S_t = suspended solids content of trade effluent

S_e = suspended solids content of raw sewage

TDS_t = total dissolved solid concentration of trade effluent

TDS_s = total dissolved solid concentration of raw sewage

Modified Mogden Formula

$$C = R + V + \left(\frac{O_t}{O_s}\right) B + \left(\frac{S_t}{S_e}\right) S$$

where,

C = total charge per m^3 of industrial effluent

R = reception and conveyance charge per m^3

V = volumetric and primary treatment cost per m^3

O_t = COD (in mg/L) of the indl.effluent after 1 hr quiescent effluent at a pH of 7.0

O_s = COD (in mg/L) of settled sewage

B = biological oxidation cost per m^3 of settled sewage

S_t = total SS in mg/L of the trade effluent at pH 7.0

S_s = total SS in mg/L of crude sewage

S = treatment and disposal of primary sludges per m^3 of sewage

R, V, O_s, B, S_s and S are assessed annually on a regional basis [7]. It can be seen that no direct charge for the oxidation of ammonia is made although the unit charge B for biological oxidation does, of course, include an element for the cost of nitrification at those partially or fully nitrifying sewage works included in the regional average.

In contrast, the biological charge in the original Mogden formula was calculated on the basis of McGlowan Strength [8] which takes into account the nitrogen content of the industrial effluent and sewage. Problems may arise in using modified Mogden formula as a basis for equitable charging for the treatment employing activated sludge process for nitrification of wastes which are high in ammonical nitrogen concentration and low in carbonaceous BOD. In such cases, the biological treatment costs incurred may greatly exceed the charges levied [9].

Fukashiba CETP FormulationF

At Fukashiba industrial wastewater treatment plant in Japan, catering Kashima Petrochemical Complex consisting of 19 core industries and 39 other factories, a specific methodology is adopted for sharing the cost burden. The quantity and quality of the wastewaters reported by the factories are called the 'approved wastewater quantity' and the 'approved wastewater quality'. Each factory has its own values for the approved wastewater quantity and quality but it may discharge less wastewater than the approved quality. Each factory has only one discharging outlet to the sewerage system where wastewater consist of a quantity rate and a quality rate (the quantity rate as reported in 1988 was 32 yen/m^3 of wastewater discharged). The quality rate is determined using the following expressions:

$$F = (B + C)/2 + S + 6N$$

where,

B = BOD_5 concentration, mg/L

C = COD concentration, mg/L

S = suspended solids concentration, mg/L

N = concentration of the N-hexane extracts, mg/L

When the F value is less than 120 mg/L, the quality rate is 10 yen/m^3 of wastewater discharged; when the F value exceeds 120 but is less than 240, the quality rate is 30 yen/m^3, where the F value exceeds 240 but is less than 360, the quality rate is 40 yen/m^3. Thus, the quality rate increases 10 yen/m^3 for the increase in F value of 120. F values are determined based on analysis of the samples taken manually by the chemists of the Fukashiba treatment plant. The number of sampling times is determined based upon the 'F load' which is the F value multiplied by the quantity of wastewater discharged daily. When the F load of a factory exceeds 500 kg/day, the factory is classified as 'A' and wastewater is sampled 5 times per month. When the F load exceeds 100 but is less than 500, the factory is classified as 'B' and the wastewater is sampled twice per month. When the F load exceeds 5 but is less than 100, the factory is classified as 'C' and the wastewater is sampled once a month. When the F load is less than 5, the factory is classified as 'D' and sampling is done once a year. Thus, the Kashima Sewerage Works and the Fukashiba Wastewater Treatment Plant are independently operated based on the charging system [10].

Borough of Glossop STP Formulation

The sewage treatment plant serving Borough of Glossop and Textile mill operating for 5 days in a week preferred to charge capital and running costs separately. Running costs are charged on a modified 'Mogden' formula, whereas the charging formula for normal industrial effluents is based on the 'Mogden' formula.

The schedule of capital charges for weekday effluents is given in Table 62.7 [11].

Table 62.7: Apportionment of Capital Costs

Description	*Cost (£)*	*Textile Effluent Proportion*	
		Per cent	*Cost (£)*
Effluent pipeline	33,347	100	33,341
Pumping station	14,111	—	—
Inlet works	5,663	—	—
Sedimentation tanks	56,030	—	—
Filters	133,280	57	75,970
Humus tanks	55,053	38	20,920
Recirculation pumping station	24,092	2.4	579
Pressing plant	142,111	9	12,790
Miscellaneous	39,576	—	—
Sub-total	503,263	28.5	143,600
Contingencies, etc.	94,878	28.5	27,040
Administrative costs	41,900	28.5	11,940
Total	640,041		182,580

Two formulae are in operation for financial apportionment:

i) To recover running costs associated with the weekday effluent

$$C = V + P\left(\frac{BOD_t - 125}{125}\right) + O\left(\frac{BOD_t}{BOD_{ss}}\right) + S\left(\frac{C\%\ BOD_t}{SS_{ms}}\right) + AK$$

where,

C = Cost of treatment per m^3

V = Capital and running costs of the inlet works and sedimentation tanks (when used)

P = Running costs of the recirculation pumping station (In the event of the V units above being used, the capital costs of this station are to be included in P)

O = Running costs of biological filters and humus tanks

S = Running costs of the sludge processing plant

A = Administration and laboratory costs

K = A utilization factor, initially unity, to be reviewed by each party at regular intervals

BOD_t = Average BOD of industrial effluent in mg/L

BOD_{ss} = Average BOD of the mixed settled sewage and industrial effluent in mg/L

SS_{ms} = Average SS of the mixed settled sewage and industrial effluent in mg/L

C% = A factor for the conversion rate of BOD to secondary sludge. Initially 40 per cent, but to be reviewed annually in the light of treatment records

ii) To recover capital and running costs associated with the weekend effluent (normaindustrial effluent charging formula)

$$C = V + O\left(\frac{COD_{te}}{COD_{ss}}\right) + S\left(\frac{SS_{te}}{SS_{cs}}\right)$$

where,

C = Cost of treatment per m^3

V = Capital and operating costs of all volumetric units

O = Capital and running costs of all oxidation units

S = Capital and running costs of all sludge units

COD_{te} = COD of industrial effluent in mg/L

COD_{ss} = COD of the settled sewage in mg/L

SS_{te} = Suspended solids in industrial effluent in mg/L

SS_{cs} = Suspended solids in crude sewage in mg/L

Sanitation Districts Surcharge Formula

The Sanitation Districts in Los Angeles county, USA, consist of 27 individual Districts, out of which 15 Districts are joined together into the joint outfall system providing a common sewerage system for approximately 750 square miles of area and 8000 industrial companies. The Sanitation Districts provide and operate wastewater treatment facilities and trunk sewers. The largest treatment facility, the Joint Water Pollution Control Plant, provides primary treatment for about 350 MGD of wastewater and is planned for conversion to secondary treatment capabilities in the near future.

The Sanitation District's main source of revenue has been *ad valorem* (property) taxes on land, improvements and certain personal property within the District's boundaries. Each individual Sanitation District establishes an *ad valorem* tax rate sufficient to pay for capital improvements, and for operation and maintenance costs of sewerage facilities needed for wastewaters generated within the District.

Beginning in 1966, the Sanitation Districts began to investigate alternate sources of revenue. Federal and State grant funds for construction of sewerage facilities were first received by the Sanitation Districts at this time. The Districts investigated user charges requiring industrial companies to reimburse the Districts for the costs of industrial wastewater treatment and disposal.

It was observed that the Districts were not receiving payments from industrial wastewater discharges proportional to the costs of providing sewerage service.

In addition to obtaining supplementary revenue from industrial companies, the Districts at that time needed to obtain more information about industrial companies and to establish regulatory authority over industrial discharges. Due to these needs, the Districts adopted an industrial waste ordinance in 1972 which provided for an industrial wastewater treatment user charge programme.

Accordingly, the Sanitation Districts established a quantity-quality surcharge structure making use of the charge parameters of total wastewater flow, COD, suspended solids and peak flow. The Sanitation Districts industrial wastewater surcharge formula is as follows:

$$\text{Surcharge} = a\,(V) + b\,(COD) + c\,(SS) + dM\,(P) - TAX$$

where,

Surcharge = Net annual industrial wastewater treatment surcharge in dollars. No refund is made if a negative number results

V = Total annual volume of flow, MG

COD = Total annual discharge of COD in 1000 lbs

SS = Total annual discharge of SS in 1000 lbs

P = Peak discharge rate over a 30 min. period, occurring between the hours of 8.00 a.m. and 10.00 p.m.

A = Average discharge rate, determined by dividing 'V' by the total annual hours of operation or working time for the industrial discharger, converted to GPM (see M)

a, b, c, d = Unit charge rates adopted annually by the individual District based upon the projected annual total costs for wastewater collection, treatment and disposal, in dollars per unit.

M = A multiplying factor accounting for increased District's costs due to high ratios of industrial discharge peak to average flow rates (P/A). Factor 'M' is obtained from the formula $M = 2.50 \log (P/A)$.

Tax = the annual ad valorem taxdes paid to the Districts during the accrual year on the land or property utilized for the generation of industrial wastewater in dollars.

For the first three years of surcharge programme, the Districts endeavoured to recapture the full industrial share of capital (exclusive of Federal grant funds) and O and M costs. For these years, the total capital outlay for construction of sewerage facilities over the future five fiscal years was estimated and distributed to the three parameters of total wastewater flow, COD and SS. The distribution was in accordance with the percentage of the total existing capital facilities of the sewerage system predominantly related to each parameter. For example, sludge digester costs are primarily related to the suspended solids parameter while activated sludge aeration tanks are primarily related to the COD parameter. The total O and M costs were estimated for the future five fiscal years and distributed among the three surcharge parameters.

The sum of the capital and O and M costs determined for the parameters of total flow, COD and suspended solids were divided by the projected five year total volume of wastewater and the total weight of COD and suspended solids to be treated by the sewerage system, in order to obtain the unit charge rates 'a', 'b' and 'c'. Table 62.8 gives the surcharge rates used for the two fiscal years 1972-73 and 1973-74.

Table 62.8: Industrial Wastewater Treatment Surcharge Rates in USA

Unit Rate	*Parameter*	*Surcharge Rates (in $)*	
		1972-73	*1973-74*
a	Volume, MG/year	83.25	104.00
b	COD, 1000 lbs/year	4.75	6.25
c	SS, 1000 lbs/year	11.00	14.25
d	Peak Flow, GPM	22.00	18.75
Flat Rate Charge	Volume, MG/year	181.00	230.00

In the latest revisions to the Districts ordinance, the total annual capital costs are now calculated as 1/30 of the total capital investment in the sewerage system during the 30-year period prior to the accrual fiscal year. The O and M costs used are those estimated for the next future fiscal year. The sum of these capital and O and M costs is distributed to flow, COD and suspended solids parameters as indicated above.

A peak flow charge was established to develop revenue from companies who make periodic short term demands upon the sewerage system that are not reflected in the annual charges for flow, COD and SS and to encourage companies to discharge the wastewaters during the periods of low flow within the District's system. The companies are not charged for peak flow discharged between the hours of 10.00 p.m. and 8.00 a.m. Industrial companies having yearly wastewater flows of 6 MG or less are permitted to file a surcharge statement using a flat rate charge equal to the charge for 'typical' industrial wastewater.

The companies with flow rates reaching a maximum of over 100 GPM or an average over 50000 GPD must provide a continuous measurement and recording of the wastewater flow. Palmer-Bowlus flumes are used for flow measurements as they offer several advantages over weirs and Parshall flumes in open channel measurements [12,13].

Chemtech Formulation

The formulation developed by Chemtech Consultants [14] is based on the proportion of the CETP usage. A fair distribution of cost for a CETP would be that where industries pay according to the load which they impose on the CETP. This load comprises not only the volumetric flow but also include the influence that chemical, physical and biological conditions of the waste have on capital and operational cost of the CETP.

The formulation is as given below:

$$CC_{(i)} = \frac{Q_{(i)}}{Q_{(T)}} \times CC_{(Q)} + \left(0.5 \times \frac{B_{(i)}}{B_{(T)}} + 0.5 \times \frac{C_{(i)}}{C_{(T)}}\right) \times CC_{(B)} = a \times CC_{(A)}$$

where,

$CC_{(i)}$ = Capital cost contribution by the ith industry in Rs.
$Q_{(i)}$ = Wastewater flow produced by the ith industry in m^3/d
$Q_{(T)}$ = Total wastewater flow to the CETP in m^3/d
$CC_{(Q)}$ = The flow dependant cost of the CETP in Rs.
$B_{(i)}$ = The BOD load of effluent from the ith industry in kg/d
$B_{(T)}$ = The BOD capacity of the CETP in kg/d
$C_{(i)}$ = The COD load of effluent from the ith industry in kg/d
$C_{(T)}$ = The COD capacity of the CETP in kg/d
$CC_{(B)}$ = The BOD and COD capacity dependant cost in Rs. of the CETP in Rs.
$CC_{(A)}$ = The cost of common facilities for the CETP in Rs.
a = The factor that is a function of the ith industry's design flow

Cost of neutralization has not been included in the formula as this operation must be a part of the individual industry's pretreatment.

Rather than complicating this formula to adjust for a connection fee, it is recommended that this be kept as a separate item to be charged according to actual cost of labour and magterials at the time of connection, distance from collection system, size of piping and existing conditions at the site.

For the recovery of operating and maintenance cost, the following formula has been suggested:

$$CM_{(i)} = \frac{Q_{(i)}}{Q_{(T)}} \times CM_{(T)} \times PF_{(i)}$$

where,

$CM_{(i)}$ = Operation and maintenance cost of the ith industry in Rs.

$CM_{(T)}$ = Total operation and maintenance cost in Rs.

$PF_{(i)}$ = The pollution factor for the ith industry, defined as follows:

$$PF = 0.5 \times \frac{BOD_{(i)} + 200}{600} + 0.5 \times \frac{COD_{(i)} + 500}{1500}$$

If $BOD_{(i)}$ and/or $COD_{(i)}$ values result in $PF_{(i)} < 1$, then $PF_{(i)} = 1$.

The cost for capital recovery and operation and maintenance is expressed as the combined figure in the calculations made for each industrial estate under the different grant schemes and plant capacities. It is, therefore, convenient to do the same for the individual industry's contribution which will allow the CETP operator to recover all costs, or in case of a private operator, all costs plus a profit.

Since both the capital and operating costs to a very large extent are dependant on only volumetric throughput and level of organic pollution, and since the proportion between these two parameters remain reasonably constant for the most of the type of design generally considered, it is possible to combine the above formulae to a single payment schedule which can be used at each estate and is as follows:

$$CO_{(i)} = \frac{Q_{(i)}}{Q_{(T)}} \times PF_{(i)} \times CO_{(T)} + A$$

where,

$CO_{(i)}$ = Total annual contribution by the ith industry in Rs.

$CO_{(T)}$ = Total annual income required by the CETP operator in Rs.

A = A minimum contribution by each industry

$$PF_{(i)} = 0.5 \times \frac{BOD_{(i)} + 200}{600} + 0.5 \frac{COD_{(i)} + 500}{1500}, \text{ for } PF_{(i)} > 1$$

$$PF_{(i)} = 1.0, \text{ for } PF_{(i)} < 1.0$$

Graduated Payments Formulation

If desired, it is possible to arrange graduated payments whereby the major (large) industries pay marginally higher cess.

Generally, the major industries are few in numbers in any industrial estate, but they contribute the largest share to the total effluent flow in an estate. A small increase in the cess for these industries will ensure that approximately 50 – 65 per cent of the industries producing less than 10 m^3/d of waste, only need to pay a minimal fixed amount for being connected to the CETP.

The Chemtech formulation for the annual contribution from each industry, CO (i), can be easily corrected by eliminating all but the minimum contribution A for small industries:

$$CO_{(i)} = \frac{Q_{(i)}}{Q_{(T)}} \times PF_{(i)} \times CO_{(T)}, \text{ for } Q_{(i)} > 10\,m^3/d$$

$$CO_{(i)} = A, \text{ for } Q_{(i)} < 10\,m^3/d$$

Apart from making the CETP economically more acceptable for the large proportion of industries in the estate, this arrangement would also result in much simplified policing and accounting procedures for the CETP operator and the authorities. It would also reduce the amount of work required to determine the wastewater production from the many small individual industries, as well as avoiding arguments between the CETP operator and the smaller industries over actual volumes of effluent.

Other payment methods benefiting the smaller industries are, of course, possible. Following illustrations for a typical Indian industrial estate explain the concept [14]:

- ☆ In instead of all industries paying 4.89 Rs./m^3 for the effluent going to the CETP, those with more than 50 m^3/d effluent paid 5.4 Rs./m^3, then those below that figure would have to pay only 3.0 Rs./m^3 to ensure the same income for the CETP operator.
- ☆ If industries with more than 50 m^3/d effluent paid 5.1 Rs./m^3 and those with 10 – 50 m^3/d paid 4.6 Rs./m^3, then the small operator with less than 10 m^3/d effluent would have to pay only 3 Rs./m^3 or a fixed amount of 7,000 Rs./year.
- ☆ If the very large industries (more than 1,000 m^3/d) paid 5.5 Rs./m^3, and the medium scale producers (50 – 1,000 m^3/d) paid 4.6 Rs./m^3, then the small industries (in this case less than 50 m^3/d) would need to pay only a fixed minimum amount.

The above illustrations show that, if required, it will not be difficult or not much more costly to the larger industries, to favour small industries to an extent that they can afford [14].

Drainage Service Charge Formulation

The formulation developed is as follows [15]:

Wastewater service charge = $V R_v + V [(BOD-250)R_B + (SS-300)R_S]$

where,

V = Wastewater volume in '000 m^3

BOD = Average BOD of wastewater, mg/L

SS = Average suspended solids of wastewaters, mg/L

R_V = Charge rate per 1000 m^3 volume

R_B = Charge rate per kg of excess BOD

R_S = Charge rate per kg of excess SS

250 = BOD standard, mg/L

300 = Suspended solids standard, mg/L

Barnard and Eckenfelder Formulation

The factors which significantly influence the cost of the wastewater treatment are the flow rate, BOD, suspended solids and the biological reaction rate in case of biological treatment facilities.

The flow rate affects the size of all processing units and, therefore, it will have the most significant influence on capital cost. BOD value will affect the size of the aeration basins and the required aeration horsepower. Higher concentrations of BOD will result in the production of more biological sludge and will, therefore, affect the sludge handling facilities. The concentration of suspended solids will primarily affect the overall sludge handling facility. A relationship developed by Barnard and Eckenfelder [16] to assess the capital cost of a combined municipal activated sludge plant is as follows:

$$\text{Cost}(\$1000) = 583\,Q^{0.84} + (110 + 37Q)\frac{S_o}{200} + (77 + 23Q)\left(\frac{SS}{200} - 1\right)$$

where,

Q = Hydraulic flow rate, in MGD

S_o = Influent raw waste BOD concentration in mg/L

SS = Raw waste SS concentration in mg/L

Using this equation, it is possible to define the net increase in cost of a waste treatment facility for various percentages of industrial wastewater. It is also possible to compute the cost of CETPs from process and cost models of each unit. The capital cost models for combined waste treatment as developed by Smith [17] and modified by Barnard [18] are given in Table 62.9. Operating cost models are also presented in Table 62.10. When considering domestic or combined wastewater treatment facilities, excess capacity is usually provided to handle peak flows and to maintain adequate treatment during periods of maintenance or breakdown. The excess capacity factors to be included in the cost analysis are shown in Table 62.11.

Table 62.9: Capital Cost Models for Domestic and Combined Wastes

Treatment Unit	*Capital Cost in 1000 Dollars*
Pretreatment Cost	$= 19 \times Q^{0.63}$
Primarhy sedimentation[b]	$= 173\ SA + 6.7\ (SA)^{0.1}$
Activated sludge	= 226 (Vol. In MG) + 67
Blower house	= 13.6 + 7.6 (CFM/1000)
Sludge return pumps	= 4.7 + 1.45 Q
Final sedimentation[b]	$= 16.2\ SA + 6.9/\ (SA)^{0.13}$
Chlorination	$= 11.6 \times Q^{0.47}$
Thickener	= SA[24.2 + 11.7/exp (SA/13.3)]
Anaerobic digester[c]	$= V\ (1.34 + 13.8/V^{0.87})$
Vacuum filters	$= 16.5 + 48\ [\text{area}\ (\text{ft})^2/100]$
Sludge incinerator[d]	$= (S/24000)\ (170 + 7.15\ S^{0.61})$
Control house[e]	$= 51.6 \times Q^{0.7}$

a: Cost updated to Sep. 1969; b: SA = Surface area in 1000 sft; c: V = Vol. In 1000 cft; d: S = Sludge production in lbs/day; e: Q = Flow rate in MGD.

Table 62.10: Operation Cost Models for Combined Wastes*

Treatment Unit	*Annual Operation Cost ($ per MGD)*
Pretreatment[a] Cost	$= 500 + 2150/Q^{0.63}$
Primarhy sedimentation	$= 909 + 2370/Q^{0.5}$
Digested sludge[b]	$= 2700 + 2500/(\text{Vol. in MG})^{0.67}$
Anaerobic digester[c]	$= 1200(V/Q)(0.048 + 0.54/V^{0.5})$
Drying beds	$= 1.2\ (S/Q)(0.21 + 29.7/S^{0.5})$
Vacuum filtration	$= 0.18(S/Q)[700/F + 0.38(2 - 0.1Q) + 0.027\ C/F]$
Incinerator	$= 1500 + 6450/Q^{0.63})$

*Source = Eckenfelder W.W., Jr. and Adams C.E., Jr., ASCE, SA 1, 1972.

a: Q = The flow rate in MGD, F = Area of the vacuum filter in 100 sq ft.;

b: V = The volume in 1000 cft, C = Capital cost in 1000 dollars;

c: S = Total sludge in lbs/day.

Cost Recovery Based on Water Consumption

Although it is considered by many that water consumption is proportional to wastewater volume with reasonable accuracy, there are two unfair elements in assuming that the capital and operating cost of the CETP are also proportional to the water consumption:

Table 62.11: Excdess Capacity Factors for Process Units*

Process	*Factor*
Preliminary	1.0
Primarhy clarification	2.0-0.008 Q
Activated sludge	1.3-0.002 Q
Aeration	1.8-0.004 Q
Sludge return	2.0-0.005 Q
Final clarifier	2.0-0.007 Q
Chlorinator	1.0
Thickener	1.5-0.004 Q
Aerobic digester	1.5-0.003 Q
Anaerobic digester	2.0-0.005 Q
Centrifuge	2.0-0.005 Q
Vacuum filter	2.0-0.005 Q
Sludge drying beds	1.0

* For flow rate, Q, ranging from 1 MGD to 100 MGD.

- ☆ Though the pollution load of the wastewater is an almost as equally important as volume of waste, it is not taken into consideration while arriving at the construction and operating cost of the CETP
- ☆ The water flow meters at industrial estates are most of the time found to be out of order, which makes the total water consumption figures more difficult to obtain and to rely on. Obviously, the water consumption figures in the industrial estates are unreliable, especially as the basis for estimation of total wastewater flow [(14)].

GBMSD Methodology

Green Bay Metropolitan Sewerage District (GBMSD) in 1970 found 25 per cent reduction in capital cost while treating municipal waste jointly alongwith paper mill wastes. The District developed a rate structure for financial apportionment in CETP. The final rate to be charged each year reflected actual costs as developed from the regular annual audit of the District's financial records. Because major users of the treatment system committed themselves to a fixed capacity, the capital portion of the user charge remained essentially constant. As operating cost depends on actual use, this portion of the user charge varied from month to month based on load. The annual user charge relating to the construction of plant facilities and sewers was developed initially on the basis of cost estimates and subsequently it was refined by using actual costs once construction was completed. Major users of the system were billed in advance by the District on a monthly basis. An estimate of total charges was made at the beginning of each month, with these charges being prorated on the basis of the flow, BOD and SS expected during that month. At the end of the fiscal year, these estimated charges were reconciled against charges for waste quantities actually discharged during the year. A refund or an additional bill was then issued to clear the books [19].

Flecksedar Methodology

Flecksedar [20] suggested apportioning of costs amongst the member industries of the CETP scheme as follows:

- ☆ As indicated below, differentiate into design and operation; where 'design' is based on the estimates set for the design, and operation in real practice, latter has to be founded on continued measurements of the actual state.
- ☆ Parts of a combined system which are exclusively attributable to one partner (*e.g.* separate primary sedimentation for a pulp mill for fiber recovery) are his financial responsibility.
- ☆ Costs of pumping stations, sedimentation tanks and conduits are split based on flow from every partner (Design/Operation).
- ☆ Investment for aeration tank is split according to loads from partners being used in design calculations of aeration tank volume (Design).
- ☆ Investment for aeration equipment is split according to every partner's contribution to aeration capacity (Design).
- ☆ Running cost for aeration is split same as done in above step but with average loads (Operation).
- ☆ Cost of thickening, sludge dewatering and sludge handling are split on the basis of excess sludge generated (Design/Operation).
- ☆ Remaining costs are split according to 'normalized ratios' of the previous categories.

NEERI's Formulation

The formula developed by NEERI for apportioning costs of CETP construction, and operation and maintenance are based on the premise that the total cost of collection, treatment and disposal depends, directly or indirectly, on wastewater flow and waste characteristics.

The apportionment of the total cost, directly attributable to each waste parameter, will be estimated by assigning the unit cost for the items, such as those listed below, to the parameters that primarily affect the design of CETP.

CETP Components	*Parameter that Primarily Affects the Cost of CETP*
Drainage	Total flow
Pumps and pumping chambers	Total flow
Screen chamber	Screenings
Grit chamber	Grit contents
Parshall flume	Total flow

Contd...

Contd...

CETP Components	*Parameter that Primarily Affects the Cost of CETP*
Equalization tank	Fluctuations in flow and characteristics
Neutralization tank	pH
Clariflocculator	Suspended and Colloidal Solids
Sand filter	Colloidal Solids
Adsorption columns	Trace organics

Some costs are not directly attributable to flow or waste characteristics, and are best assigned as fixed overheads. Examples of such costs are:

- Capital, operation and maintenance costs for water supply in CETP
- Capital and maintenance costs for land, building and roads
- Costs of plant supervisors and trained operating staff
- Cost of green-belt development

Depending upon the level of wastewater treatment proposed for end use, the above basis will be modified, and the mode of payment ascertained.

The users share will be estimated on the basis of the loading parameters. For example, the cost assigned to the COD can be allocated to each user's contribution of COD.

The formula derived, on this principle, is of the form:

$$TC = TC_v + TC_{COD} + TC_{SS}$$

where,

TC = Total annualized cost of waste treatment facility

TC_v = Total cost assigned to flow

TC_{COD} = Total cost assigned to COD

TC_{SS} = Total cost assigned to suspended solids for a particular industry, if

F_v = Fraction of total flow contributed by the industry

F_{COD} = Fraction of total COD contributed by the industry

F_{SS} = Fraction of total suspended solids contributed by the industry

Then, the total cost allocated to the industry, is

$$TC_i = F_V\ TC_V + F_{COD}.TC_{COD} + F_{SS}.TC_{SS}$$

or

$$TC_i = \left(\frac{V_i}{V}\right) TC_V + \left(\frac{COD_i}{COD}\right) T_{COD} + \left(\frac{SS_i}{SS}\right) TC_{SS}$$

where,

V_i = Flow contributed by the industry, cum/d

V = Total flow, cum/d

COD_i = COD contributed by the industry, kg/d

COD = Total COD, kg/d

SS_i = Suspended solids contributed by the industry, kg/d

SS = Total suspended solids, kg/d

Proposed Cost Apportionment

For CETPS in NCT by NEERI

The formulae, based on wastewater flow and characteristics normally used for cost apportioning of CETPs, are not directly applicable to CETPs in NCT as:

- ☆ The survey of industrial typology in NCT by NEERI in April 1996 reveals that there are significant number of non-polluting/dry units in the industrial estates.
- ☆ Most of the water polluting units are in small scale sector, and are spatially distributed in the industrial estates.
- ☆ Aforereferred formulae are applicable for biological treatment of wastewater, whereas the COD and heavy metals are the primary pollutants emanating from industries in industrial estates of NCT under consideration. Thus the options for design and, hence the cost apportionment, in CETPs of NCT are:
    - Strict adherence to the 'Polluter Pays Principle' requiring thorough routine monitoring of wastewater from water polluting industries. This option requires wastewater to be transported through tankers to the CETP sites.
    - Design and construction of new drainage systems/upgradation of existing drainage systems, and to design CETPs to treat entire wastewater from industrial estates and apportionment of costs to 'all industries' on individual plots in the industrial estate (s).

  Accordingly, the proposed cost apportionment formulae are:

Cost Apportionment based on 'Polluter Pays Principle'

$$C_i = \left(\frac{AT_i}{\Sigma AT_i}\right) \times C_s$$

where,

C_i = Annual cost burden from waste treatment to be shared by water polluting industrial unit i

C_s = Total annual cost burden from waste treatment for the entire industrial estate

AT_i = Annual turnover of water polluting industrial unit I, Rs.

ΣAT_i = nnual turnover of all water polluting industries served by the CETP, Rs.

Alternatively,

$$C_i + \left(\frac{Q_i}{\Sigma Q_i}\right) x\ C_s$$

where,

C_i = Cost to be shared by water polluting industrial unit I

C_s = Total cost of CETP in the estate

Q_i = Wastewater flow from polluting industrial unit i, cum/d

ΣQ_i = Total wastewater flow treated in CETP, cum/d

Cost Apportionment to all units Existing in Industrial Estates

$$C_{p(i)} = \left[\frac{Po_{p(i)}}{\Sigma Po_p} x\ 0.4 + \Sigma \frac{WS_{p(i)}}{\Sigma WS_p} x\ 0.3 + \frac{Lab_{p(i)}}{\Sigma Lab_{p(i)}} x\ 0.2 + \frac{La_{p(i)}}{La_p} x\ 0.1\right]$$

$$x\ C_s + C_{f(i)}$$

where,

C_p = Cost to be shared by units located on plot (i)

C_s = Total cost to be raised by the industrial estates

$Po_{p\,(i)}$ = Total horsepower of pumps installed on tubewells in plot (i), hp

ΣPo_p = Total horsepower of pumps installed on tubewells in the estate, hp

$WS_{p\,(i)}$ = MCO water supply to plot (i), kilolitres/month

ΣWS_p = Total MCD water supply to the estate, kilolitres/month

$La_{p\,(i)}$ = Area of plot (i), sq.m.

ΣLa_p = Total area of plots in the estate, sq.m.

$Lab_{p\,(i)}$ = Number of labours working in plot (i)

ΣLab_p = Total number of labours working in the industrial estate'

$C_{f\,(i)}$ = Additional fixed cost to be paid by plot owner (i) depending upon the

Nature of activaties carried out in the plot, such as:

Activity Carried Out in Plot	*Addl. Fixed Cost, C_p, Disbursed by the Owner of Plot in the Industrial Estate, Rs.*
A. Highly polluting activities	10,000
Pickling	
Electroplating/anodizing	
Chemical manufacturing	
Pharmaceutical manufacturing	
Soap and detergent manufacturing	
Paint/varnish/dye manufacturing	
Rubber/Plastic reprocessing	
Similar industries as per DPCC classification	
B. Medium polluting activities	5,000
Cloth Dyeing and printing	
Food processing	
Meat processing	
Foam manufacturing	
Automobile Service stations	
Laundry and dry-cleaning	
Similar industries as per DPCC classification	
C. Low polluting activities	2,000
Wire drawing	
Activities resulting in only cooling water generation, *e.g.* Cold and hot rolling	
Printing press	
Similar industries as per DPCC classification	

Industries, where a functional ETP exists meeting prescribed standards, will be treated at par with dry industries and in such cases, only land and labour coefficients of Equation 17 will be operative.

In view of the typology and small/medium scale of operation of industrial units in various industrial estates in NCT, as also the findings of the survey conducted by NEERI team in April, 1996 on the state of drainage system; it is proposed to upgrade the existing drainage systems/design new drainage systems; and to design CETPs to treat entire quantity of wastewater generated by the industrial estates. The costs will be apportioned among the plot owners employing Equation 17. The endeavour in the design is on producing wastewater from CETPs in industrial estates that is amenable for reuse in proposed green-belts, and for cooling, flushing, and other usages in the industrial estates thereby conserving water, and recovering part of O and M costs.

It is estimated that the order of magnitude amount to be paid by individual plot owners would range from Rs.5000 to Rs.50,000 to meet 20 per cent share of the capital cost of CETPs as the initial instalment of industries' contribution. Repayment of loan share by the individual plot owner will be worked out on the same basis.

Watson *et al.* Recommendations

It is apparent that the regulations regarding user charges and industrial cost recovery provides a maximum flexibility for the establishment of equitable charges for industrial plants connected to municipal sewer systems.

There are obvious costs, such as administrative, billing and collection, that have little to do with the quantity and quality of discharge. The large user, therefore, should not be required to pay user costs for such items on the same basis as the small ones. Furthermore, the guidelines ignore use of the basic and long used concept of rate-making for the water and sewerage fields, that treatment capacity and its operation decrease in cost per unit volume as the total volume to be handled increase. Therefore, in the interest of complete sound equity for all users of the system, there is a justification for the use of a sliding scale in many cases [(21)].

References

1. Subramanian V., 'Planning and development of combined effluent treatment facility for small industries – A case study', *IAWPC Tech. Annual*, XII, 1985.
2. Kulkarni V.S., Saraswati N., and Khanna P., 'Development and implementation of clean technologies of industrial production', Invited paper – presented at *IAEM Annual Conference*, 1990.
3. 'Current practices in municipal treatment of industrial wastes' – Chemical Industry Committee of Ohio River Valley Water Sanitation Commission – *Sewage and Industrial Wastes*, Vol. 29, No. 6, June 1957.
4. Malz F., 'Water Pollution – Institutions and their financing – chemical analytical aspects', *Progress in Water Tech.*, Vol. 7, No. 5, 1975.
5. Cheda P.V., 'Wastewater management for tannery complex at Jajmau, Kanpur', *M.Tech. Project*, ESE group, IIT, Bombay, 1982.
6. Roman M., 'Economic aspects of water protection against pollution in Poland', Proc. IDF Seminar on Dairy Effluents, Doc. 104, IDF, Brussels, 1978.
7. Yorkshire Water Authority, 'Code of practice for the control of trade effluent draining to publicsewers', 1976.
8. McGowan G., 'Royal commission on sewage disposal', 5th report, Appendix IV, 1-9 Cmt. 4282, HMSO, London, 1908.
9. Hall I.R. and Paine R.E., 'The treatment of coal carbonization wastewaters in admixture with sewage', *Water Pollution Control*, Vol. 80, No. 1, 1981.
10. Okawa Y., Shinozuka S., Ota R. and Matsui S., 'Experience of 16 years operation and maintenance of the Fukashiba industrial wastewater treatment plant of the Kashima Petrochemical Complex – Operation and maintenance', *Water Science and Technology*, Vol. 20, No. 10, 1988.

11. Jones A.D., 'The treatment of high proportions of a textile effluent in admixture with sewage', *Water Pollution Control,* Vol. 73, No. 5, 1974.

12. Parkhurst J.D., Kremer J.G., and Chu J.P., 'Municipal industrial waste program nets $ 8.9 million in two years', *Water and Sewage Works,* April 1976.

13. Parkhurst J.D., Kremer J.G. and Chu J.P., 'Development of an industrial waste program in a major sewerage agency', *30th Industrial Waste Conference,* Purdue University, 1975.

14. World Bank Industrial Pollution Control Projects – 'Feasibility assessment of common treatment facilities in India', Report prepared by M/s Chemcontrol, Copenhagen, Denmark, Vol. 1, March 1991.

15. 'Operation of wastewater system – *Water and wastewater technology* – SI version', Chapter 12, Ed. Mark J. Hammer, John Willey and Sons Inc., New York.

16. Eckenfelder W.W. Jr., and Adams C.E. Jr., 'Design and economics of joint wastewater treatment', *JASCE (San. Engg. Dn.), SAI,* February 1972.

17. Smith R., 'Cost of conventional and advanced treatment of wastewaters', U.S. Deptt. of Interior, FWQA, Cincinnati, Ohio, 1968.

18. Barnard J., Thesis presented to Vanderbilt University at Nashville, Tenn. In partial fulfillment of the requirements for the degree of *Ph.D.*, 1971.

19. Barton G.A., Byrd J.F., Faulkender C.R., and Moss O.H., 'Joint treatment of pulping and municipal wastes', *JWPCF,* Vol. 47, No. 5, 1975.

20. Fleckseder H., 'Preliminary studies on combined or separate treatment (as experienced with effluents from industries processing wood)', *Prog. in Water Tech.*, Vol. 12, No. 3, 1980.

21. Watson K.S., Goldsmith H.I., and Pico R.F., 'Joint treatment versus pretreatment of food processing wastes', *JWPCF,* Vol. 45, No. 12, 1973.

22. NEERI's State-of-the-Art Report on *Common Effluent Treatment Plant,* October 1991.

Chapter 63

Utilisation of Fly ash in Pesticide Dusting Powder Formulation and as Wasp Repellent

S.S. Bagchi [1] *and V.S. Kale* [2]

[1]*Reliance Infrastructure, D-3 MIDC Butibori, Nagpur*
[2]*Yawalkar Pesticides Ltd., Kamptee Road, Nagpur*

ABSTRACT

Methyl parathion 2 per cent dusting powder (D.P) is an organophosphorus pesticide recommended for controlling chewing and sucking type of pests of agricultural crops it kills insect by contact, stomach and respiratory action. Methyl parathion 2 per cent dusting powder is formulated as per the norms of Central Insecticide Board (CIB). Methyl parathion is cholinesterase inhibitors therefore fumigation is to be prevented during manufacturing and applications. In this current process, white clay is pulverized and it is then blended with soap stone powder and after which liquid technical grade pesticides are dispersed on this blended material.The problem is related to the dispersion process. The nebuliser, which is used to spray liquid technical pesticides on pulverized clay powder, gets choked frequently. This is due to the evaporation of solvent of technical pesticide, which results in bigger droplet causing agglomeration of absorbent. This hampers the process and finally affects the product quality. To overcome with the problem of agglomeration, the study was conducted by using fly ash as a dispersant.In the formulation of methyl parathion 2 per cent dp and soapstone powder was replaced completely. By using fly ash up to 50 per cent it was found to be a suitable dispersant to solve the problem of agglomeration and moreover its very cost effective as its saves 40 per cent on electricity, soapstone powder,crusher wear and tear, time and manpower.Wasp colonies are common in power plant, wasp stung causes pain, burning, itchiness, redness,swelling and even allergy. Pain will last for a few hours. Displacement of wasp colonies is often required, Fly ash particle size below 45 (μ)micron has wasp repellent properties and fly ash is a cheap source can be used to displace the wasp colonies.

Keywords: *Alkalinity, Bulk density, Methyl parathion, Dusting powder, Dispersant, Fly ash, Wasp repellent.*

Introduction

Fly Ash Mission (govt. of India) has isolated many thrust areas of fly ash utilization, one of them is agriculture related studies and application. Bituminous fly ash is a multinutrient carrier and soil conditioner.[4] Present study is undertaken to give suitable dispersant to pesticide formulation industry replacing costly soapstone powder, and gainful utilization of fly ash. Methyl parathion 2 per cent dp is selected for study as this pesticide is very common. Moreover it is one of the cheapest and largest selling dustable powder. Methyl parathion is an ogrganophosphorus pesticides recommended for controlling chewing and sucking type of pests and mites contains 2 per cent (w/w) methyl parathion as active ingredient and 98 per cent (w/w) processed carriers. It is a ready to use dust formulation 15-25 kgs of dosage per hectare is recommended for different types of crops.

Manufacturing Process

The formulation of 2 per cent methyl dp is done as per the norms of Central Insecticide Board (CIB). The major steps are as follows:

Step I Pulverization of white clay

Step II Blending of soapstone powder with clay powder

Step III Technical Pesticide addition mixture of clay powder and soapstone powder through nebulisation process.

In the step (III) the nebuliser get chocked frequently due to the evaporation of solvent of technical pesticides, hence bigger droplet causes agglomeration of absorbent, which hampers the process,increases the time period and finally affect the product quality.To overcome this problem of agglomeration the study was conducted using fly ash as a dispersant by replacing soapstone powder.

Material and Methods

Characterization of fly ash was done and dry fly ash was collected from silo of Koradi Thermal Power Station and analysis is carried out as per IS 6940: 1982.Fly ash of Koradi TPS is of silicious type and dry fly ash from last row is preferred because of its particle size. Formulation of methyl parathion dp is prepared in 1kg lab batch. 25gm (w/w) technical grade methyl parathion (purity 80 per cent) is added to 200gm, 300gm, 400gm, 500gm and 600gm of fly ash and clay blended mixture to make it 1kg sample and simultaneously 1kg sample of methyl parathion is prepared by current formulation method. Analysis of finished product is carried out and compared with IS specification IS6940: 1982.

Current formulation as per Indian standard.

1. Technical Grade Material Methyl Parathion (80 per cent) 2.5 per cent
2. Adjuvant
   a) Mineral absorbent Clay 10 per cent
   b) Mineral carrier Soapstone powder 87.5 per cent sample are numbered as follows:

Name	Kind of ingredient	Per cent (w/w)
Sample A — Current formulation		
10 per cent (w/w) clay + 87.5 per cent (w/w) soapstone powder+2.5 per cent (w/w) Technical		
Sample B — 97.5 fly ash (w/w) +2.5 per cent (w/w) Technical		
Sample C — 10 per cent (w/w) fly ash +87.5 per cent (w/w) clay +2.5 per cent (w/w) Technical		
Sample D — 20 per cent (w/w) fly ash + 77.5 per cent (w/w) clay +2.5 per cent (w/w) Technical		
Sample E — 30 per cent (w/w) fly ash + 67.5 per cent (w/w) clay +2.5 per cent (w/w) Technical		
Sample F — 40 per cent (w/w) fly ash +57.5 per cent (w/w) clay +2.5 per cent (w/w) Technical		
Sample G — 50 per cent (w/w) fly ash +47.5 per cent (w/w) clay +2.5 per cent (w/w) Technical		
Sample H —— 60 per cent (w/w) fly ash +37.5 per cent (w/w) clay +2.5 per cent (w/w) Technical		
Sample I —— 70 per cent (w/w) fly ash +27.5 per cent (w/w) clay +2.5 per cent (w/w) Technical		
Sample J —— 97.5 per cent (w/w) clay +2.5 per cent (w/w) Technical		

Results and Discussion

To act as a carrier or diluent, the material must be homogenous, free flowing and devoid of lumps while fineness should be less than 75μ, preferably between 50-80μ.[1] Bulk density should not be decreased more than 60 per cent after compaction as it directly affects the flowability of the dusting powder and further, should absorb less moisture. Inert carriers or fillers should not catalyze the decomposition of pesticide.[1] Hence following tests were carried out to test the suitability of fly ash as a carrier: (1) Particle size (2) Bulk Density (3) AlkalinityIt was observed that addition of 100 per cent fly ash would not serve the purpose and the agglomeration problem remains the same; hence the formulation was not selected. Requirement of adjuvant for methyl parathion 2 per cent DP is as per IS 8960: 1978 and analysis of fly ash used in formulation is shown in Table 63.1.

Sl.No. 4 and Sl.No. 5 is the additional test carried out for the future stability of the product as it has been observed that more per cent of unburnt carbon in ash will absorb more moisture and it will increase the alkalinity/acidity of product, which result in active ingredient degradation. Fly ash and methyl parathion technical batches are kept constant in entire formulation. Specification of Methyl Parathion Technical is presented in Table 63.2.

Table 63.1: Fly Ash Properties for Formulation

Sl.No.	*Characteristics*	*Requirement as per IS 8960: 1978*	*Observed Values of Fly Ash*
1.	Per cent Material passing through 75 micron Sieve	90	92.44 ± 0.52
2.	Bulk Density after compacting per cent	Not to exceed the value obtained before compacting by more than 60 per cent	51.15±0.41
3.	Alkalinity (as NaoH) per cent	0.1	0.018 ± 0.0011
4.	Unburnt carbon per cent		0.50
5.	Moisture per cent		0.20

Table 63.2: Specification of Methyl Parathion Technical from Manufacturer's Analytical Report[8]

Specification	*Value*
Content of Methyl Parathion, per cent (w/w)	80.2
Acidity (as H_2SO_4), per cent by mass	0.2
Insoluble substance in acetone, per cent by mass	0.3
Content of free paranitrophenol and paranitrophenol caused by hydrolytic impurities, per cent by mass	1.2
Water content, per cent by mass	0.18
Insoluble substance in acetone, per cent by mass	0.3
Molecular Formula	$C_8H_{10}NO_5PS$
Molecular Weight.	263.21

Analysis of finished product is presented in Table 63.3.

It was observed that the mean value of per cent increase in bulk density of samples C, D, E, F and G with 10 per cent, 20 per cent, 30 per cent, 40 per cent and 50 per cent fly ash w/w were complying the Indian Standard and samples H, I, B with 60 per cent, 70 per cent and 97.5 per cent fly ash were not suitable for formulation of Methyl parathion 2 per cent dusting powder. It has been observed that increase in fly ash per cent will decrease the bulk density of the product which is an important parameter in this product, but it prevents agglomeration and hence the shelf life. From the above data (Table 63.3) it is observed that upto 50 per cent dry fly ash from last row of ESP is optimum.

Stability Study

A stability study was conducted after six months and again after one year of stored samples in ambient conditions, to assess the degradation of product in the presence or absence of fly ash. Sample B (97.5 per cent fly ash) was discarded as the methyl parathion technical did not retain on the fly ash. Results of six monthly

Table 6.3: Analysis of Finished Product

Sample Name	Description	Material Passing through 75μ Sieve (per cent)	Bulk Density g/cc of Sample Bd-b	Bulk Density g/cc of Sample Bd-a	Per cent Increase in from BD-b	Alkalinity as NaOH per cent by mass	Methyl Parathion Content per cent by Mass (Average of five analysis)	Mean value ±Sd	Per cent Increase in Bd- b after One Year
Sample–A	Free flowing powder	94.50	0.520	0.680	30.76	0.018	1.98-2.02	2.03±0.02	32.7
Sample–B	Free flowing powder	92.30	0.780	1.36	74.35	0.019	1.39-1.42	1.41±0.02	67.1
Sample–C	Free flowing powder	93.75	0.600	0.740	23.33	0.019	2.0-2.03	2.02±0.01	23.7
Sample–D	Free flowing powder	93.55	0.620	0.790	27.41	0.018	2.0.-2.1	2.06 ±0.02	27.4
Sample–E	Free flowing powder	93.30	0.650	0.930	43.07	0.017	2.0-2.08	2.02 ±0.02	4.35
Sample–F	Free flowing powder	93.10	0.660	0.980	48.48	0.017	1.98-2.01	1.99 ±0.01	48.42
Sample–G	Free flowing powder	92.85	0.690	1.10	59.42	0.018	1.99-2.05	2.0±0.01	59.50
Sample–H	Free flowing powder	92.70	0.710	1.18	66.19	0.020	2.02-2.10	2.05±0.02	66.50
Sample–I	Free flowing powder	92.40	0.720	1.22	69.44	0.019	2.02-2.12	2.04±0.02	66.50
Sample–J	Free flowing powder	94.6	0.580	0.690	17.9	0.019	1.99-2.05	2.01±0.02	18.2

Bd-a: Bulk Density after compaction; Bd-b: Bulk Density before compaction.

stability analyses have shown that the product did not degrade and the per cent content was same. However, degradation was observed after one year. Stability study after one year indicates that there was decrease in percentage purity in all the samples and it was in the range of 3.5 to 7 per cent but it was at par with the controls. The stability data revealed that Methyl parathion 2 per cent Dusting Powder remains stable for one year in ambient conditions of the laboratory (Table 63.4).

Table 63.4: Percentage Increase of Bulk Density of Samples A to J

Sl.No.	*Sample Name*	*Mean Value of Per cent Increase of Bulk Density*
1.	A (Current formulation)	30.19
2.	B (fly ash 97.5 per cent)	76.27
3.	C (fly ash 10 per cent)	24.51
4.	D (fly ash 20 per cent)	27.12
5.	E (fly ash 30 per cent)	43.98
6.	F (fly ash 40 per cent)	48.46
7.	G (fly ash 50 per cent)	57.85
8.	H (fly ash 60 per cent)	65.99
9.	I (fly ash 70 per cent)	68.93
10.	J (clay 97.5 per cent)	19.61

Field Trials

Field trials were conducted in *Jatropha curcas* nursery. Application was done on 40 plants which was under study.*Jatropha curcas.* was also planted across the border of the nursery which is considered as control and no application of methyl parathion was done. It was observed that 50 per cent of the plants were found under termite attack in the first year. 10 per cent of the plants were attacked by termite in the second and third year. But no attack was observed in soil applied with methyl parathion dusting powder. No adverse change in the vegetative growth and yield of the crop was seen when it was applied with current formulation sample A. Sample C (10 per cent Fly ash), Sample D (20 per cent fly ash), Sample B (100 per cent fly ash) was used in the Jatropha nursery trials.

Economy in Use of Fly Ash

Fly ash is cheaper than any other material and if the formulation company is situated within 50 km of Power plant.All the experiments are carried out in M/s Yawalkar Pesticides Pvt. Ltd., Nagpur.Company has projected major saving using fly ash in pesticide formulations are shown in Table 63.5.

Fly Ash as Insect Repellent

Fly ash below 45μ was tried on wasps (Vespa orientalis). It is necessary to study the behaviour of the wasp very closely for application to be effective. The wasps belong to the order Hymenoptera. Social wasps that are commonly found in Koradi are Vespa orientalis. The former is a wasp with yellow and reddish or brown bands. They build enormous combs on trees, in the unused portion of buildings, especially near toilets or bathrooms, beneath water valves in power plants (Figure 63.1), but they avoid direct sunrays. Vespa covers all its combs with a envelope that leaves a passage all around, adequate for movement and enabling the workers to have easy access to every part of the nest.[5] They are beneficial to man as they kill many destructive insects but if they build their nests in place where they become a hazard to human comfort, then they need to be controlled. If the nest is in the garden and on auxiliary where children play they can cause nuisance by inflicting painful stings. A wasp kills its enemy like spiders, flies, crickets or beetles by the poison that it injects with the sting to paralyze or kill its enemy.

Table 63.5: Economical Aspect

Sl.No.	*Area*	*Per cent Saving Considering Current Cost of Production*
1.	Electricity	40
2.	Crusher wear and tear	25
3.	Soapstone powder/ Natural resource	100
4.	Health and safety of worker	Lesser Exposure to pesticide during production as agglomeration problem is solved

Treatment of fly ash was made in the day by wearing a face mask. The fly ash dust was directed closely into the nest opening, which is very important. One puff of fly ash weighed about 18-19 g. Two/three strong puffs of ash filter through the nest and repel the wasp within few minutes. The treatment was done in the residential

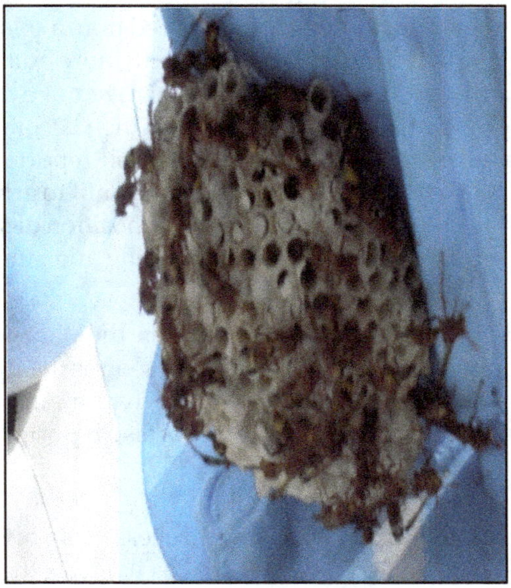

Figure 63.1: Wasp Colonies Beneath Valve.

Figure 63.2: After Fly Ash Application.

premises of Koradi power plant and in the plant beneath the valves, where the nests are built.

It was observed that dry fine dust (below 45µ) of fly ash adhered on the surface of the walls when the wall was moist due to pozzolanic reaction between fine fly ash and lime on the wall. As long as the ash was there, no wasp returned to the nest even after a year (Figure 63.2).It has also been observed that fresh fly ash serves as a better repellent than old fly ash as fresh fly ash has a typical odour of metal oxides especially sulphurous oxides. 10 per cent mortality was observed from a colony of approximately 25 wasps during application. As they get exposed to more quantity of ash, ash gets stuck on their body and they instantaneously want to remove the same. During this act their mandibles get affected due to silica which then becomes unfit to grind the food particles. Narayanasamy[6] and Arputha Sankari[7] reported that mandibles of insects treated with fly ash revealed that they become unfit to grind the food particles subsequently. Fly ash is toxic mainly due to its composition of silica as SiO_2 at 60-65 per cent. This results in the cessation of insect feeding and subsequently the insect dies of hunger.

In normal course, wasps always first moisten the part with saliva and then chew the fibers thoroughly to a fine pulp. Hence as soon as the fly ash is sprayed they exhibit similar action and dislike the taste. Hence fly ash can be a very good wasp repellant. Asthama and allergy in human being is also reported if stung by wasp.

Conclusion

Application of fly ash in formulation process has solved the problem of agglomeration, but it is restricted to 50 per cent as it affects the bulk density of the finished product.The dry fly ash replaces the costly soapstone powder completely without affecting the quality of the finished product and therefore it is concluded that it can be used as dispersant in formulation of 2 per cent methyl parathion dp. Field study of the product it is concluded that the change in adjuvant not only fulfills all the requirements, but it is also very cost effective. Moreover it was observed that fly ash particle size below 45 micron has very good wasp repellent properties.

References

1. Central Insecticide Board and Registration Committee Guidelines (1997).
2. IS 6940: 1982. Methods of test for pesticide and their formulation, Bureau of Indian Standards, New Delhi.
3. IS 8960: 1978. Specification for Methyl Parathion dusting powders, Bureau of Indian Standard, New Delhi.
4. Matte D.B, Kene D.R. 1995. Studies on utilization of fly ash from Thermal Power Station in Agriculture.
5. Mani, M.S., 1995. *Insects*, 2nd edition, National Book Trust, New Delhi, India.
6. Narayanasamy, P. and Gnanakumar D., 1989. A Lignite fly ash-a nonpolluting substance for tackling pest problems, Progress in pollution research, Bangalore, pp. 201-206.
7. Arputha, S and Narayanasamy, P., 2007. Bio-efficacy of fly ash–based herbal pesticides against pests of rice and vegetables, *Current Science*, Vol. 92, No. 6, pp. 811–816.
8. Huayang Agrochemical Group, 2005. Methyl parathion (Technical grade) specification, China.

Chapter 64

Application of Fluidized Bed Reactor for Recovery of Chemical from Pickling Wastewater

T. Nandy and S.N. Kaul

CSIR-NEERI, Nagpur

ABSTRACT

This paper represents the application of full-scale fluidized bed technology with recourse to recovery of value-aided chemicals from the spent liquor generated from steel pickling processes in the steel manufacturing industry, achieving Zero Effluent Discharge [ZED]. Optimization of the process variables of a full-scale fluidized bed reactor for the recovery of hydrochloric acid [HCl] and iron oxide [Fe_2O_3] are discussed.

Keywords: *Pickling: Waste pickle liquor, Hazardous, Fluidized bed, Resource recovery, Hydrochloric acid, Iron oxide.*

Introduction

Pickling is the process of chemically removing oxides and scales from the surface of hot-rolled steel by the action of water solutions of inorganic acids. Depending on the type of product being processed, various acids such as sulfuric, hydrochloric, nitric, hydrofluoric, and phosphoric may be utilized either individually or in combination, in both batch and continuous processes. Until 1950, sulfuric acid accounted for more than 90 per cent of the pickling acid: however, in recent years an increasing number of steel mills have shifted to hydrochloric acid offers several advantages over sulfuric acid as follows: [Rennhack 1974; Azad 1976; Negro *et al.*,

1994]; [a] produces a more even and cleaner surface, [b] requires lower liquid temperatures, and [c] can be regenerated.

The pickling process includes a series of hydrochloric acid solutions at varying concentration and temperatures, and hot rolled steel is passed through it. This process removes fine scale, and bright the clean surface is produced. Iron oxide dissolved in acid are present in the form of insoluble chlorides, and spent acid is discarded which forms wastewater stream commonly referred to as "Waste Pickle Liquor". The pickled steel is then rinsed with water by sequential immersion in rinse tanks to remove all acid and dissolved iron. The rinse water is also discarded as wastewater alters certain washings. The following equations illustrate the chemical reactions occurring during the pickling process.

$$Fe_2O_3 + 6\,HCl \rightarrow 2\,FeCl_3 + 3H_2O \quad (1)$$

$$FeO + 2\,HCl \rightarrow FeCl_2 + H_2O \quad (2)$$

$$2\,FeCl_3 + H_2 \rightarrow 2\,FeCl_3 + 2HCl \quad (3)$$

Waste pickle liquor from the pickling bath contains free hydrochloric acid and iron chlorides, which are considered as a source of environmental pollution. Around 0.13Kg. of the pickle liquor is generated per kg. Of steel processed. Because of their pollutant content and corrosiveness, these liquids appear in the list of hazardous wastes in all legislations [Negro et. Al, 1994]. The discharge of untreated waste pickle liquors into surface water causes water pollution due to its high chloride and iron contents, besides lowering the pH of the receiving stream. Dissolved oxygen of water is also lowered due to hydrolysis of iron salts to iron hydroxides. Discharge into land results in building up of iron and chlorides in soil, making it unfertile, and further polluting the groundwater.

To combat environmental pollution due to discharge of pickling liquors, two main options are available, which include end-of-the-pipe treatment, and recovery of value-added chemicals leading to zero effluent discharge.

End-of-the-pipe treatment of waste pickle liquor consists of neutralization, and/or deep-well injection. Neutralization of waste pickle liquor using lime and aeration is the most commonly adopted treatment process in various steel industries. Ferrous iron in the spent acid gets precipitated as ferrous hydroxide during neutralization, and subsequently converted to ferric hydroxide during aeration. Ferric hydroxide sludge separated is then dewatered through a vacuum filter or sludge-drying beds, where it is oxidized during drying. The neutralized wastewater is evaporated in open solar evaporation ponds. Usually, a large land area is required for solar drying of the neutralized wastewater. Also, an increase in chloride concentration attenuates the evaporation process. Removal of accumuln pressure lated salts causes problems in its disposal due to its hygroscopic nature. The overflow of wastewater from ponds may also contaminate the ground and surface water during monsoons (Gurnham, 1965; Kumar *et al.*, 1998: Kumar *et al.*, 1999; Nandita *et al.*, 2000].

Deep-well injection of waste pickle liquor is another method of treatment and disposal of pickling liquors. This method has now become obsolete due to severe environmental problems associated with it. Continuous disposal of pickle liquor causes sealing of permeable strata of the wells by deposition of reaction products, thereby reducing the capacity of the wells. The possibility of serious pollution of groundwater also cannot be ruled out [Rudolfs, 1997].

Resource recovery techniques that include roasting, pressure oxidation, ion exchange, and oxi-precipitation processes are also available for recovery of value-added chemicals from waste pickle liquors [Negro *et al.*, 1994]

The iron oxide obtained in roasting processes can be considered as a high quality raw material due to its high purity with a chloride content of 0.05-0.1 per cent and can be used in different industries, for example, as components for ferromagnetic materials [ferrites and high power magnets], as raw material for the production of iron dust [electrodes, use in chemical sector, etc.] and as an additive in the manufacture of different products [magnetic strips, colorings, abrasives, bricks, glass, cosmetics, etc.] In pressure oxidation processes, the capital and operational costs ae very high, and the purity of regenerated acid is only 18-20 per cent. Ion exchange process does not extract the ferrous iron in recoverable form, and environmental pollution problems are not fully solved. The oxi-precipitation process can only be used for liquors when the pickling agent has been recovered. This process obtains similar iron elimination rate than other processes, and yields an iron by product for its use, without further transformation, in other industries like electronic, painting, etc. [Negro *et al.*, 1994].

This paper highlights the application of a full-scale fluidized bed reactor achieving Zero Effluent Discharge [ZED] with recourse to recovery of value-added chemicals [hydrochloric acid and iron oxide] from the waste pickle liquor and optimizing the operational parameters.

Materials and Methods

Application of a full-scale fluidized bed reactor system was assessed and a steel coils, to achieve zero effluent kg/d of hot-rolled steel coils, to achieve zero effluent discharge of waste pickle liquor with recourse-to-resource recovery of value-added chemicals. The reactor system was optimized through evaluation of the reactor performance by varying the operating parameters, which included the reactor temperature and the waste pickle liquor feed rate, and also the optimization of air velocity for fluidization of the reactor bed.

Details of Fluidized Bed Reactor System

The full-scale fluidized bed reactor system understudy consisted of the following unit processes.

1. Fluidized bed reactor of size 2 m diameter and 2.5 m height.
2. Cyclone, including dust re-circulation system
3. Venturi, in which hot waste gases are cooled down from approximately 85°C to 100° C. During this process, the normal pickling acid is strongly concentrated due to evaporation of water.
4. Separator, in which the waste gases coining from the venturi are separated from the venturi liquid.

5. Absorber, including scrubber, in which the hydrochloric acid is recovered by absorption of gaseous HCl in water.
6. Fume stack, which sucks off the gases developed in the reactor and keeps the entire system below atmospheric pressure. The residual inert is led through the stack in to the atmosphere.

Figure 64.1 shows the schematic of the full-scale fluidized bed reactor system under study, and the details are presented elsewhere [NEERI Technical Report, 2000]

Process Reaction

The fluidized bed media consisted of iron oxide particles, which are kept in a turbulent state by the combustion gas in the reactor; the spent liquor is pumped via a venturi concentration circuit. The high temperature maintained in the fluidized bed reactor results in the conversion of $FeCl_2$ into Fe_2O_3 and HCl. Thus, the reverse of the pickling reaction takes place in the reactor as follows:

$$2\,FeCl_2 + 2\,H_2O + \tfrac{1}{2}O_2 \rightarrow Fe_2O_3 + 4HCl \quad (4)$$

About 95 per cent of the reaction takes place on the surface of oxide particles and only 5 per cent in the gap. While the in oxide reacting on the surface of the oxide particles contributes to the growth of oxide particles, the iron oxide formed in the gas phase generates fine iron oxide dust, which is carried out of the reactor with the combustion gases, in the cyclone, this dust is again removed from the gas and sent back to the reactor. The iron oxide formed is constantly drawn off by means of a vibrating cooling trough. This facilitates maintaining the constant height of the fluidized bed in the reactor.

The hot reaction gases from the reactor are cooled in the venturi to approximately 100°C. The finest dust particles carried with the gases are captured at the same time making use of dissipated heart to evaporate the spent liquor. A split stream is taken out of this venturi circulation system and conveyed to the reactor.

The gas current passes to the absorber and the hydrochloric gas is absorbed adiabatic ally by water in a reversible flow, achieving a solution 16-17 per cent hydrochloric acid.

Experimental Evaluation

Waste pickle liquor generated from steel processing was analyzed for pH, acidity, total iron, chlorides, free hydrochloric acid, and other parameters as per *Standard.*

Methods [American Public Health Association, 1995]

Similarly, rinse water from the process was also analyzed for various parameters. Table 64.1 presents the physicochemical characteristics of waste pickle liquor and rinse water used for the experimental study.

Propane gas was used as the fuel, and air was used as the oxygen supplement and also to maintain the bed in a fluidized condition. The propane gas requirement was in the range of 49.8 – 52.8m^3/h.

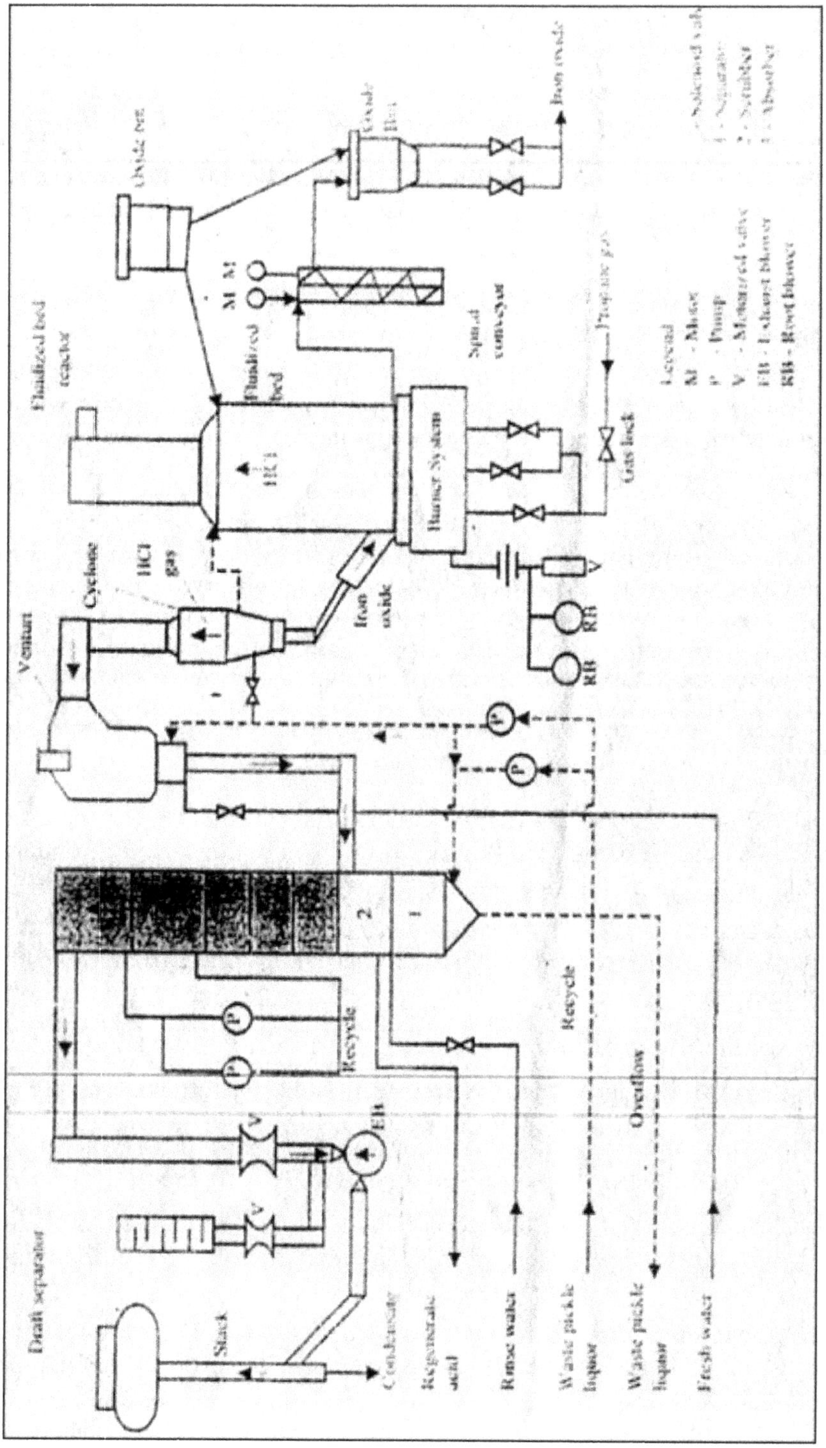

Figure 64.1: Schematic of the Full Scale Fluidized Bed Reactor System.

Table 64.1: Physico-chemical Characteristics of Waste Pickle Liquor and Rinse Water

Parameters	*Concentration**	
	Waste Pickle Liquor	*Rinse Water*
pH	<1	1–2
Acidity	21,000 – 25,000	15,000 – 20,000
Chlorides	88,000 – 100,000	42,000 – 50,000
Free HCl	3.3 – 4.2	0.6 – 0.9
Total iron	1,50,000 – 1,80,000	37,500 – 40,000
Total dissolved solids	90,000 – 110,000	45,000 – 60,000
Suspended solids	400 – 500	125 – 250
COD	26,500 – 28,400	5,500 – 6,000
Specific gravity	1.28 – 1.35	1.03 – 1.05

* All values are expressed in mg/l, except pH and specific gravity.

To optimize the rector system with respect to temperature for recovery of acid, the studies were initially carried out by varying the temperature at a constant pickle liquor feed rate. The temperature was varied between 400 and 1,000 C, while the pickle liquor feed rate was kept at design value of 1,600 kg/h. [200 kg/h, m^3 of reactor volume]. The rinse water was used to absorb the HCl gases generated in the reactor, and the feed rate of rinse water was maintained at 2,400 kg/h [300 kg/$m^{3,}$h].

Optimizing the reactor temperature studies were carried out to assess the optimal pickle liquor feed rate at the optimum operating temperature achieved. The feed rate was varied between 900 and 1700 kg/h [112.5–212.5 kg/m^3• h] and the temperature was maintained in the range of 850 to 900° C.

To assess the effect of airflow rate on expansion of the bed and pressure gradient across the bed, studies were carried out by varying the airflow rate from 200 to 1.900 m^3/h. The pressure at various heights of the reactor was measured using pressure gauges installed > Fluidized systems were extensively used because of the high rete of heat transfer and uniform temperature within the bed, and the heat transfer coefficient was calculated based on the expression as depicted hereunder [Dow and Jakob]

$$hD/k = 0.55\,[D/L]^{0.65}\,[D/d]^{0.17}\{\,(1\text{-}e)\,Ps\,Cs/epCp\}^{0.25}\,(DG/\mu A)^{0.80} \quad (5)$$

Results and Discussion

Effect of Remperature on Recovery of HCl and Fe_2O_3

Conversion reaction of $FeCl_2$ to Fe_2O_3 and HCl occurs at temperatures above 450°C; hence, the reactor performance was assessed at temperatures varying from 400 to 1000°C, and the results are presented in Table 64.2. The effect of temperature on recovery of acid and iron oxide is depicted in Figure 64.2. Perusal of the figure

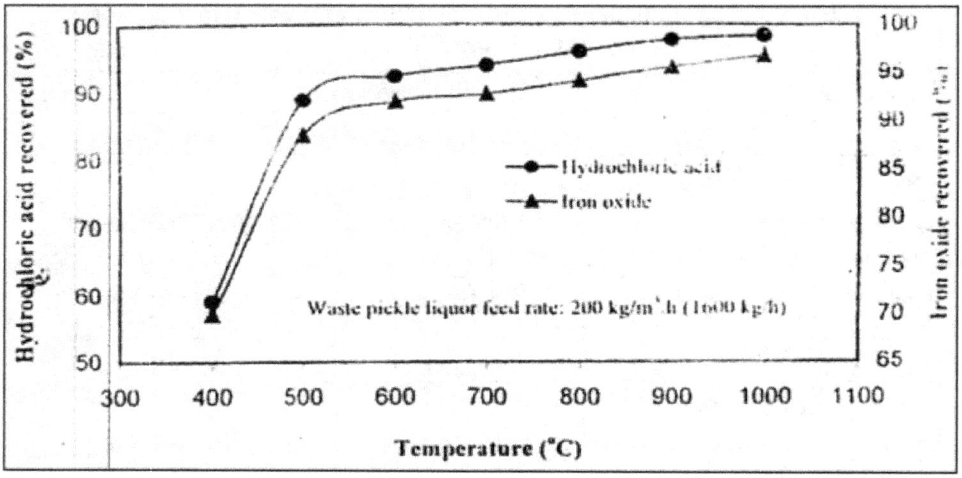

Figure 64.2: Effect of Fluidized Bed Reactor Temperature on Recovery of Hydrochloric Acid and Iron Oxide.

indicates that hydrochloric acid and iron oxide recovery of 88.8 per cent (112.5 kg/m³ • h) and 88.6 per cent (14.2 kg/m³• h) respectively, are achieved at a temperature of 500°C with a waste pickle liquor feed rate of 1,600 kg/h (200 kg/m³• h). The increase in acid recovery is marginal (only 9.6 per cent), with the increase in temperature from 500 to 1,000°C at a constant feed rate. Also, in case of iron oxide, the recovery is only around 8.2 per cent, with an increase in temperature from 500to 1,000°C. Operating the reactor system at temperatures below 500°C, a decrease in recovery of both acid and iron oxide is observed.

Table 64.2: Effect of Reactor Temperature on Recovery of Hydrochloric Acid and Iron Oxide at Constant Pickle Liquor Feed

Rector Temperature (ºC)	*Value-Added Chemicals Recovered*			
	Hydrochloric Acid		*Iron*	*Oxide*
	kg/h	*kg/m³.h*	*kg/h*	
400	600	75.0	90	11.25
500	900	112.5	114	14.25
600	930	116.2	118	14.75
700	950	118.7	120	15.00
800	970	121.2	122	15.25
900	990	123.7	124	15.50
1000	1000	125.0	125	15.62

Waste pickle liquor feed: 200 kg/m³. h (1600 kg/h).

Value-Added Chemicals from Waste Pickle Liquor

Studies indicate an optimum operating reactor temperature of 500°C in terms of recovery of acid and iron oxide. However, temperatures in the range of 800-900°C are required to yield a dense grain of iron oxide resisting attrition, the reason being that under turbulent conditions, the granules of iron oxide developed are so small in size (in the range of 0.5-0.7mm) that they are transported by the gas stream out of the reactor. Hence it is necessary to agglomerate these granules to such a size that they are retained in the reactor in a fluidized condition. The agglomeration of iron oxide presupposes a molten state in the course of the reaction, and therefore, temperature in the reactor needs to be maintained above the melting point of ferrous chloride, *i.e.*, above 675°C. However, studies indicated temperatures in the range of 800-900°C to obtain a dense grain of iron oxide in the size of 4-5mm to retain them in the reactor in a fluidized condition. Thus operating the reactor system at temperatures of 800-900°C, recovery of HCl and Fe_2O_3 in the range of 96-97.7 per cent (121.2-123.7 kg/m^3 • h) and 94.2 – 95.6 per cent (15.2 – 15.5 kg/m^3 •h), respectively, are achieved with a pickle liquor feed rate of 200 kg/m^3 • h.

Effect of Waste Pickle Feed Rate on Recovery of HCl and Fe_2O_3

The fluidized bed reactor was also assessed by varying the waste pickle liquor feed rate operating the system at a constant temperature range of 850-900 °C. Table 64.3 presents the performance of the system at varying feed rates. The effect of waste pickle liquor feed rate on the recovery of acid and iron oxide, as presented in Figure 64.3 depicts a steady recovery of around 99 per cent hydrochloric acid and 98 per cent iron oxide with an increase in the feed rate from 112.5 to 187.5 kg/m^3• h. However, a decrease in the recovery of both acid and iron oxide is observed upon further increasing the feed rate.

Table 64.3: Effect of Waste Pickle Liquor Feed on Recovery of Hydrochloric Acid and Iron Oxide at Consult Reactor Temperature

Waste Pickle Liquor Feed		*Value-Added Chemicals Recovered*			
		Hydrochloric Acid		*Iron Oxide*	
(kg/h)	*Kg/m³·h*	*(kg/h)*	*Kg/m³·h*	*(kg/h)*	*Kg/m³·h*
900	112.5	640	80.00	80	10.00
1,050	131.2	750	93.75	93	11.62
1,150	143.7	810	101.25	102	12.75
1,300	162.5	900	112.50	113	14.12
1,500	187.5	1030	128.75	129	16.12
1,600	200.0	990	123.75	124	15.50
1,700	212.5	950	118.75	117	14.62

Reactor temperature: 850-900ºC.

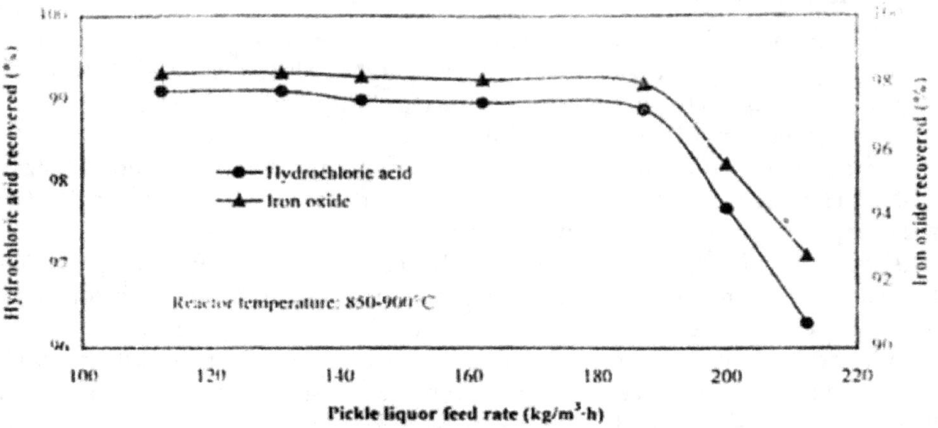

Figure 64.3: Effect of Waste Pickle Liquor Feed on Recovery of Hydrochloric Acid and Iron Oxide at Constant Reactor Temperature.

The data on the performance of the system at optimum operating conditions are presented in Table 64.4. A feed ratio of 187.5 kg/m³• h (1500 kg/h) ami reactor temperature in the range of 850–897°C was found optimum with recovery of HCl and Fe_2O_3 in the range of 125-128.7 kg/m³• h (1,000 – 1030 kg/h) and 15.9 – 16.2 kg/m³• h (127-130 kg/h), respectively. Power consumption (average) in the process was found to be 0.0015 kWh per kg of HCl recovered.

Table 64.4: Performance of Fluidized Bed Reactor at Optimum Operating Conditions

Day	*Waste Pickle Liquor Feed*	*Reactor Tempe-rature*	*Value-Added Chemicals Recovered**				
			Hydrochloric Acid		*Iron Oxide*		*Energy Consumption*
	(kg/m³ h)	*(º C)*	*kg/h*	*(kg/m³ h)*	*kg/h*	*(kg/m³ h)*	*(kWh/kg HCl recovered)*
1	187.5	853-893	1030	128.7	130	16.25	0.00155
2	187.5	855-864	1020	127.5	130	16.25	0.00152
3	187.5	854-884	1010	126.2	129	16.12	0.00142
4	187.5	854-897	1000	125.0	127	15.87	0.00156
5	187.5	850-867	1020	127.5	130	16.25	0.00153
6	187.5	856-869	1000	125.0	129	16.12	0.00168
7	187.5	853-892	1000	125.0	128	16.00	0.00144
8	187.5	867-889	1030	128.7	129	16.12	0.00151
9	187.5	858-883	1020	127.5	128	16.00	0.00141
10	187.5	851-892	1010	126.2	129	16.12	0.0158

* Data presents average of 8 h monitoring each day.

Value-Added Chemicals from Waste Pickle Liquor

Effect of Air Velocity on Fluidization of Bed

The plot of superficial velocity of air induced and pressure gradient across the bed as monitored is shown in Figure 64.4. The initial part of the curve indicates the fixed bed, and as the velocity increases, the pressure gradient passes through a maximum value and then reduces within increase in velocity, because the porosity of the bed increases. This declining part of the curve indicates the fluidized bed. The decrease in pressure gradient continues until the velocity is high enough for transport of the material to take place. At this point, the pressure gradient starts to increase again, which indicates the transport of the bed.

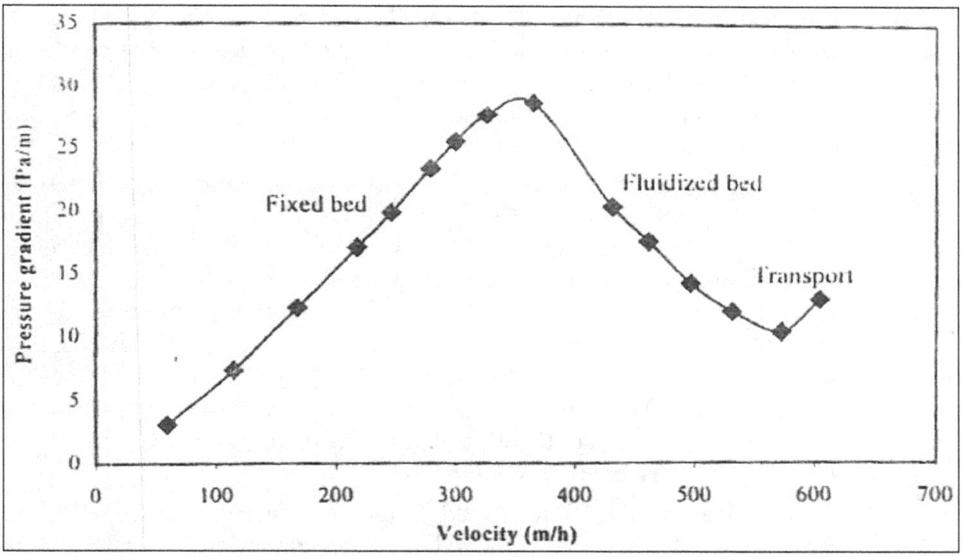

Figure 64.4: Plot of Superficial Velocity of Air vs. Pressure Gradient in Reactor Bed.

In the case of gas-solid fluidized systems, two separate phases – a dense phase and a lean phase – are formed, and a condition of "aggregative fluidization" is produced. Froude Number ($u^2/g \bullet d$, where u is the velocity of air flow at which fluidization takes place, d is the diameter of the particles, and g is the acceleration due to gravity) gives a criterion from which the type of fluidization taking place can be predicted. At values greater than unity, aggregative fluidization takes place (Coulson and Richardson, 1991).

In the present study, the Froude Number determined ranges between 4 and 6 indicating aggregative fluidization. Due to intimate mixing and high degree of turbulence, heat transfer within the bed was found to be rapid, and heat transfer coefficient determined based on equation (5) is 305 $W/m^2 \bullet K$, and therefore uniform temperature was quickly attained throughout the system.

Conclusions

The application of a full-scale fluidized bed reactor for recovery of value-added chemicals from waste pickle liquor achieved an efficiency of 99 per cent in terms of acid (HCl) recovery, achieving Zero Effluent Discharge (ZED). In addition, iron oxide (Fe_2O_3) was also recovered with very high purity (chloride content of 0.05 – 0.1 per cent). The recycling of recovered acid has resulted in the reduction of HCl requirement in the pickling section, achieving savings of 55.3 – 60.2 per cent in a steel plant of a capacity of 1000 X 10^3 kg/d of hot-rolled steel coils. Based on the studies, the following conclusions are drawn:

1. The reactor temperature of 500°C was found optimum for the recovery of hydrochloric acid and iron oxide. However studies indicated that operating temperatures in the range of 800–900°C is necessary to obtain dense grain of iron oxide in the size of 4-5 mm so as to retain them in the reactor in fluidized condition, and hence, the reactor temperature must be maintained between 800 and 900 °C to obtain a dense grain of iron oxide resisting attrition.
2. Steady recovery of both hydrochloric acid and iron oxide of around 99 and 98 per cent respectively, was achieved with an increase in the feed rate of waste pickle liquor from 112.5 to 187.5 kg/m^3·h (900 1500 Kg/h). However, recovery of both acid and iron oxide dropped upon further increasing the feed rate. A feed rate of 187.5 kg/m^3· (1500 kg/h) and temperatures in the range of 850-900°C were found optimum with recovery of HCl and Fe_2O_3 of 125-128.7 kg/m^3·h (1000 – 1030 kg/h) and 15.9 – 16.2 kg/m^3 (127-130 kg/h), respectively. The strength of the recovered acid was found to be in the range of 15.5 – 17 per cent. Average power consumption was found to be 0.0015 kWh per kg of HCl recovered.
3. The Froude Number determined ranges between 4 and 6 indicating "aggregative fluidization" achieved in the reactor and heat transfer within the bed was found to be rapid, as indicated by a very high heat transfer co-efficient (305 W/m^2 · K), thereby attaining quick uniform temperature throughout the system.

Nomenclature

h = heat transfer coefficient (W/m^2 • K)
k = thermal conductivity of the air (W/m • K)
d = diameter of the particle (m)
D = diameter of the reactor (m)
L = depth of the bed (m)
E = voidage of the bed
P_s = density of the solid (kg/m^3)
P = density of the air (kg/m^3)
C_s = specific heat of the solid (J/kg • K)

C_p = specific heat of the air constant pressure (J/kg • K)

μ = viscosity of the air (Pa • s)

G = mass rate of flow of air (kg/s), and

A = cross-sectional area of the bed (m^2).

References

1. AMERICAN PUBLIC HEALTH ASSOCIATION, AMERICAN WATER WORKS ASSOCIATION. Water environment federation. (1995). *Standard Methods for the Examination of Water and Wastewater.* 19th ed, Washington, DC.
2. AZAD, H.S. (1976). *Industrial Wastewater Management Handbook* New York; McGraw-Hill Book Co.
3. COULSON, J.M., and RICHARDSON, J.F. (1991). *Chemical Engineering* Vol. 4, London: Pergamon Press.
4. DOW, W.M. and JAKOB, M. (1951). Heat transfer between a vertical tube and a fluidized air solid mixture. *Chem. Engg. Prog.* 47, 120-126.
5. GURNHAM, C.F. (1965). *Industrial Wastewater Control. New York*: Academic Press.
6. KUMAR, S.M., CHARE, N.Y., VAIDYA, A.N., AND BAL A.S. (1998). Recovery of acid from pickling liquors, *Environ. Eng. Sci.* 15 (4) 259-263.
7. KUMAR, S.M., VAIDYA, A.N., and BAL, A.S. (1999). Treatment options for the management of waste pickling liquor. *J/ntl Assoc. Environ. Manage 26*. 35-39.
8. NANDITA, D., KUMAR, S.M., CHARE, N.Y., VAIDYA., A.N. and BAL., A.S. (2000). Recovery of Chemicals; from waste liquor – A case study. *J. Av. Wr. Environ. Manage.*, 27, 189-194.
9. NEERI technical report (2010). *Waste Management in a Steel Industry*, Nagpur, India.
10. NEGRO, C, LATTORREE, R. DUFOUR, J., FORMOSO. A., and LOPEZ M.F. (1994). Treatment and recovery of pickling liquors. *J. Environ. Sel Helth* A29 (9), 1899-1926.
11. RENNHACK R. (1974). Novel developments in the regeneration of acid pickling liquors and other chloride bearing solutions. In P.E. Wulfinghoff, E., *Wastes – Solids Liquids and Gases*, New York; Chemical Publishing Co., p. 311-319
12. RUDOLFS W. (1997). *Industrial Wastes – Their Disposal and Treatment*. Bikner: Allied Scientific Publishers, India.

Chapter 65

Optimization of Aerated Lagoon System Treating Municipal Wastewater

P. Nema, S.N. Kaul and S.K. Goyal

CSIR-NEERI, Nagpur

ABSTRACT

A sewage treatment plant comprising of anaerobic and aerobic lagoon system treating domestic wastewater was studied in detail. The BOD removal efficiency achieved was in the range of 35-40 per cent against the design efficiency of 80 per cent. The paper highlights the possible causes for low efficiency, and suggests various measures to achieve the designed efficiency through the concept of BOD removal kinetics using dispersed flow conditions instead of mix flow.

The study reveals that about 85 per cent BOD removal efficiency can be achieved by modifying the hydrodynamics of the wastewater treatment system. Incorporation of additional lagoon system required for economical flow arrangements shall enhance the capacity of the system by about 60 per cent without any additional expenditure in terms of power input per unit volume of wastewater.

***Keywords**: Sewage treatment plant, Aerobic lagoon, Hydrodynamics, Dispersion number.*

Introduction

Water is a precious and critical resource. Water resources are depleting fast at an alarming rate worldwide due to continuous growth in population and to meet the matching requirements of the increased population. Besides, drinking purposes, water is extensively used in industries, irrigation, and many other activities. Daily per capita consumption of water is about 180-300 litres, of which about 80-85 per cent returns as wastewater. A city with 1 million population generates about 150 MLD wastewater. Thus a huge quantity of wastewater is generated from domestic sources, therefore, treatment of domestic wastewater and reuse of this treated water can become a potential water resource for non-potable purposes, such as irrigation and industry. Simultaneously degradation of natural water bodies/land can be prevented and public health problem could be mitigated to a large extent. Further, discharge of untreated sewage can lead to human health problems due to spread of disease-causing bacteria and viruses.

Usually, the domestic wastewater treatment scheme consists of preliminary treatment (or pre-treatment) processes involving screening, shredding and grit removal, followed by biological secondary treatment processes, consisting (aerobic and/or anaerobic processes.

A STP having aerobic biological treatment process of treating domestic wastewater was studied in detail essentially with the objective of its assessment. The BOD removal efficiency achieved was observed in the range of 35-40 per cent against the design efficiency of 80 per cent. The paper highlights the possible causes for low efficiency, and suggests measures to achieve the designed efficiency. The concept of BOD removal kinetics for the dispersed flow type conditions has been applied to arrive at the obtainable removal efficiency.

Materials and Methods

Study Region

A 75 MLD sewage treatment plant is operational in a major city in India located at 23° 1′ latitude and 70° 37′ longitude. The present population of the city is about 3.5 million, with an area of about 234 km^2 and population density of about 15000 persons per km^2. The present water supply to the city is about 440 million litres per day (MLD), of which more than 70 per cent is derived from groundwater. There are presently two treatment plants (with two more under construction) for the treatment of wastewater generated from the city.

The sewage treatment plant comprises of screen and grits removal followed by biological treatment consisting of a series combination of anaerobic lagoon and facultative aerated lagoon with settling zone. The STP receives only domestic waste from the surrounding populated area.

Description of STP

Two mechanically cleaned screens with an opening of 15 mm are provided in parallel. Belt conveyors are provided to remove the screenings. Screened sewage flows through two detritors designed for a surface overflow rate of 1008 $m^3/m^2/d$, detention time of one minute and velocity of flow through the chamber as 0.15 cm/s. The degritted wastewater is equally distributed to two anaerobic lagoons; each of size 85m x 181m x 6.5m. The design detention period in anaerobic lagoon is one-day with the designed maximum BOD loading of 0.28 kg BOD/m^3/d.

The anaerobic lagoons are followed by two aerated lagoons each of size 85m x 298m x 5m with a detention period of two days. Each aerated lagoon is provided with four, 50 HP mechanical surface aerators. A settling basin is provided in each aerated lagoon as an integrated unit for settling of suspended solids of the effluent from the aerated lagoon. The settling basins are of the size 85m x 74m x 6.5 m each and have a design detention period of 6 hrs. The treated effluent from each lagoon discharges into a common mixing chamber and finally let out into a river. The schematic flow diagram of the treatment plant is depicted in Figure 65.1.

Sample Collection and Analysis

Relevant plant design data for the plant was collected from the concerned authority. Wastewater flow was measured using parshell flume at the STP, and 24 hourly flow composite samples were collected at 2 hrs intervals at the inlet and outlet of the aerobic lagoons (two parallel separate units). The samples were analyzed for relevant performance parameters *viz.* BOD_5, COD and TSS, as well as other anionic constituents following the standard procedures. The wastewater samples were collected four times during summer season; March-May 2000.

Results and Discussion

Performance of Aerated Lagoon System

The average wastewater flow rate ranged from 47 to 67 MLD, with an overall average of 54 MLD. The designed average flow rate for the STP is 75 MLD. The plant performance data with respect to BOD_5, COD and TSS in influent and effluent is presented in Table 65.1.

Table 65.1: Performance of Aerated Lagoons: Summarized Data

Parameter	*Lagoon-1*		*Lagoon-2*	
	Range	*Average*	*Range*	*Average*
BOD				
Influent, mg/L	108-153	134	102-152	133
Effluent, mg/L	65-98	84	68-93	83
Reduction, per cent	35-40	37	33-42	38
COD				
Influent, mg/L	180-256	226	177-251	220
Effluent, mg/L	91-172	145	98-165	140
Reduction, per cent	33-38	36	28-39	36
TSS				
Influent, mg/L	96-158	120	95-153	116
Effluent, mg/L	72-80	76	76-88	79
Reduction, per cent	20-54	37	19-51	32
pH			—	
Influent, mg/L	7.3-8.0	—	7.7-8.2	—
Effluent, mg/L	7.6-8.3	—	7.8-8.2	—

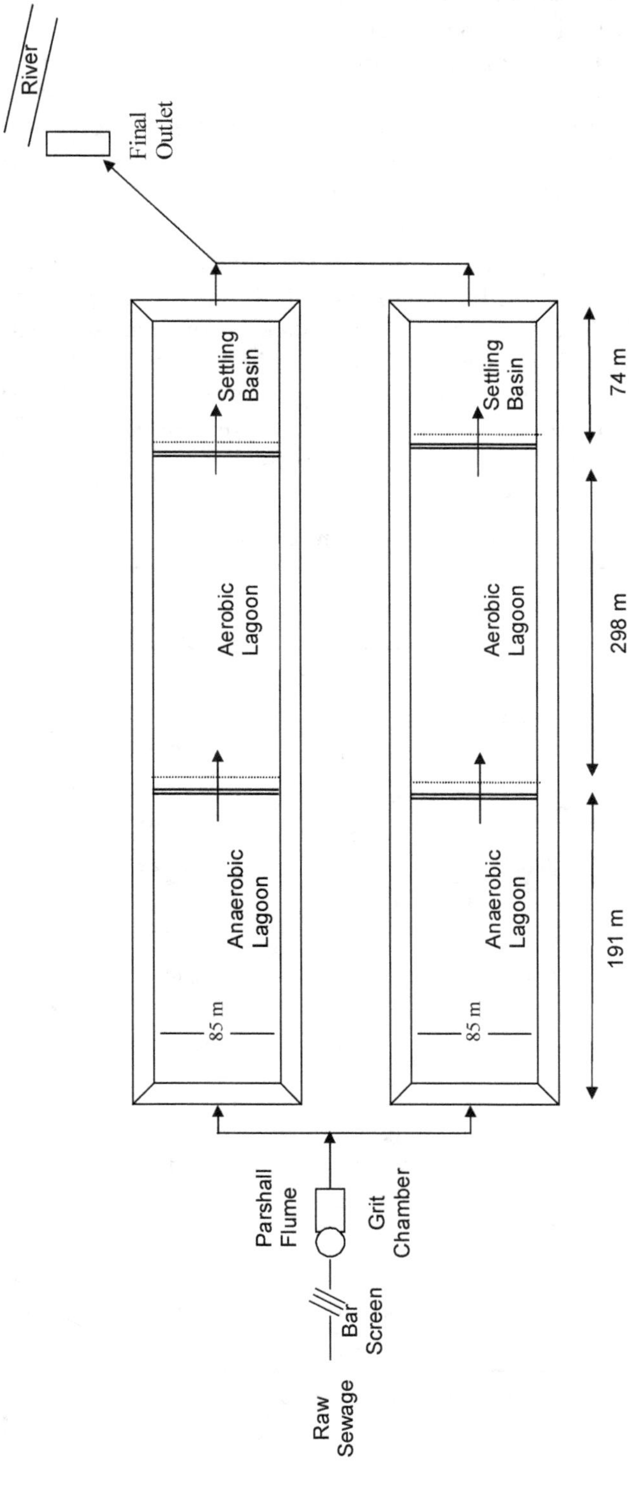

Figure 65.1: Schematic Diagram of Existing STP.

In the lagoon system-1, influent BOD was observed in the range of 108-153 mg/L (average 134 mg/L), whereas the effluent BOD varied from 65 to 98 mg/L with an average of 84 mg/L, thus giving BOD reduction between 35 per cent and 40 per cent (average 37 per cent). Reduction in COD and TSS were found in the range of 33-38 per cent (average 36 per cent) and 20-54 per cent (average 37 per cent) respectively. Similarly in the lagoon system-2, average reduction in BOD, COD and TSS were of the same order, *i.e.* 38 per cent, 36 per cent and 32 per cent respectively. Performance of both the lagoon systems was found almost similar.

Status of nutrients and anions in the influent and effluent to the aerated lagoon system 1 and 2 is presented in Table 65.2. No significant variation was observed in the concentration of NH_3-N, TKN, NO_3-N and SO_4, however, sulfides showed a reduction of more than 55 per cent, and chlorides increased marginally by about 3 per cent. MLSS concentration for such system should be in the range of 50-150 mg/L, whereas, it was observed 53-104 mg/L with an average of 82 mg/L. The MLVSS concentration ranged from 21 to 45 mg/L with an average of 36 mg/L in the zone in close proximity of the aerators.

Table 65.2: Status of Nutrients and Anions in the Aerated Lagoons

Parameter	*Lagoon-1*		*Lagoon-2*	
	Range	*Average*	*Range*	*Average*
NH_3-N				
Influent, mg/L	14-22	19	14-21	19
Effluent, mg/L	7-24	17	10-22	18
TKN				
Influent, mg/L	19-25	22	20-21	20
Effluent, mg/L	20-27	23	18-22	20
NO_3				
Influent, mg/L	0-4.9	1.2	0-2.2	0.6
Effluent, mg/L	0-4.5	1.1	0-3.4	0.8
SO_4				
Influent, mg/L	136-152	140	116-150	136
Effluent, mg/L	125-165	142	120-162	140
Sulphide				
Influent, mg/L	3.2-4.0	3.6	3.3-4.0	3.6
Effluent, mg/L	1.4-1.6	1.6	1.4-1.7	1.5
Reduction, per cent	50-62	56	48-63	58
Chloride				
Influent, mg/L	410-570	510	390-570	492
Effluent, mg/L	470-560	522	430-550	505

Based on 4 data sets collected during March-May 2000.

Average detention time observed was 3.8 days as against the design value of 2 days. The average surface overflow rate (SOR) was found to be 1.0 $m^3/m^2/d$ as against the design SOR of 3.0 $m^3/m^2/d$. The lagoons showed a BOD reduction of about 40 per cent against the designed value of about 80 per cent indicating that the aerated lagoon had been performing at approximately half of their design efficiency for BOD removal. In spite of higher detention time, lower SOR and lower BOD loading than the design values, the aerated lagoons were not found to perform at the designed efficiency.

The inadequate power supply, improper maintenance and operation of the aerators and failure of the aerators to perform at its rated efficiency have an equal share in the below par performance of the aerated lagoon. Agglomeration of all these factors appears to be major causative factors for the poor performance of the aerated lagoons.

The analysis of the data indicates that the power provided to achieve the desired effluent BOD was considerably lower than the required. Absence of DO throughout the lagoon area supported this fact. Therefore, a careful review of the installed power in the lagoons is warranted along with assessment of the actual oxygenation capacity of the aerators under prevailing conditions rather than going by their rated capacity for power input computation.

Aerators Performance vis-à-vis Power Requirement

Currently 4 aerators of 50 HP capacity with designated oxygen transfer rate of 1.22 kg/KW are installed in each lagoon. Oxygen and power requirement for the achieved and required BOD reduction efficiency is estimated, as given in Table 65.3. The total power requirement for 37 per cent BOD removal (as achieved in the present case) would be about 121HP with the available aerators and their oxygenation capacity (with 70 per cent O_2 transfer rate under actual field conditions). This implies that the existing aerators (total 200 HP capacity) are delivering only about 120 HP. In order to achieve effluent BOD of 20 mg/L to meet the regulatory requirements, about 85 per cent reduction in BOD is required, which in turn would require about 278 HP of aeration power. To meet this requirement either supplemental power may be provided or altogether additional existing aerators can be replaced by four new aerators of 50 HP, but with high oxygen transfer rate (about 2 kg/KW). Further, continuous operation of aerators will be a pre-requisite to achieve the desired BOD removal efficiency.

Proposed Modification in the Treatment System

In the existing system, BOD at the inlet to the anaerobic lagoon ranged between 144 and 245 mg/L (average 195 mg/L), and the BOD reduction achieved ranged between 25-38 per cent (average 31 per cent). In aerobic lagoon, average BOD reduction was about 37 per cent, whereas settling basin also showed about 9 per cent (range 3-15 per cent) reduction. The BOD concentration in the treated wastewater was observed in the range of 50-95 mg/L with an average of 75 mg/L. Therefore, in order to meet the regulatory requirement of 20 mg/L BOD, further treatment becomes necessary. This could even be achieved by enhancing the efficiency of aerated lagoon by installing additional aerators.

Table 65.3: Power Requirement Calculations

Case I: Existing System

Input BOD load = 134 mg/L x 10^{-3} kg/m^3 x 37500 m^3/d = 5025 kg/d = 209.4 kg/hr

Rated O_2 transfer rate of aerator = 1.22 kg/KW

Actual O_2 transfer rate under field conditions (aerator efficiency)* = 70 per cent

Achieved efficiency of aerated lagoon system = 37 per cent

Then for 37 per cent BOD removal efficiency

O_2 requirement = 0.37 x 5025 = 1859 kg/d = 77.5 kg/hr

Power requirement = 77.5/ (0.7 x 1.22) = 90.7 KW = 121 HP

For 87 per cent BOD removal efficiency

O_2 requirement = 0.85 x 5025 = 4271.2 kg/d = 178 kg/hr

Power required = 178/ (0.7 x 1.22) = 208.4 KW = 278 HP

Case II: Replacement by efficient aerators

Rated O_2 transfer rate of aerator = 2.0 kg/KW

Assuming actual O_2 transfer rate under field conditions = 70 per cent

Then for 85 per cent BOD removal efficiency

Power requirement = 178/ (0.7 x 2) = 127.1 KW = 170 HP

Estimation of O_2 transfer rate under field conditions

$$N = No\ [Cwalt - Cl)/Cs20] \times 1.024^{T-20}\ (\alpha)$$

where,

N = kg O_2/kw H transferred under field conditions

No = kg O_2/kw H transferred in water at 20 °C and zero DO (rated aerator capacity)

Cwalt = Oxygen saturation conc. for water at given temp. and altitude, mg/L (7.54 for the city under study)

Cl = Operating oxygen conc., mg/L; usually 1 mg/L

T = O_2 transfer correction factor for water, (25 °C)

Cs25 = 8.38 mg/L and

α = Usually 0.8-0.85 for wastewater

Since installation of additional/new aerators shall require substantial recurring expenditure in terms of electrical power, partial modification in the existing system based on BOD removal kinetics is explored, as discussed here.

Optimization of Aerated Lagoon System Based on BOD Removal Kinetics

The design of aerated lagoons for BOD removal is based on first order kinetics and the complete mix hydraulics model. The model commonly used is:

$$S = So/\ (1 + K.t)$$

where,

So = Influent BOD (mg/L)

S = Effluent BOD (mg/L)

K = Temperature dependent substrate removal rate constant, (per day)

K_{20} = 0.276 per day at 20 °C

q = Temperature coefficient (1.035)

K = $K_{20} \times q^{(T-20)}$

T = Temperature of water (°C)

t = Total detention time in system (d)

Using the above equation, the substrate removal rate has been calculated for the present operating conditions (liquid temperature 25 °C) of the aerated lagoons. Values of K and K.t have been computed for individual observations and summarized in Table 65.4. The value of K is found to vary from 0.14 to 0.21 with an average of 0.162 d^{-1}, which is much lower than the reported value of 0.3 d^{-1}. The average value of K.t is found to be about 0.6 (K = 0.3 per day, and t = 2.0 d), which is also very-very low. The reported is about 6, which results about 95 per cent BOD removal efficiency.

Table 65.4: Kinetic Constants of Sewage in Aerated Lagoons

Date	*BOD Inf. (So), mg/L*	*BOD Eff. (S), mg/L*	*HRT (d)*	*K= (So-S)/ S.t (per day)*	*K.t (Dimension-less)*
		Lagoon -1			
Set I	126	77	4.3	0.148	0.636
Set II	108	65	4.2	0.158	0.662
Set III	150	98	3.8	0.140	0.531
Set IV	153	95	3.0	0.204	0.611
Average	134	84	3.8	0.162	0.610
		Lagoon–2			
Set I	130	75	4.3	0.170	0.733
Set II	102	68	4.2	0.119	0.500
Set III	148	95	3.8	0.147	0.558
Set IV	152	93	3.0	0.211	0.634
Average	133	83	3.8	0.162	0.606

Arceivala (1) has also reported overall BOD removal rate constant between 0.6 and 0.8 per day at 20 °C for facultative type lagoon with 3-12 days detention time, resulting in 70-90 per cent efficiency. However, in the present study, the value of K as well as t are found considerably low.

Normally, the mixing conditions in an aerated lagoon are neither plug flow type nor completely mix type. The mixing conditions prevailing are between the two, and fall under dispersed flow conditions representing a non-ideal situation. For which

Fick's second law equation is used, wherein diffusion coefficient is replaced by dispersion coefficient (D). Hence, the change in concentration is given by:

$$\frac{\partial S}{\partial t} = D.\frac{\partial^2 S}{\partial X^2} \quad \frac{\partial S}{\partial t} = D.\frac{\partial^2 S}{\partial X^2}$$

where,

S = Concentration of substrate, (mass/volume)

D = Axial or longitudinal coefficient (L^2/t)

X = Distance L in direction of main flow

Taking a steady state differential materials balance at any cross section in the reactor, Levenspiel (2) obtained the following general equation for any reactant following the nth order kinetics:

$$D.\frac{\partial^2 S}{\partial X^2} - U.\frac{\partial S}{\partial t} - KC^n = 0$$

where,

U = Mean flow velocity along the reactor, L/t

K = Removal rate constant, per unit time

n = Order of the reaction

The above equation was further solved by Wehner and Wilhem (3) for the first order kinetics, and equation obtained was

$$\frac{S}{So} = \frac{4\,a\,e^{1/2d}}{(1+a)^2.e^{a/2d} - (1-a)^2.e^{-a/2d}}$$

where,

a = $(1+4\,K.t.d)^{1/2}$ (dimensionless)

d = D/UL = $D.t/L^2$ (dimensionless)

D = Axial or longitudinal coefficient (L^2/t)

U = Mean flow velocity (L/t)

L = Length of axial travel

t = Theoretical detention time (V/Q)

K = Substrate removal rate constant, per unit time

So and S = Initial and final BOD concentrations (mg/L)

The value of D/UL varies widely from zero to infinity. When D/UL approaches zero, it follows plug flow conditions, and when D/UL approaches infinity, it represents a completely mix flow regime. Hence, the above equation is applicable over the whole range of D/UL values.

In practice, the typical range of D/UL values observed in different types of treatment reactors lies between 0.1 and 4.0. Of course, some reactors may show values of less than 0.1 and much greater than 4.0 (even 100). However, there is no ideal plug flow or ideal completely mixed flow system. When a system shows D/UL value of about 0.2 or less, it is considered to be approaching plug flow, while at 3.0-4.0, it is considered close to well mixed or complete mixing.

Use of Wehner-Wilhem equation is facilitated by reference to substrate (BOD) removal efficiency charts. Figure 65.2 shows one such chart between dimensionless product K.t against percentage of remaining substrate (S/So) and percentage of substrate removed (1-S/So). It is evident from the figure that:

- ✰ A completely mixed or even a reasonably well mixed system (D/UL > 4) is incapable of giving more than 90-97 per cent efficiency for K.t values less than 20,
- ✰ For a given K.t value, the efficiency increases as D/UL decreases, namely, plug flow type system always tend to give higher efficiency than well mixed systems and,
- ✰ Very high efficiencies of greater than 99 per cent, as are often required for bacterial removal can be reached if the system approaches plug flow conditions. Thus D/UL is as important design parameter as the substrate removal rate K is.

Considering the dispersed flow type conditions, the value of dispersion number (D/UL) has been estimated for the present system. There are two aerated lagoons, each of 85 m width (W), 298 m length (L), and 4 m liquid depth (d) with 2 days detention time. Since no baffle is provided in the lagoon, the empirical correlation for dispersion coefficient (D) used is, D = 16.7 W, where W is width of lagoon, *i.e.* 16.7 x 85 = 1419.5 m^2/h.

For the given conditions of flow rate (design and observed) and using the geometry of aerated lagoon, the value of D/UL has been computed for each day's observation. Further, for the estimated value of K.t, theoretical BOD removal efficiency has been read from the Wehner-Wilhen chart. All these values are given in Table 65.5. The BOD removal efficiency from the Wehner-Wilhen chart (extrapolated) is found to be around 40 per cent, which is close to the actual values observed from the sampled data. For the present design data, the value of D/UL and K.t are computed as 0.77 and 1.4 respectively, for which the BOD removal efficiency would be in the range of 60-65 per cent, not as initially envisaged to be about 80 per cent.

Therefore, in order to increase the BOD removal efficiency to about 80 per cent or even more, the basic design parameters need to be modified, either by enhancing the K.t value and/or by reducing the D/UL value. The value of product, K.t can be enhanced either by increasing K or t. Rate constant K can be enhanced by improving the aeration conditions. Increase in detention time means lower flow rate, which for the existing conditions may not be practical.

For the theoretical detention time of 2.7 days, the value of dispersion number (D/UL) is 1.04 for the given aerated lagoon, and for achieving 85 per cent BOD removal

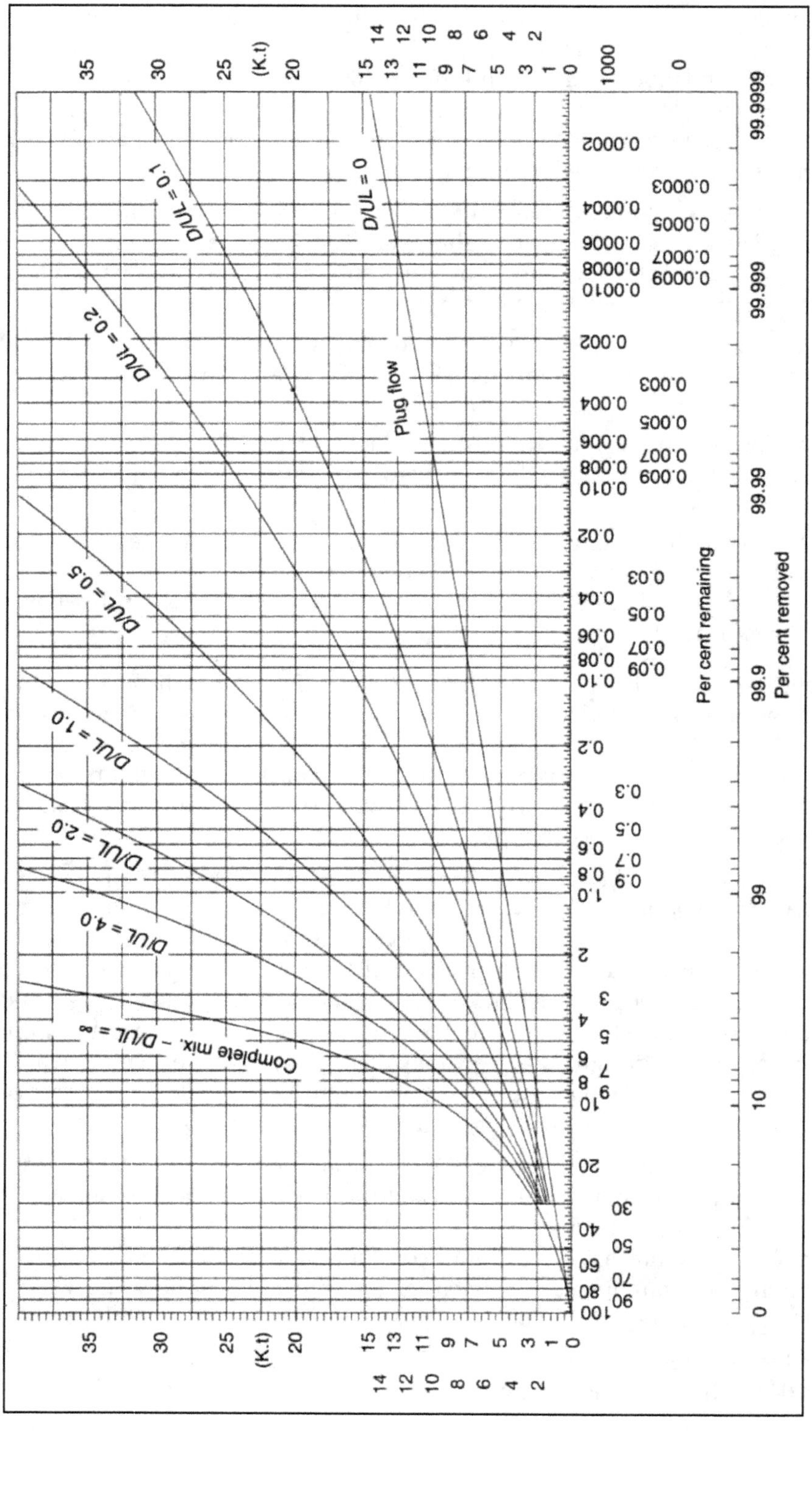

Figure 65.2: Determination of Substrate Removal Efficiency using the Dispersed Flow Model (Wehner-Wilhem Equation).

efficiency, the value of K.t from Wehner-Wilhen chart is estimated to be 4.5. For the designed detention time of 2.7 days, the value of rate constant (K) would be 1.66. Thus the value of rate constant need to be increased by about 10 times by installing aerators of proper capacity and operating round the clock delivering required oxygen input. Need for additional aerators have been discussed in earlier section.

Table 65.5: K.t, D/UL and BOD Removal Efficiency (Based on Wehner-Wilhem Chart)

Date	*Flow Rate (m^3/d)*	*HRT (d)*	*K.t (Dimension-less)*	*D/UL= $D.t/L^2$ (Dimension-less)*	*BOD Removal Efficiency (per cent)*
Observed					
			Lagoon -1		
Set I	23500	4.3	0.636	1.65	< 50 per cent
Set II	24000	4.2	0.662	1.61	Actual values not
Set III	27000	3.8	0.531	1.46	readable from the
Set IV	33500	3.0	0.611	1.15	chart (Figure 65.2)
Average	27000	3.8	0.610	1.46	
Lagoon–2					
Set I	23500	4.3	0.733	1.65	< 50 per cent
Set II	24000	4.2	0.500	1.61	Actual values not
Set III	27000	3.8	0.558	1.46	readable from the
Set IV	33500	3.0	0.634	1.15	chart (Figure 65.2)
Average	27000	3.8	0.606	1.46	
Designed	**37500**	**2.0**	**1.4***	**0.77***	**60-65 per cent ***

* Computed value.

Therefore, as an alternate, the dispersion number needs to be reduced, which can be obtained by providing a baffle along the length of the lagoon. In this case, the dispersion number shall reduce to 0.2, for which the value of K.t is about 3.4. Hence, about 85 per cent BOD removal efficiency can be achieved for 2.7 days detention time with the K value of would be 1.26. However providing a baffle in the existing lagoon system shall require considerable changes in the wastewater inlet and/or outlet arrangements, and hence may not be an attractive option.

Alternatively, an additional lagoon system can be added with inlet flow arrangements, and a baffle can be provided in the existing complete lagoon system. A schematic diagram of the proposed system is shown in Figure 65.3. The new lagoon could be of the same dimensions as the existing one with baffle [*i.e.* 40 m width x (191+298+74) m length x 4 m liquid depth). The new lagoon system shall act as anaerobic lagoon, and the complete existing system (anaerobic and aerobic) shall act as aerobic lagoon with relocating of the aerators. The existing settling basin and out let system can be retained/slightly modified. After converting the existing anaerobic

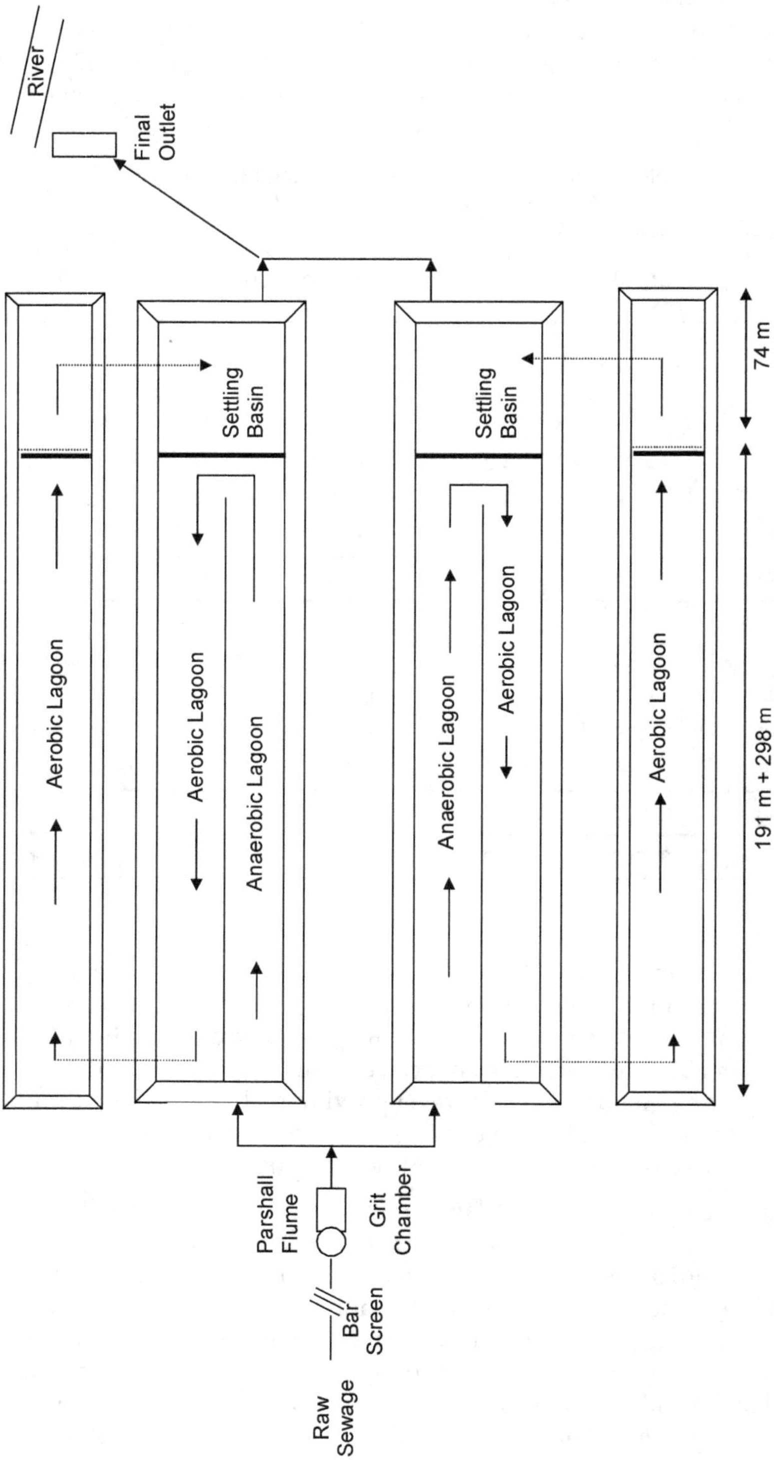

Figure 65.3: Schematic Diagram of Proposed Modification in STP.

and aerobic lagoon with baffle, the value of dispersion number would reduce to about 0.1, and for achieving 85 per cent efficiency, the value of K.t would be about 3.0. For 2.7 days detention time, the value of K would be about 1.11 per day. Therefore the proposed modification shall have twin advantage of:

- ☆ High BOD removal without any additional cost of power input per 1000 m^3 of wastewater
- ☆ Provision for treatment of additional wastewater (capacity expansion from 75 MLD to 120 MLD).

Conclusion

The bio-kinetics approach to improve the plant performance in terms of BOD removal would require adjustment in values of dispersion number and K.t. This could be achieved by providing an additional unit of aerated lagoon and a length-wise baffle in the existing aerated lagoon system. These measures will have cost implications. However, the capital cost will be only one time cost, whereas the resultant improved efficiency in terms of effluent quality would be a long-term benefit with only marginal change in the energy requirement for operation of the system. Thus, the recurring cost will not increase substantially.

The other benefit accrued from the suggested modifications would be an effluent quality in compliance with the regulatory standards and thus suitable for discharge in the river. The treated effluent will help supplement groundwater recharge, as also could be utilized as a source for water for irrigation and even industrial/recreational requirements for down stream areas with adequate treatment.

References

1. *Wastewater Treatment for Pollution Control*, by S.J. Arceivala, Tata Mc-Graw Hill Publishing Company Ltd., New Delhi, 1998.
2. *Chemical Reaction Engineering* by Levenspiel O., John Wiley and Sons, New York, 1962.
3. Wehner J. and Wilhem R. (1956), *Chem. Engg. Sci.*, Vol. 6, 89.

Chapter 66

A Study on Groundwater Quality in the Surrounding Area of a Sewage Treatment Plant

S.N. Kaul, P. Nema, S.K. Goyal and C.G. Mehta

CSIR-NEERI, Nagpur

ABSTRACT

A detailed investigation into the groundwater quality in the surrounding areas of two sewage treatment plants was carried out. Water samples were collected from different groundwater sources, *viz.* hand pumps and bore wells located in the region and analyzed for their physico-chemical and bacteriological parameters. Most of the samples were found to exceed the standard limits for drinking water parameters with respect to total hardness, TDS, sulfates, chlorides, fluorides, calcium and magnesium. Further, some water samples were also found bacteriologically contaminated. This paper presents the groundwater quality data, and highlights various possible causes of contamination of groundwater sources, and suggests remedial measures, both preventive as well as curative, to arrest the contamination.

Introduction

Groundwater has become major source of water in the country to meet the agricultural, animal, industrial and domestic (including drinking) needs. In India, groundwater resources feed more than 50 per cent of the total area under irrigation. However, the quality of the groundwater available is a concern for its sustainable and effective use. During the last few decades, disposal of effluents from industries and municipal sewage and their percolation to groundwater has adversely affected the quality of the groundwater. Increased use of chemical fertilizers and pesticides further aggravates the condition. The deterioration in quality of groundwater is creating several health problems to the users, as this is a major and easily available source for human consumption in most of the villages and urban areas (Bhatnagar and Sharma, 2002).

A study on assessment of groundwater quality in a suburb area of a major city in western India located at 23° 1′ latitude and 70° 37′ longitude, where groundwater is the prime source of drinking water supply was carried out. The present population of the city is about 3.5 million, with an area of about 234 km^2 and population density of about 15000 persons per km^2. The present water supply to the city is about 440 million litres per day (MLD), of which more than 70 per cent is met from groundwater. The groundwater sources comprise of about 300 bore wells scattered all over the city. A sewage treatment plant of 180 MLD capacity exists in the vicinity of the study area selected for groundwater quality assessment. The groundwater sources in the area include hand pumps and bore wells.

The possible causes that can lead to contamination are sewage overflow, leakage in sewer lines or final discharge of treated wastewater into the receiving water body/ land. Residents in the nearby area of the STP, which constitutes study area also, raised their concern over the potential threat of groundwater contamination because of their proximity to the STP. Keeping in view the known health implications, a detailed investigation was carried out for assessment of water quality in order to determine the extent of contamination, if any and to suggest necessary effective precautionary and remedial measures for preventing groundwater contamination.

Methodology

A water quality monitoring programme was designed and sampling locations were fixed covering a distance up to 1.5 km distance from the STP taking into consideration the distance that the enteric bacteria can traverse. Since, groundwater is the main source of water supply in the region, samples were taken from hand pumps (7nos.) and bore wells (3 nos.), which are used by the people for domestic as well as farming purposes. Sampling locations for groundwater quality assessment are depicted in Figure 66.1, whereas details of these sources are furnished in Table 66.1.

Samples were collected during Oct.-Nov. (representing post-monsoon/winter season) and May-June (representing summer season), following the WHO guidelines (WHO, 1985), and were analyzed for the relevant physico-chemical and bacteriological parameters following the standard procedures (APHA, 1992; Mara, 1968).

The groundwater sources *viz.* hand pumps and bore wells are located within a distance of 200 to 1500 m from the STP. The hand pumps were installed during 1970-1985, whereas the bore wells were installed relatively later. The depth of hand pumps was in the range of 15 to 26 m, whereas bore wells were more than 137 m deep. Reconnaissance survey of the study area revealed location of soakpit/septic tanks in the vicinity of some of these groundwater sources. Practices of defecation in open and existence of unsewered channels were observed in the village areas.

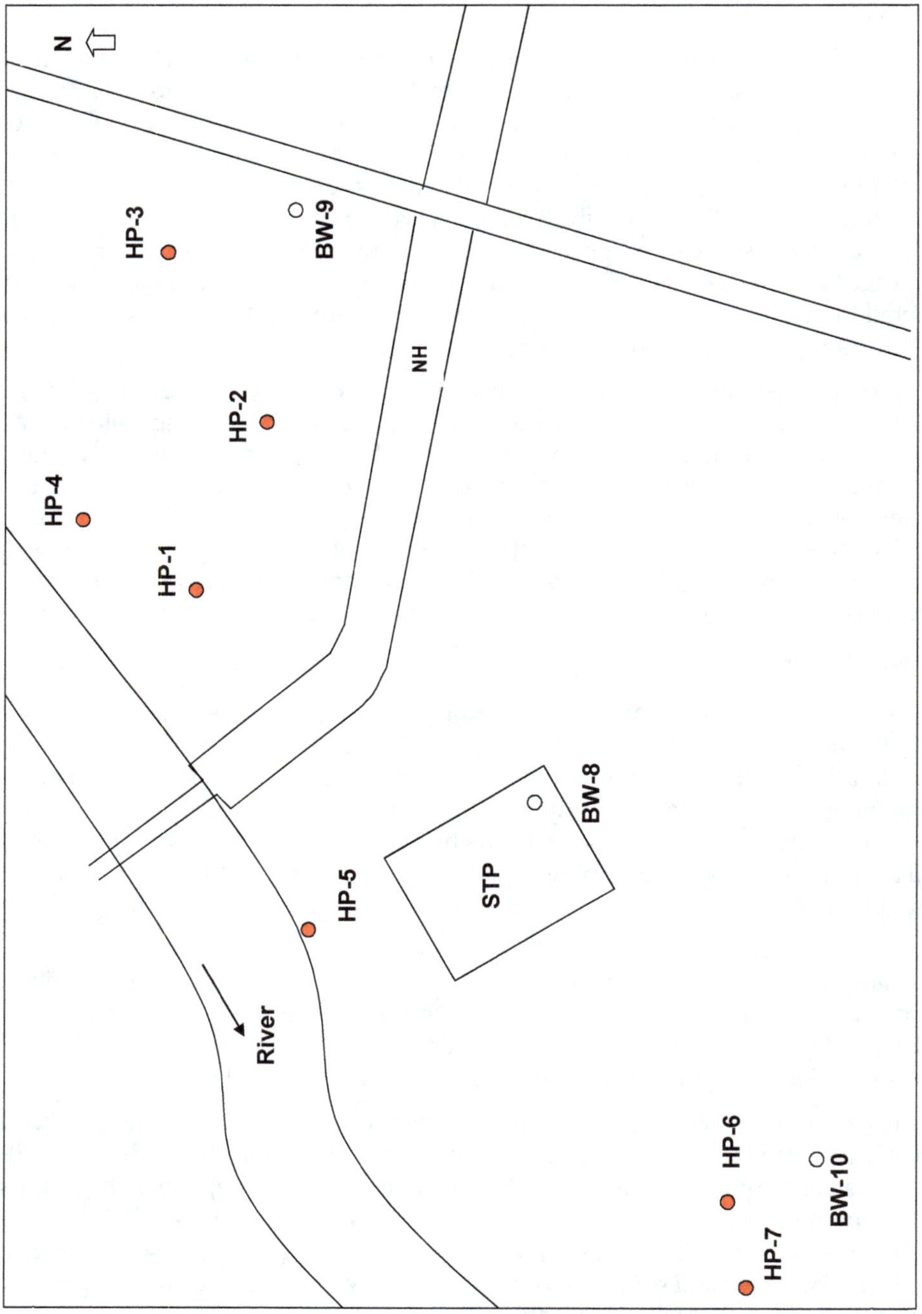

Figure 66.1: Map Showing Groundwater Sampling Locations around STP.

Table 66.1: Details of Groundwater Sampling Locations

Sampling Location	*Distance (m)*	*Direction*	*Depth (m)*	*Year of Commissioning*	*Monitoring in Season*
			Hand pumps		
1	1000	NE	23	—	PM/W and S
2	1100	NE	23	1970	PM/W and S
3	1500	NE	23	—	PM/W and S
4	1500	NE	23	1975	PM/W and S
5	500	N	26	1985	S
6	1000	SW	—	—	S
7	1100	SW	15	1982	S
			Bore wells		
8	200	E	130	—	PM/W and S
9	1100	SW	167	2000	PM/W and S
10	1500	E	—	—	PM/W and S

PM/W: Post-monsoon/Winter; S: Summer.

Results and Discussion

Physico-chemical Quality of Groundwater Sources

Physico-chemical characteristics of groundwater samples collected from different sources (hand pumps and bore wells) in the surrounding areas of STP during post-monsoon/winter and summer seasons are summarized in Tables 66.2 and 66.3 respectively. Season and source-wise comparison of groundwater quality parameters with the drinking water quality standards (IS 10500: 1995) is presented in Figure 66.2. Per cent exceedence of various parameters, *viz.* TDS, total hardness, chlorides, sulfates, nitrates, calcium and magnesium ions in the water samples collected from different hand pumps and bore wells are presented in Figure 66.3. A perusal of the data presented in these tables and figures highlights the following salient points:

- ☆ In hand pumps water samples, average concentration of total alkalinity, total hardness, TDS, chlorides, sulfates, calcium and sodium were found more in summer season than post-monsoon/winter, however in bore wells water samples, average concentration of different parameters were found more or less same in both the seasons.
- ☆ Comparison of water samples collected from hand pumps and bore wells, in general indicated higher concentration in shallow water sources (hand pumps) as compared to deep water sources (bore wells) in both the seasons indicating that the shallow water sources are more affected than the deep water sources.
- ☆ Concentration of nitrates was found well within the permissible limits in all the samples in both the seasons, maximum being 29.5 mg/L in one hand pump sample.

Table 66.2: Source-wise Physico-chemical Quality of Groundwater Around STP: Post-monsoon/Winter

Parameter	*1*	*2*	*3*	*4*	*5*	*6*	*7*	*Avg.*	*8*	*9*	*10*	*Avg.*
Source →	*HP*	*HP*	*HP*	*HP*	*HP*	*HP*	*HP*	*HP*	*BW*	*BW*	*BW*	*BW*
Number of samples	5	5	5	5	ND	ND	ND	5	5	5	5	5
pH	6.6-7.1	6.7-7.3	6.5-7.1	6.6-7.1	—	—	—	—	6.0-7.4	7.0-7.6	6.4-7.6	—
Alkalinity (as $CaCO_3$)	702	665	627	565	—	—	—	640	360	380	582	441
Total Hardness (as $CaCO_3$)	706	536	410	490	—	—	—	536	302	538	417	419
Ca-Hardness (as $CaCO_3$)	431	320	263	285	—	—	—	325	215	333	225	258
Mg-Hardness (as $CaCO_3$)	275	241	149	205	—	—	—	218	85	202	192	160
Chlorides (as Cl^-)	930	525	316	906	—	—	—	669	350	449	392	397
Sulphate (as SO_4^{--})	252	124	163	226	—	—	—	191	112	92	134	113
Nitrate (as NO_3^-)	11	18	17	17	—	—	—	16	11	12	11	11
Fluoride (as F^-)	3.3	2.4	2.9	3.4	—	—	—	3.0	2.2	2.4	2.5	2.4
Calcium (as Ca^{++})	172	128	105	114	—	—	—	130	86	133	90	103
Magnesium (as Mg^{++})	66	58	36	49	—	—	—	52	20	81	46	49
Sodium (as Na^+)	572	384	408	568	—	—	—	483	291	245	305	280
Potassium (as K^+)	1.0	1.5	0.8	1.5	—	—	—	1.2	1.1	1.2	1.6	1.3
TDS	2680	1839	1975	2540	—	—	—	2258	1370	1314	1346	1343
COD	9	7	10	5	—	—	—	8	7	10	—	8

All parameters except pH are in mg/L.

Table 66.3: Source-wise Physico-chemical Quality of Groundwater Around STP: Summer

Parameter	*1*	*2*	*3*	*4*	*5*	*6*	*7*	*Avg.*	*8*	*9*	*10*	*Avg.*
Source →	*HP*	*HP*	*HP*	*HP*	*HP*	*HP*	*HP*	*HP*	*BW*	*BW*	*BW*	*BW*
pH	7.6-7.8	7.6-8.2	7.6	7.4-8.9	7.9-8.2	7.6-8.4	8.0-8.1	—	7.5-7.6	7.6-7.7	7.5	—
Alkalinity (as $CaCO_3$)	790	745	690	635	950	868	865	792	465	490	510	488
Total Hardness (as $CaCO_3$)	710	595	475	610	383	556	490	546	335	535	410	427
Ca-Hardness (as $CaCO_3$)	400	335	270	335	290	448	400	354	220	330	290	280
Mg-Hardness (as $CaCO_3$)	310	260	205	275	93	108	80	190	115	205	120	147
Chlorides (as Cl^-)	895	515	305	855	600	820	890	697	345	440	380	388
Sulphate (as SO_4^{--})	293	236	187	305	360	440	496	331	116	104	150	123
Nitrate (as NO_3^-)	15	7	11	15	8.1	29.5	7.7	13	11	9	12	11
Fluoride (as F^-)	3.5	2.4	3.1	3.7	3.2	1.6	2.2	3	2.3	2.5	2.6	2
Calcium (as Ca^{++})	160	134	108	134	116	180	179	142	88	132	116	112
Magnesium (as Mg^{++})	74	62	49	66	22	26	19	46	28	49	29	35
Sodium (as Na^+)	530	385	350	510	667	830	898	596	272	230	348	283
Potassium (as K^+)	0.7	1.3	1.5	1.2	22	27	30	12	1.2	1.3	1.7	1
TDS	2513	1726	1727	2594	2457	3102	3338	2494	1272	1217	1510	1333
COD	7	7	12	9	48	64	23.5	24	9	6	—	8

All parameters except pH are in mg/L.

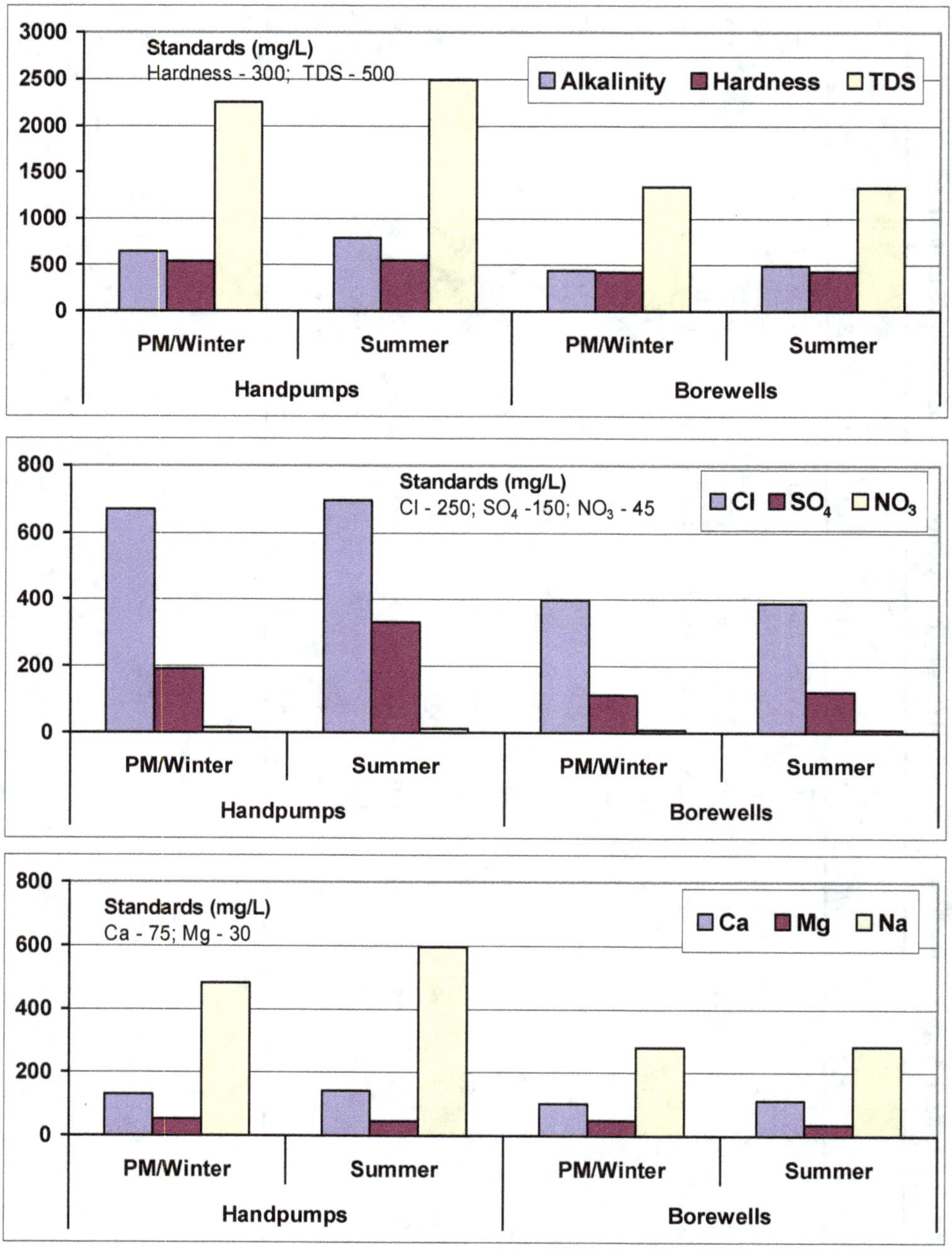

X - axis : Parameter; Y - axis : Concentration, mg/L

Figure 66.2: Average Hydro-chemical Characteristics of Groundwater Samples.

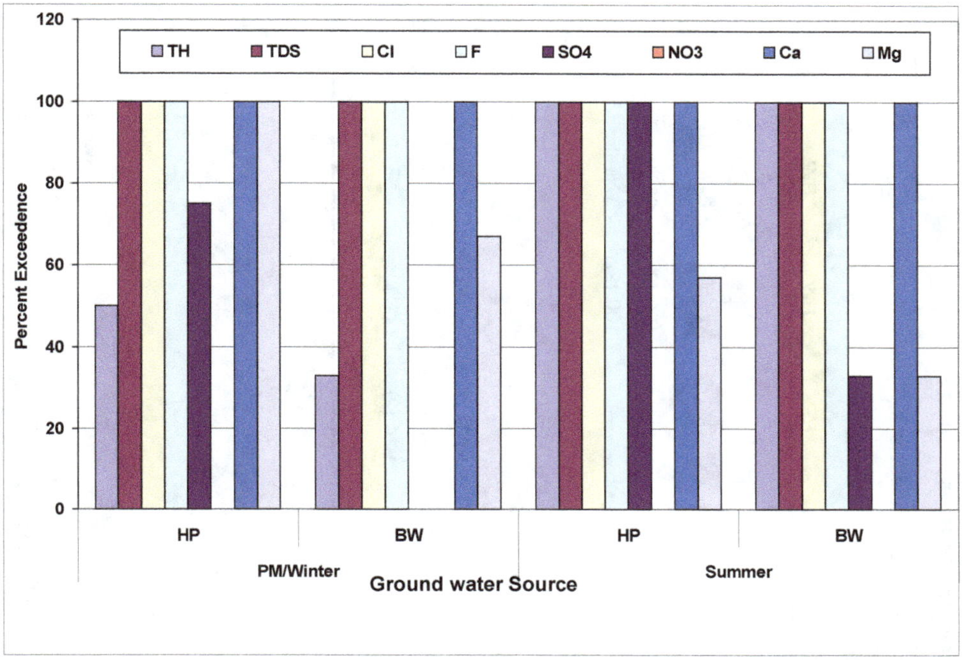

Figure 66.3: Percent Exceedence to the Standards.

- Average concentration of sodium was found considerably high (596 mg/L) in the hand pump water samples as compared to the samples of bore wells (283 mg/L) in summer.
- Average fluoride concentration was found to be exceeding the desirable limit of 0.6-1.2 mg/L in all the groundwater samples in both the seasons.
- All the samples showed chemical oxygen demand (COD). Average COD concentration during summer was higher in hand pumps (24 mg/L), as compared to 8 mg/L in bore wells indicating some organic pollution, whereas during post-monsoon/winter COD levels were observed around 8 mg/L.

Bacteriological Quality

Bacteriological quality of water samples collected from different groundwater sources during post-monsoon/winter and summer seasons are summarized in Table 66.4. The indicator organisms chosen were total coliforms, fecal coliforms, E. coli and fecal streptococci, presence of which in water indicates bacteriological contamination. The salient observations are summarized as under:

- Most of the water samples collected showed presence of total coliforms. Presence of fecal coliforms and E. Coli was also observed in number of samples, whereas, fecal streptococci were found in only 2 samples.

Table 66.4: Bacteriological Quality of Groundwater Sources around STP

Parameter	*Maximum Count (CFU/100ml) and Frequency (No. of samples out of total samples collected)*									
	1	*2*	*3*	*4*	*5*	*6*	*7*	*8*	*9*	*10*
Source →	*HP*	*HP*	*HP*	*HP*	*HP*	*HP*	*HP*	*BW*	*BW*	*BW*
Post-monsoon/Winter										
Number of samples	5	5	5	5	—	—	—	5	5	5
Total coliforms	Nil	7 (3)	13 (3)	25 (2)	—	—	—	TNC (2)	33 (5)	20 (2)
Fecal coliforms	Nil	Nil	5 (1)	18 (1)	—	—	—	100 (2)	25 (2)	15 (2)
E. coli	Nil	Nil	Nil	15 (1)	—	—	—	100 (2)	Nil	1 (2)
F. streptococi	Nil	Nil	Nil	1 (1)	—	—	—	Nil	Nil	1 (1)
Summer										
Number of samples	2	2	2	2	6	6	2	2	2	2
Total coliforms	Nil	1 (1)	15 (2)	Nil	TNC (6)	TNC (6)	TNC (2)	Nil	12 (2)	Nil
Fecal coliforms	Nil	Nil	5 (2)	Nil	TNC (6)	45 (6)	42 (2)	Nil	1 (1)	Nil
E. coli	Nil	Nil	2 (1)	Nil	TNC (6)	25 (6)	14 (2)	Nil	Nil	Nil
F. streptococi	Nil	Nil	Nil	Nil	Nil	Nil	Nil	Nil	Nil	Nil

Method used: Pore plate technique

CFU/100 ml

mb.Air.Samp.Pt.

- ✰ Very high count of TC, FC and E.Coli was found in three of the hand pumps, located in the downstream of treated effluent discharge. All the six sets of sample showed high bacteriological contamination.

Bacteriological quality of water samples collected from different hand pumps and bore wells in the study region indicated groundwater contamination. Existence of STP in the area could be a possible source of groundwater contamination. Raw sewage overflow, leakage in sewer lines, land application of sludge and partially treated wastewater and also septic tanks in the region can cause contamination of groundwater sources. A brief account of mode of transport of wastewater leading to groundwater contamination is described here.

Raw Sewage Overflow

Storm water systems in urban areas are sometimes combined with sanitary sewer systems en-route to STP. Excessive storm water can cause this combined system to overflow leading to flow of untreated sewage into waterways, resulting in contamination.

Leakage from Sewer Lines

Leakage may occur due to transportation of untreated raw sewage through various pipes/open channels on its way to STP and further final discharge to river body. Sewer leaks can occur from tree root invasion, soil slippage, seismic activity, loss of foundation due to washout, flooding and sewage back up, among other events (USEPA, 1990). When leaking sewer lines are located deep underground below the biologically active portion of the soil, the sewage can enter groundwater directly. This can result in the introduction of chlorides, microorganisms, organics, trace metals and other chemicals that may cause disease and foul taste and odors in drinking water (Rail, 1989).

Systematic inspection of sewer lines, exclusion of hazardous waste, and adherence to modern construction and maintenance specifications are necessary preventative measures for protection of groundwater sources from sewer leaks (USEPA, 1973).

Land Application of Partially Treated Wastewater and Municipal Sludge

Sludge, the residue of the chemical, biological, and physical treatment of municipal and industrial wastes can be applied to land as fertilizer or as fill. Sludge usually contains concentrated organic matter, nitrogen, inorganic salts, heavy metals, and bacteria. It is a common practice to use partially treated wastewater for fertilization, irrigation, and water supply recharge as an alternative to direct discharge into waterways. Movement and percolation of wastewater through the soil, biologically and physically removes biodegradable substances, pathogenic organisms, and inorganic substances (Rail, 1989).

Further, application of chemicals on agricultural lands, lawns and farm fields to kill weeds and insects and to fertilize the plants, may lead to groundwater contamination, as these chemicals may get washed particularly during rainy season.

Septic Tanks

Individual septic tank systems, those are not connected to a city sewer system, particularly in rural or in semi-urban areas, can be serious contamination sources. Septic tanks are mainly designed to slowly drain away human waste underground at a slow, harmless rate. An improperly designed, located, constructed, or maintained septic tank can leak bacteria, viruses, and household chemicals, causing serious problems of groundwater contamination.

Further, site specific environmental factors around the septic tank and leachfield such as soil properties, water table location, subsurface geology, climate, and vegetation may also affect the quality and quantity of released wastewater (USEPA, 1990).

Action Plan/Remedial Measures

Based on the findings of the study, the following action plans / remedial measures are suggested:

- ☆ Sources of contaminated water used for potable purposes as identified by the study should be red-marked and sealed. Alternate means of safe water supply to these areas must be found.
- ☆ In order to arrest possible seepage of sewage into the groundwater, the bottom and the sides of the treatment plant units should be made impervious employing suitable means *viz.* clay consolidation, HDPE sheets and/or chemical sprays etc.
- ☆ A water quality-monitoring programme should be carried out on regular basis and necessary preventive measures should be taken to ensure that the residents get safe water.
- ☆ As an additional safeguard, periodic health check-ups of the residents should be carried and necessary measures should be taken to prevent water borne diseases.
- ☆ The affected population in the vicinity of the treatment plants should be made aware about the risk associated with the use of contaminated water. Information should also be imparted to the residents regarding simple methods of disinfection using disinfectants like chlorine in tablet/liquid form.

Conclusion

The quality of water samples collected from different groundwater sources (hand pumps and bore wells) located in the surrounding areas of sewage treatment plant, indicated contamination. Contamination of drinking water sources by sewage can occur from raw sewage overflow, septic tanks, leaking sewer lines, land application of sludge and partially treated wastewater.

Contamination could be caused by seepage of sewage from the lagoons. Continuous discharge of treated sewage into the river could be the possible cause of groundwater contamination in this area. Use of untreated sewage for irrigation and its infiltration into the groundwater appeared to be the causative factor. Further, inadequate sanitation facility, proximity of groundwater sources to septic tanks/soak pits could also be a source of bacteriological contamination. However, the exact source of contamination could not be pinpointed.

The findings of the study indicate that a detailed hydro-geological study of the area with recourse to predictive modeling for assessment of water quality is warranted in order to establish the exact cause of groundwater contamination. The hydro-geological studies would comprise of quantitative/qualitative assessment of water reserves, its pathways (flow directions) and the identification of potential sources of contamination. The predictive modeling would also provide the future scenario on the basis of the present data and this will help concerned authorities in planning their future strategies.

References

1. APHA, AWWA, WEF, 1992. *Standard Methods for Examination of Water and Wastewater*, 18th ed., American Public Health Association, Washington, D.C.
2. Bhatnagar P.R. and Sharma B.R., 2002. Groundwater Pollution through Agricultural Practices and Agro-Industries in India. IWMI-TATA Water Policy Research Program, Annual Partners' Meet 2002. (Web Article).
3. Imhoff, K., 1982. *Handbook of Urban Drainage and Wastewater Disposal*, John Wiley and Sons, New York.
4. Mara D.D., 1968. *Bacteriology for Sanitary Engineers*, John Wiley and Sons, New York.
5. Metcalf and Eddy, 1991. *Wastewater Engineering*, Treatment, Disposal, and Reuse, Tata McGraw-Hill, Inc., New Delhi.
6. Rail, Chester D., 1989. *Groundwater Contamination: Sources, Control and Preventive Measures*. Technomic Publishing Company Inc., Lancaster PA, pp. 37-40.
7. Todd D. K., 1980. *Groundwater Hydrology*, John Wiley and Sons, New York.
8. USEPA, 1990. *EPA Handbook, Vol.1: Groundwater and Contamination*. Office of Research and Development, Washington DC.
9. US EPA. Water Programs, Guidelines Establishing Test Procedures for Analysis of Pollutants. Federal Register, 38 (199): 28758-28760. Oct 16, 1973.
10. WHO, 1985. *Guidelines For Drinking Water Quality*, Vol. 3.

Chapter 67

Wastewater Treatment and Management through Natural Systems

T. Nandy and S.N. Kaul

CSIR-NEERI, Nagpur

ABSTRACT

Wastewater treatment and management processes depend on natural responses, *viz.* gravitational forces for sedimentation or on natural components, *viz.* biological organisms, and these are often supported by complex array of energy intensive mechanical equipments. However, natural waste treatment systems are designed specifically to utilize natural resources to the maximum possible degree while obtaining the intended wastewater treatment or management goal. These systems cost less to build and to operate, and require less energy than mechanical treatment alternatives. Natural systems though typically include pumps and piping for wastewater conveyance, however, does not depend exclusively on external energy sources to maintain the major treatment responses.

An attempt has been made in this paper to address the various natural waste treatment systems explored for renovation of wastewater alongwith the detailed functioning, performances anticipated and achieved, and the design considerations.

Keywords: *Natural systems, Wastewater treatment, Management, Aquatic systems, Stabilization ponds, Water hyacinth, Duckweeds, Wetlands, Terrestrial systems.*

Introduction

Wastewater management has received considerable attention in recent years to meet the stringent regulations of the statutory bodies in addition to water scarcity.

Research in this field has resulted in development of various treatment technologies by applying various concepts, *viz.* physical, chemical and biological.

Amongst the various treatment routes available for renovation of wastewater, natural systems, also form a key component. The main underlying principle of operation of this system is its dependence on natural components for achieving the intended purpose. Natural systems explored so far for effective wastewater treatment comprise mainly aquatic, wetland and terrestrial concepts. These systems depend on natural, physical, chemical, and also unique biological processes, which highlights significance of each of the treatment process.

Aquatic Treatment Systems

The major treatment responses in the aquatic treatment system, are biological components. These systems are further subdivided to distinguish between lagoon and/or pond systems, depending on microbial life and lower forms of plants and animals, in contrast to aquatic systems, which also utilize higher plants and animals.

Wetland Treatment Systems

Wetlands are identified as lands with water table at or above the ground surface to maintain saturated soil conditions, and the growth of related vegetation. The wetlands used for renovation of wastewater include pre-existing marshes, swamps, strands, logs, peat lands, cypressdomes, and systems especially constructed for wastewater treatment.

Terrestrial Treatment Systems

The basic terrestrial concepts include, slowrate, rapid infiltration, overland flow and on-site which are dependent on physical, chemical and biological reactions on and within the soil matrix. In addition, the slow rate and overland flow also require presence of vegetation as a major treatment component.

Generally, wastewater constituents of concern include organics, pathogens, metals, and nutrients. In all the afore-referred systems, the natural components which provide critical reactions and responses include bacteria, protozoas, algae, vegetation (aquatic and terrestrial), and the soil, while the responses involved include a range of physical, chemical and biological reactions. However, the major concern of management in these processes include the potential for travel of contaminants with groundwater, the risk of leakage from ponds and other aquatic systems, groundwater mounding beneath the system, the need for drainage and maintenance of design flow conditions. The fundamental relationships amongst the various hydraulic parameters, *viz.* permeability, groundwater flow velocity, aquifer transmissivity, dispersion and retardation have been detailed elsewhere [(1, 2, 3, 4, 5)].

In addition, the use of natural systems demand the necessary understanding. Herremoes [(6)] indicate that it is a frequent conception that "low technology" approaches are simple and easy solutions that are "naturally" better than high technology. However, often the opposite holds true as the low technological natural systems indeed being simple from a technical point of view, yet they may be very complex from an ecological point of view and difficult to operate.

Therefore, a sequential approach include potential feasibility and the land requirement for treatment, with identification of possible sites based on technical and economic factors, followed by detailed field investigations, identification of the most cost-effective alternative, and development of the criteria needed for final design. The conservative methods for area estimation for the various natural treatment systems to assess their suitability are presented in Table 67.1.

Table 67.1: Conservative Design Parameters and Area Estimation for various Natural Treatment Systems

Sl.No.	Type of Pond	Design Conditions	Area Estimation
1.	Oxidation pond	Depth: 1 m Detention – 30 days Organic loading rate – 90 kg/ha.d Effluent BOD – 30 mg/L Effluent SS > 30 mg/l	A = kQ A – Area in hectare k – 3.2×10^{-3} Q – Flow in m^3/d
2.	Facultative ponds in cold climate	Depth: 1.5 m Detention – 80 days Organic loading rate – 16.8 kg/ha.d Effluent BOD – 30 mg/L Effluent SS > 30 mg/l	A = kQ A – Area in hectare k – 1.68×10^{-2} Q – Flow in m^3/d
3.	Facultative ponds in warm climate	Depth: 1.5 m Detention – >60 days Organic loading rate – 56 kg/ha.d Effluent BOD – 30 mg/L Effluent SS > 30 mg/l	A = kQ A – Area in hectare k – 5.1×10^{-3} Q – Flow in m^3/d
4.	Controlled discharge ponds	Depth: 1.5 m Detention – 180 days Effluent BOD – 30 mg/L Effluent SS > 30 mg/l	A = kQ A – Area in hectare k – 1.63×10^{-2} Q – Flow in m^3/d
5.	Partial mix aerated ponds	Depth: 2.5 m Detention – >50 days Organic loading rate – 100 kg/ha.d Effluent BOD – 30 mg/L Effluent SS > 30 mg/l	A = kQ A – Area in hectare k – 2.9×10^{-3} Q – Flow in m^3/d
6.	Secondary hyacinth ponds	Depth: 1.5 m Detention – >50 days Organic loading rate – 30 kg/ha.d Effluent BOD – 30 mg/L Effluent SS > 30 mg/l Temperature–>10°C	A = kQ A – Area in hectare k – 9.5×10^{-3} Q – Flow in m^3/d
7.	Advanced secondary hyacinth ponds	Depth: <1 m Detention – >6days Organic loading rate –1000 kg/ha.d Effluent BOD–<10 mg/L Effluent SS- <10 mg/l	A = kQ A – Area in hectare k – 9.5×10^{-4} Q – Flow in m^3/d
8.	Tertiary hyacinth ponds	Depth: <1 m Detention – >6 days Organic loading rate – 50 kg/ha.d Effluent BOD–<10 mg/L Effluent SS- <10 mg/l	A = kQ A – Area in hectare k – 7.1×10^{-4} Q – Flow in m^3/d

Contd...

Table 67.1–*Contd...*

Sl.No.	Type of Pond	Design Conditions	Area Estimation
9.	Constructed wetlands	Depth: 0.3 m Detention – >7days Organic loading rate – 100 kg/ha.d Effluent BOD–<10 mg/L Effluent SS- <10 mg/l	A = kQ A – Area in hectare k – 4.3 x 10^{-3} Q – Flow in m^3/d
10.	Overland flow	Depends on length of operating season Detention time – 1-2 hr Effluent BOD < 10 mg/L Effluent SS = 10 mg/L Total N < 10 mg/L Total P < 6 mg/L Hydraulic loading -15 cm/week	$A = 3.9 \times 10^{-4} (Q_m + 0.05\, q_c t_s)$ $Q_m = q_c + t_s . q_c / t_a$ Q_m – Av. monthly design flow, m^3/month q_c – Av. monthly flow in community, m^3/month t_s – Number of months storage is required t_a – Number of months in operating season
11.	Slow rate systems	Hydraulic loading -15 cm/week Effluent BOD <2 mg/L Effluent SS <1 mg/L	$A = 6.0 \times 10^{-4} (Q_m + 0.03\, q_c t_s)$
12.	Rapid infiltration systems	Effluent BOD < 5 mg/L Effluent SS <2 mg/L	$A = k\, Q_m$ $k = 5 \times 10^{-5}$ Q_m–Av. monthly design flow to land treatment side, m^3/month

The various natural treatment systems have been discussed highlighting the performances being anticipated and achieved, and are detailed in this paper. The expressions/equations/relationships presented in the paper provide accurate interrelations amongst various components/variables/parameters; however, use of only compatible SI units for all relevant parameters must be ensured.

Pollutants Removal Technique through Natural Treatment Systems

Organic loading in terms of biochemical oxygen demand (BOD) is found to be the limiting design factor for pond, aquatic and wetland systems, the basis being maintenance of aerobic conditions in the upper column in the unit, and the sources of dissolved oxygen being surface reaeration and photosynthetic oxygenation. However, BOD is seldom the limiting design factor for land treatment processes. Factors, *viz.* nitrogen, metals, toxics, hydraulic capacity of soils, etc. control the design, and therefore the system almost never approaches the upper limits for successful biodegradation of organics.

Most of the plants in these processes do not remove the BOD directly, however, they serve as hosts for a variety of attached growth organisms, and the microbial activity is primarily responsible for the organic decomposition.

Suspended solids though not a limiting factor for design, improper management of solids within the system can result in process failure. The removal of suspended solids in pond system depends primarily on gravity sedimentation though algae can be of concern in some situations. Sedimentation and entrapment in the microbial growths are both contributing factors in hyacinth, wetland, and the overland-flow processes. Filtration in the soil matrix is the principal mechanism for slow rate and rapid infiltration systems. The design features and anticipated performance of aquatic treatment systems and wetlands are detailed in Tables 67.2 and 67.3, respectively [7, 8, 9, 10].

Volatilization, adsorption, and then biodegradation are the principal methods for removing trace organics which are resistant to biological decomposition in natural treatment systems. Volatilization occur at the surface of ponds, wetlands and rapid infiltration basins, in the water droplets from sprinkles used in land treatment and from the liquid films in overland-flow systems. Adsorption occurs primarily on the organic matter in the treatment system which is in contact with the wastewater. Microbial activity then degrades the adsorbed matter.

The basic kinetic equation used for determination of concentration of contaminant at any time t through volatilization is given as,

$$\frac{C_t}{C_o} = e^{-(k_{vol})(t)/(y)}$$

where,

C_t = Concentration at time t, mg/L

C_o = Initial concentration at t = 0, mg/L

k_{Vol} = Volatilization mass transfer coefficient, cm/hr

Y = Depth of liquid, cm

Sorption of trace organics to the organic matter present in the treatment systems is the primary physico-chemical mechanism of removal [11]. The emperical equation used to determine the sorption rate is as follows [12]:

$$\log K_{oc} = (1.00)\,(\log K_{ow}) - 0.21$$

where,

K_{oc} = Sorption coefficient expressed on an organic carbon basin = K_{sorb}/O_c

K_{sorb} = Sorption mass transfer coefficient, cm/hr

O_c = Percentage of organic carbon present in the system

K_{ow} = Octanol water partition coefficient

Jenkins *et al.* [13] estimated the sorption of trace organics on an overland flow through first order kinetics with rate constants defined as follows:

$$k_{sorb} = \frac{B_3\, K_{ow}}{Y(B_4 + K_{ow})m^{1/2}}$$

where,

k_{sorb} = Sorption coefficient, h^{-1}

m = Molecular wt. of organic chemical, g/mol

B_3 = Coefficient specific to treatment system (0.730 for overland flow system studied)

y = Depth of water on overland flow slope

K_{ow} = Octanol water partition coefficient

B_4 = Coefficient specific to treatment system (170.8 for overland flow system studied)

The rate of movement of organic compounds in soil has been found to be a function of velocity of the carrier water, organic content of the soil, octanol-water partition coefficient for organic compound and physical properties of the soil system. The movement velocity during saturated flow in the soil system can be estimated using the following equation [13]:

$$V_C = \frac{K\,G}{n - 0.63\,\rho\,O_C\,k_{ow}}$$

where,

V_c = Velocity of organic compound, m/d

K = Saturated permeability of soil, m/d in vertical or horizontal direction

G = Hydraulic gradient of flow system, m/m (1 for vertical flow and $\Delta H/\Delta L$ for horizontal flow)

n = Porosity of the soil

r = Bulk density of soil, g/cm^3

O_c = Organic content of soil, per cent

k_{ow} = Octanol water partition coefficient

Nutrients of importance include nitrogen, phosphorus and potassium, and are of dual concern, since their control is necessary to avoid adverse health or environmental effects, but the same are essential for the performance of natural biological treatment systems. Removals are achieved either by nitrification/denitrification, plant uptake, adsorption, sludge deposition or volatilization of ammonia.

Pathogens present in wastewater may pose the potential risk of impact on groundwaters from both aquatic and land treatment systems, the contamination of crops or infection of grazing animals and off site loss of aerosolized organisms. However, Reed [14] has shown that natural aquatic wetland and land treatment systems provide effective control of pathogens. Ponds and wetland system achieve removal through natural die-off, predation, sedimentation and adsorption, other factors being detention time and temperature. Removal of bacteria and viruses in land treatment systems is due to a combination, desiccation, adsorption, radiation and predation.

Waste Stabilization Ponds

Waste stabilization ponds are most appropriate where land is inexpensive, climate favourable and a simple method of treatment is desired not requiring equipment and operating skills [15]. The ponds are usually low cost, low maintenance, and also a sustainable method of wastewater treatment and therefore have been employed widely for treatment of wastewater for over a long time. Current refinements in design enabled ponds to be approached as reactors, affected by reaction kinetics and flow patterns. Performance criteria have also been widened to include removal of micro-organisms and nutrients, besides BOD [16].

Types of Ponds

Waste stabilization ponds can be classified by dominant type of biological reaction, duration and frequency of discharge, extent of treatment ahead of pond or arrangement among cells. The basic classification, however, depends on the dominant biological reactions occurring in the pond, *viz.*

- ☆ Aerobic
- ☆ Anaerobic
- ☆ Facultative

Aerobic waste stabilization ponds are shallow ponds of about 0.3 m depth or less, and are designed so as to maximize light penetration and also ensure growth of algae through photosynthetic action. Aerobic conditions are maintained throughout the depth of the pond at all times. Such ponds are also useful where ultimate harvesting of algae is desired, however, their use in wastewater treatment has been scanty.

Anaerobic waste stabilization ponds require no dissolved oxygen for microbial activity as the anaerobic and facultative organisms utilize oxygen indirectly from compounds, *viz.* nitrates and sulphates as their hydrogen acceptors, and form end-products methane and carbon dioxide. These ponds, therefore, accept higher organic loadings, and operate without algal photosynthesis. Light penetration being of least priority, the ponds are built deeper maintaining about 3.0 to 4.0 m depth.

Facultative waste stabilization ponds are partly aerobic and partly anaerobic, these conditions being achieved by providing about 1.0 to 2.0 m depth, thus favouring algal growth alongwith the growth of facultative micro-organisms. These ponds being predominantly aerobic during the sunshine hours and for some hours of the night, turn anaerobic, in the few remaining hours especially at the pond bottom zone. Benthal deposits are generally anaerobic beyond the first few millimeters from the solids-water interface. Most of the existing waste stabilization ponds all over the world are of the facultative type with varying degrees of aerobicity and anaerobicity. The schematics representation of a typical facultative pond is shown in Figure 67.1 [16].

The sequence of installation of these ponds in a treatment scheme classify the ponds further as:

- ☆ Primary stabilization ponds

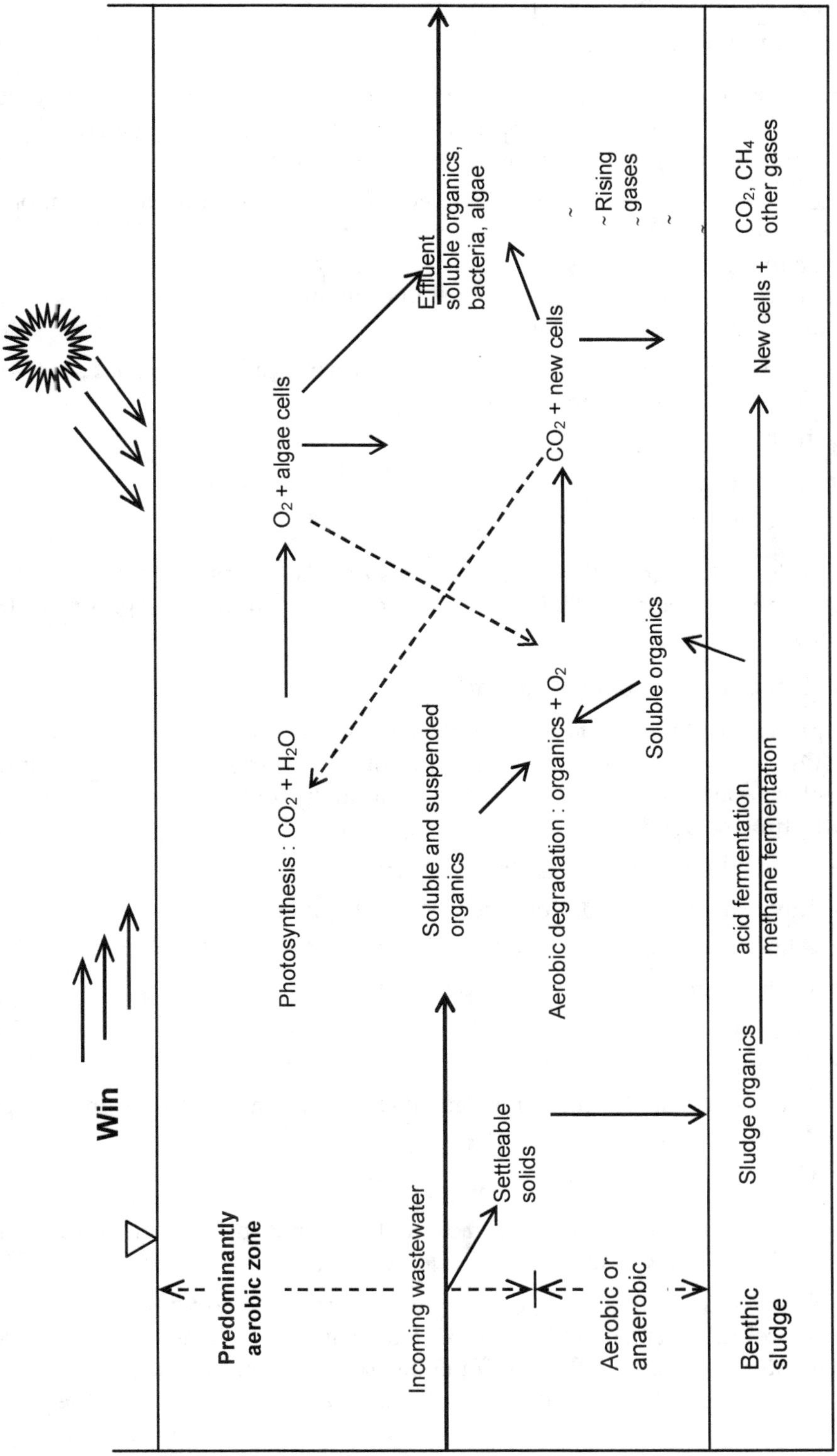

Figure 67.1: Schematic Representation of Processes in a Stabilization Pond.

- Secondary stabilization ponds
- Maturation ponds.

Ponds receiving untreated wastewater are referred to as primary waste stabilization ponds. Ponds receiving primary treated or biologically treated effluent for further treatment are termed secondary waste stabilization ponds.

In the maturation ponds, the effluent pretreated either in ponds or conventional treatment plants are exposed for a further period of time with the main objective of achieving natural bacterial die-off to desired levels. The ponds being lightly loaded in terms of organic matter, they constitute an economically feasible alternative to disinfection by chlorine in tropical climates [16].

The performance assessment of these ponds are basically carried out based on removal efficiencies achieved in terms of,

- BOD
- Micro-organisms
- Nutrients.

In addition to the above, the performance also needs to be assessed in the light of possible nuisance development *(viz.* sulphide odours, mosquitoes) and groundwater pollution.

Factors Affecting Pond Ecosystem

The principal members of abiotic component in the pond ecosystem comprise oxygen (O_2), carbondioxide (CO_2), water, light and nutrients, while the biotic component includes algae, bacteria, protozoa and a variety of other organisms. Various factors affecting the pond ecosystem are detailed hereunder:

- Fluctuations in wastewater characteristics
- Environmental factors, *viz.* radiation, light and temperature
- Algal growth patterns, and their diurnal and seasonal variations
- Bacterial growth patterns and decay rates
- Hydraulic transport patterns
- Evaporation and seepage
- Solids settlement, liquefaction, gassification, upward diffusion and sludge accumulation
- Gas transfer at interface.

Besides the afore-referred factors, algae must be retained in the pond for sufficient time to complete its growth and multiplication through binary fission.

The observed growth rate for algae in ponds usually ranges widely depending on temperature and other factors. Studies carried out in UK by Toms *et al.* [17] revealed the growth rate in the range of 0.05 to 0.7 per day. The observed rate of 0.48 per day, indicate that the algae must remain for 1/0.48 days (a little over 2 days) in the pond otherwise cell washout would occur. Therefore, provision of minimum detention period of 2 to 2.25 days for successful algal culture needs to be ensured based on the growth rate studies.

Since bacterial growth rates are much faster than algae, the algal cell washout time controls the minimum detention period to be provided in the pond.

Process Mechanism

The principle components of stabilization ponds include solar radiation and presence of algae. Basic mechanism of photosynthesis aids in algal growth converting solar energy to chemical energy in the organic form. Studies indicate that about 6 per cent of visible light energy can be converted to algal energy [18,19]. The chemical energy contained in algal cell, determined using bomb calorimeter, averages to around 6000 calories per gram of algae on ash free basis.

Since visible radiation penetrating the water surface is usually limited to 4000-7000A, depending on the "sky clearance factor" for the area under consideration, the average visible radiation (VR_{av}) received can be estimated as:

$$VR_{av} = VR_{min} + [(VR_{max} - VR_{min}) \times \text{sky clearance factor}]$$

where,

VR_{min} = Minimum visible radiation when sky is cloudy

VR_{max} = Maximum visible radiation when sky is clear

Concurrently with algal production, oxygen is being produced in accordance with the following equation:

$$106\,CO_2 + 16NO_3 + HPO_4 + 122H_2O + 18H^+ \rightarrow \underset{\substack{\text{Algae}\\(1)}}{C_{106}H_{263}O_{110}N_{16}P_1} + \underset{\substack{\text{Oxygen}\\(1.3)}}{138O_2}$$

Since on weight basis the oxygen production is 1.3 times the algal production, to maintain aerobic conditions at all times in the pond, it is necessary that ultimate BOD of wastewater does not exceed oxygen produced in algae photosynthesis thereby ensuring presence of dissolved oxygen in the effluent at all times.

Neel *et al*[20] suggested the ratio of average radiation/ultimate BOD loading to be above 0.38 to maintain dissolved oxygen in the pond effluent. However, Indian experiences indicate ratio above 0.42 as desirable. Figure 67.2 presents the relative depths of aerobic and anaerobic zones in facultative ponds at day and night [16]. Thirumurti[21] has depicted that the dissolved oxygen in the effluent diminishes as radiation reduces in intensity, however, BOD removal rate progresses undiminished until finally light becomes limiting.

It has been found that as the BOD load on a pond is increased, the depth of the aerobic zone tends to diminish both during day and night hours. With increased BOD load, relatively more algal growth is promoted, and the algal biomass, both gives and consumes oxygen. It gives oxygen only during daylight hours when photosynthesis is possible while consumes oxygen all the while in respiration. Increased algal biomass may lead to loss of oxygen in non-photosynthesis hours

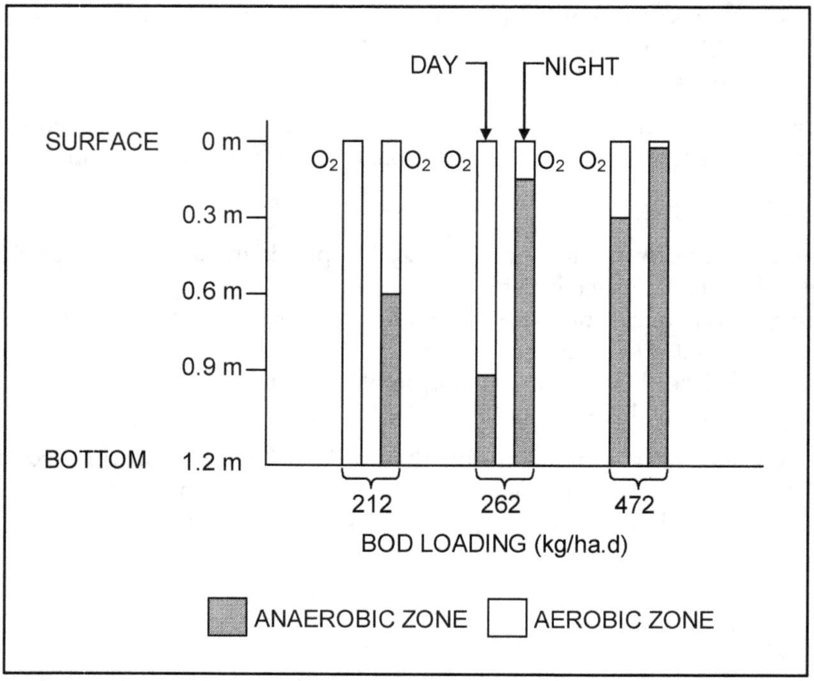

Figure 67.2: Depths of Aerobic Zones at different Loadings of Ponds [16].

thereby resulting in anaerobic conditions and sulphide formation. Further, loading the pond maintaining the ratio of average radiation/ultimate BOD loading in the range of 0.42 to 0.55 resulted in reasonably achieving nuisance free conditions [16].

The detention period required to be provided in ponds to achieve desired degree of BOD or coliform removal can be determined using the dispersion flow model [16]:

$$S = S_o \left[\frac{4a\,e^{1/2d}}{(1+a)^2 e^{a/2d} - (1-a)^2 e^{-a/2d}} \right]$$

where,

a = $\sqrt{(1 + 4K_p td)}$

d = D/UL

D = Dispersion coefficient, m^3/hr

U = Flow velocity, m/hr

L = Length of flow, m

K_p = Substrate removal rate per unit time.

After analysing the pond data treating domestic sewage at US, Canada, Thailand including India, Arceivala [16] arrived at the following relationship:

$K_{p\,(20°C)} = 0.132 \log (L) - 0.169$

The value of $K_{p\,(20°C)}$ has been found to vary between 0.10-0.15 depending on pond load. Correction to be applied to the $K_{p\,(20°C)}$ for temperature change is given below:

$K_{p\,(T°C)} = K_{p\,(20°C)}\,(1.035)^{(T-20)}$

In addition to organic matter removal, a certain amount of sulphide formation also often occurs in facultative ponds. However, excessive formation may lead to problems of malodours, and even inhibition of algal activity especially with high sulphate containing wastewater being carried out by bacteria, which utilise sulphate as a hydrogen acceptor in their catabolic reactions, *viz. Desulphovibric sp.* The favourable conditions imply dissolved oxygen concentration less than about 0.16 mg/L, (practically anaerobic condition), and temperature above 15°C.

Gloyna and Espino [22] studied sulphide production in laboratory scale stabilization ponds under different loading conditions and developed an empirical relation (at 25°C) as given below:

$$S = (mg/L) = [0.0001058\,(kg\,BOD_5/ha.day) - 0.001655\,(detention, day) + 0.0553] \times SO^{=}_4\,(mg/L)$$

Disappearance of algae has been observed by Gloyna and Espino [22] at sulphide concentration of about 6.5 to 8.4 mg/L. Sulphide formation may, become a severe constraint while adopting high BOD loadings and short detention times in warm climates. Oxidation reduction potential (ORP) measurements have been used to detect early formation of sulphides in ponds as relatively small increase in sulphide concentration cause large decrease in ORP [15].

Microorganisms removal is basically achieved in the ponds through enhancement of die down rates as reported by Marais [23] and adapted by Arceivala [16], and the following die off rate has been established:

$K_b = 1.0 - 1.2$ per day at 20°C.

For temperature other than 20°C, the rate can be obtained by the following expression:

$K_{b\,(T°C)} = K_{b\,(20°C)}\,(1.19)^{(T-20)}$

The alternative strategies to achieve higher removals of micro-organisms include the following:

- ☆ Appropriate pond geometry
- ☆ Pond in series
- ☆ Increased detention period (by increasing depth)
- ☆ Providing maturation pond
- ☆ Chlorination.

The models developed for total nitrogen removal achieved in the pond system given as hereunder have been detailed elsewhere [24,25,26]:

$$N_e = N_o \exp(-k_t [t+60.6 (pH-6.6)])$$

$$N_e = N_o \left[\frac{1}{1 + t(0.000576T - 0.00028) \exp[(1.080 - 0.042T)(pH - 6.6)]} \right]$$

where,

N_e = Effluent total nitrogen, mg/L

N_o = Influent total nitrogen, mg/L

k_T = Temperature-dependent rate constant, $d^{-1} = k_{20} q^{(T-20)}$

q = 1.039

T = Water temperature, °C

k_{20} = 0.0064

Typical pH values are reported elsewhere [25], or can be estimated using the following equation:

$$pH = 7.3 \exp [0.0005 (ALK)]$$

ALK – expected influent alkalinity, mg/L

Sludge accumulation has been estimated at 0.03-0.05 m^3 per capita per annum as per Gloyna [15]. Indian studies with four ponds have shown the accumulation to be 0.08 m^3 per capita per annum as the average after four years of accumulation [16]. Therefore, sludge cleaning needs to be planned once in 5-10 years. Due to the manurial value, the dried sludge has been highly favoured for use by farmers as soil conditioner.

Land requirement, being of prime importance in the application of stabilization ponds, include the water surface area to be provided in addition to about 15-30 per cent excess to accommodate embankments, roads, etc. Overall land requirement has been found to be 1-3 m^2 per capita in warmer areas to 3-12 m^2 per capita in colder areas.

Systems using Aquatic Plants/Animals

These aquatic systems utilize aquatic plants or animals as a component in wastewater renovation. The major biological components include floating plants, fish and other animals, planktonic organisms, and submerged plants. The treatment responses are due to either direct uptake of material by the plants or animals and by the presence of these biota altering the physical environment in the system, or as in the case of water hyacinth, the plant roots acting as the host substrate for attached microbial organisms, which provide significant degree of treatment [27]. However, specific environmental requirements must be maintained for their successful use, and in most cases a regular harvest is necessary to ensure optimum performance.

Floating Plants

Basically aquatic plants have same basic nutritional requirements as plants growing on land. The plants with greatest potential for wastewater treatment includes water hyacinths, duckweeds, pennywort and water ferns. Information on some of the critical environmental requirements of these plants has been presented in Table 67.4 [28]. Amongst these, hyacinths, pennywort and duckweeds have only been studied for wastewater treatment in pilot and full scale systems.

Water Hyacinth

Hyacinths are the most productive photosynthetic plants in the world. *Eichhornia crassipes* as being known, it is characterized as perennial, freshwater aquatic macrophyte (water-tolerant vascular plant) with rounded, upright, shiny green leaves and spikes of lavender flowers. It has been estimated that about 10 plants could produce 600000 more during an eight month growing season, and completely cover an area of 0.4 hectare of natural freshwater surface. Wolverton and McDonald [29] have estimated productivity of 140 tonnes/ha.annum (dry weight) for hyacinths grown in wastewater ponds. Due to the very rapid growth, hyacinths are serious nuisance problem, however, the same attributes are advantageous for wastewater treatment.

Since water hyacinth systems are capable of removing high levels of BOD, suspended solids, metals, nitrogen and significant levels of trace organics, application of these plants have been explored, and utilized for wastewater treatment. The dry-weight composition of water hyacinth plants removed from wastewater systems are detailed in Table 67.5 [30].

The characteristic feature of these plants are that they can survive and grow in anaerobic conditions, since oxygen is transmitted from the leaves to the root mass. The attached biological growth on the root mass is similar to trickling filter and rotating biological contactor (RBC) slimes, however, the oxygen source (from the roots) is near the center of the mass rather than on outside. Bacteria, fungi, predators, filter feeders, and detritovores have been reported to be present in large numbers on and among the plant roots. Typical performance data of this system with various loadings at different locations are given in Table 67.6 [29, 31].

Performance of this system with respect to BOD removal efficiency can be related directly to the density of the plant cover, and the depth of effluent in the system. For effluent depths of 1-2 m, BOD loading of about 6.7×10^{-4} kg/kg of wet plant mass per day has been recommended by Wolverton [31] when facultative pond effluent was applied to the hyacinth cells. At 80 per cent coverage of dense plants on the water surface, loading of 140 kg/ha.d BOD has been recommended by Wolverton and McDonald [29].

Removal of suspended solids is achieved through entrapment in the plant root zone, and by gravity sedimentation in the quiescent water beneath the surface mat of hyacinth plants. Another major contribution to solids control apart from above is the suppression of algae growth, since the hyacinth plant shades the water surface, and prevents passage of sunlight to the water beneath.

Plant uptake, ammonia volatilization, and nitrification/denitrification all contribute to removal of nitrogen in hyacinth systems. Nitrogen removal in excess of plant uptake levels observed indicate nitrification/denitrification, as the nitrifier organisms flourish attaching to the hyacinth roots which provide oxygen, while

Table 67.2: Design Features and Anticipated Performance for Aquatic Treatment Systems (7,8,9)

Systems	*Treatment Goals*	*Typical Criteria*			*Organic Loading (kg/ha.d)*	*Effluent Characteristics (mg/L)*
		Climate Requirement	*Detention Time (d)*	*Depth (m)*		
Oxidation pond	Secondary	Warm	10-40	1-1.5	40-120	BOD 20-40 TSS 80-140
Facultative pond	Secondary	None	25-180	1.5-2.5	22-67	BOD 30-40 TSS 40-100
Partial-mix aerated pond	Secondary, Polishing	None	7-20	2-6	50-200 TSS 30-60	BOD 30-40
Storage, and controlled – discharge ponds	Secondary, storage polishing	None	100-200	3-5	*	BOD-10-30 TSS 10-40
Hyacinth ponds	Secondary	Warm	30-50	<1.5	<30	BOD <30 TSS <30
Hyacinth ponds	Advanced Treatment, with secondary input	Warm	>6	<1	<50	BOD <10 TSS <10 TP <5 TN <5

* First cell in system designed as a facultative or aerated treatment unit.

Table 67.3: Design Features and Anticipated Performance for Wetland Treatment Systems [7,8,10]

Systems	*Treatment Goals*	*Typical Criteria*			*Organic Loading (kg/ha.d)*	*Effluent Characteristics (mg/L)*
		Climate Requirement	*Detention Time (d)*	*Depth (m)*		
Natural marshes	Polishing, advanced treatment with secondary input	Warm	10	0.2-1	100	BOD 5-10 TSS 5-15 TN 5-10
Constructed wetlands						
- Free water surface	Secondary, advanced treatment	None	7-15	0.1-0.6	200	BOD 5-10 TSS 5-15 TN 5-10
- Subsurface flow	Secondary, advanced treatment	None	3-14	0.3-0.6	600	BOD 5-40 TSS 5-20 TN 5-10

adjacent microsites and the benthic layer provide the anaerobic conditions and the carbon sources needed for dentrification [32]. The growth rate of about 220 kg/ha.d (dry weight) accounts for about 10 kg/ha.d of nitrogen.

Table 67.4: Floating Aquatic Plants for Wastewater Treatment [45]

Common Name, Scientific Name	*Temperature (°C)*		*Maximum Salinity Tolerance (mg/L)*	*Optimum pH*
	Desirable	*Survival*		
Water hyacinth *Eichhornia crassipes*	20-30	10	800	5-7
Water fern *Azolia caroliniana* *Azolia filculoides*	>10	5	2500	3.5-7
Duckweed *Spirodela trisculca* *Lemna trisculca* *Lemna obscura* *Lemna minor* *Lemna gibba* *Wolfia* spp.	20-30	5	3500	5-7

Table 67.5: Composition of Hyacinth Plants Grown in Wastewater [45]

Constituent	*Per cent of Dry Weight*	
	Average	*Range*
Crude protein	18.1	9.7-23.4
Fat	1.9	1.6-2.2
Fiber	18.6	17.1-19.5
Ash	16.6	11.1-20.4
Carbohydrate	44.8	36.9-51.6
Kjeldahl nitrogen (as N)	2.9	1.6-3.7
Phosphorus (as P)	0.6	0.3-0.9

Pilot-scale experiments with hyacinths and other aquatic plants in shallow containers (53 cm deep) indicate that overall nitrogen removal follow first-order reaction rate [33]. The nitrogen removal observed as function of plant density and temperature is given by the following equation and rate constants [34]:

$$\frac{N_e}{N_0} = \exp(-kt)$$

where,

N_e = Total nitrogen in system effluent, mg/L

N_0 = Total nitrogen applied wastewater, mg/L

Table 67.6: Performance of Hyacinth Wastewater Treatment Systems [45]

Hyacinth System	BOD (mg/L)		SS (mg/L)		Total N (mg/L)		Total P (mg/L)	
	Influent	*Effluent*	*Influent*	*Effluent*	*Influent*	*Effluent*	*Influent*	*Effluent*
Single, facultative cell, 122 cm deep, 2 ha, detention time 54 d, hydraulic loading 240 m^3/ha.d, organic load 26 kg/ha.d BOD.	110	7	97	10	12	3.4	3.7	1.6
Single, facultative cell, 173 cm deep, 3.6 ha, detention time 67 d, hydraulic loading 260 m^3/ha. d, organic load 44 kg/ha.d BOD.	161	23	125	6	–	–	–	–
Two-cell aerated pond, 183 cm deep, 0.3 ha, detention time 7 d, hydraulic loading 3570 m^3/ha.d, organic loading 179 kg/ha.d BOD.	50	14	49	15	–	–	–	–
Four-cell facultative pond, 85 cm deep, 0.06 ha, detention time 4.5 d, hydraulic loading 109 m^3/ha.d, organic loading 89 kg/ha.d BOD.	46	6	91	8	7.7	3.3	7	5.7
Five-cell facultative pond, 38 cm deep, 0.5 ha, detention time 11 d, hydraulic loading 378 m^3/ha.d, organic loading 113 kg/ha.d BOD.	13	3	–	3	22.4	1.0	11	3.6

k = Rate constant, dependent on temperature and plant desntity, d^{-1}
t = Detention time in system, d.

Temperature and plant density	k (d^{-1})
Summer Months	
Mean temperature 27°C ± 1°C	
Plant density, kg/ha (dry weight)	
3920	0.218
10230	0.491
20240	0.590
Winter months	
Mean temperature 14°C ± 4°C	
Plant density, kg/ha (dry weight)	
4190	0.033
6690	0.023
20210	0.184

Phosphorus removal through plant uptake has been achieved exceeding 30-50 per cent of the phosphorus present in municipal wastewaters [34]. Maximum plant uptake of phosphorus required supplemental nitrogen fertilization, since the nitrogen to phosphorus ratio of 6: 1 significantly varied with that present in wastewaters. The nitrogen deficiency therefore, result in improper utilization of the available phosphorus without additional nitrogen.

Phosphorus removal potential in hyacinth basin for over 80 per cent plant coverage and regular harvests can be estimated using the following relations:

$$L_p = (9353)\left(\frac{P_e - 0.778P_o}{P_o - P_e}\right)$$

where,

L_p = Hydraulic loading, limited by phosphorus removal, m^3/ha.d
P_e = Phosphorus concentration required in system effluent, mg/L
P_o = Phosphorus concentration in influent to hyacinth basins, mg/L

Hyacinth systems are also capable of high levels of metal removal achieved through the mechanism of chemical precipitation and adsorption on substrate, and on plant surfaces in addition to plant uptake. In addition, removal of trace organics has also been achieved through hyacinth systems, primarily due to volatilization and decomposition of the compounds by bacterial action apart from plant intake.

The design criteria for secondary treatment with hyacinth ponds are detailed in Table 67.7 [35]. The major operational concerns with hyacinth ponds include control of mosquitoes and odours, vegetation management, sludge removal, plant harvest and the disposal or utilization of the harvested materials and sludge. However, the beneficial reuse of harvested materials has been semidrying hyacinth, followed by utilization as soil conditioner/fertilizer. Utilization of hyacinth with sludge in anaerobic reactor has also been reported achieving high quality methane generation in cost effective processes [36].

Table 67.7: Design Criteria for Secondary Treatment with Hyacinth Ponds

Parameter	*Criteria*
Effluent requirements	BOD<30 mg/L, SS<30 mg/L
Wastewater input	Untreated
Organic loading:	
- Entire system surface	50 kg BOD/ha.d
- First cell in system	100 kg BOD/ha.d
Water depth	<1.5 m
Maximum area, single basin	0.4 ha
Total detention time	> 40 d
Hydraulic loading	> 200 m^3/ha.d
Water temperature	> 10°C
Basin shape	Rectangular, L: W>3: 1
Influent flow diffusers	Recommended
Mosquito control	Necessary
Harvest schedule	Seasonal or annual
Multipe cells	Essential, 2 sets of 3 basin, each recommended

Duckweed

Duckweed genera *Lemna* sp., *Spirodela* sp., and *Wolffia* sp., are small, green, freshwater plants with a leaflike frond, a few millimeters in width and short root, usually less than a centimeter in length have all been studied for pollutant removal, and used in wastewater treatment systems. Being the smallest and the simplest of the flowering plants, duckweed have been found to have very high reproduction rates [37].

Duckweed plants are more cold tolerant than hyacinths, and are found throughout the world. A minimum temperature of 7°C has been reported as the practical limit for growth of duckweeds [38]. The density of the plants at the water surface depends on temperature, availability of nutrients and frequency of harvest. The density on wastewater pond has been reported to be in the range of 1.2 kg/m^2 to 3.6 kg/m^2 wet weight.

Renovation of wastewater is mainly achieved through removal of BOD, suspended solids, nitrogen, phosphorus and metals. The major factors responsible for BOD and suspended solids removal are similar to that reported for facultative stabilization ponds. These plants create the environment for treatment, but contribute very little directly to the removal of BOD. Wolverton and McDonald have reported on the performance of a duckweed-covered basin [39].

However, removal of suspended solids from duckweed covered basins have been more effective in comparison with that from a conventional stabilization pond due to lack of algae and improved quiescent conditions under the surface mat. The final suspended solids concentration of 14 mg/L has been reported [39].

Plant uptake followed by removal by harvest is the major pathway for nitrogen removal in a duckweed basin, without supplemental aeration or alternative treatments. Studies indicate that an annual harvest of 22 tonnes/ha would remove about 880 kg/ha of nitrogen. Since duckweed plants have essentially no root zone, the nitrification/denitrification reactions, as in a hyacinth system, cannot occur. The environment in the liquid beneath the floating duckweed material being anaerobic, the availability of sufficient carbon source to support dentirification is very unlikely. Therefore, supplemental carbon source or separate dentirification reactor becomes necessary as a post treatment.

Removal of about 16 per cent phosphorus can be achieved through annual harvest of duckweed. However, phosphorus removal requirement in excess needs to be met with chemical precipitation or separate treatment.

Unlike hyacinth systems, plant uptake of metals has been reported to be less in these systems. Metal concentrations measured in the tissue growing on the municipal wastewater comprised of zinc 180 mg/kg, copper <26 mg/kg, lead <86 mg/kg, chromium <52 mg/kg and nickel <86 mg/kg.

Submerged Plants

Use of submerged aquatic macrophytes for treatment of wastewater has been studied in the laboratory, greenhouse and in pilot-scale field study by McNabb [36]. Table 67.8 provides details on some of the submerged freshwater plant that have been studied for use in wastewater treatment.

However, the utilization of this system is limited by the turbidity of wastewater, which may prevent light transmission to the plants to support photosynthetic activity and aerobic environment. Full scale application necessitates very shallow depths to ensure adequate light penetration, and contact between the plants and wastewater. However, algae growth being favoured under such conditions, additional process step for algal separation needs to be provided.

Aquatic Animals

Aquatic animals have also been considered for use in wastewater treatment, *viz.* daphnia, brine shrimp, and a wide variety of fish, clams, oysters, and lobsters [36,40,41,42]. The primary pollutants being removed mainly comprise of suspended solids and nutrients. Fish species used in wastewater treatment are detailed elsewhere [43]. However, the viability of this treatment system on the basis of cost benefits for wastewater renovation has not been explored for full scale application.

Table 67.8: Submerged Aquatic Plants with Potential for Wastewater Treatment

Common Name, Scientific Name	*Characteristics*
Pondweed, *Potamogeton* sp., *P. amplifolius* (most studied type)	Has both floating and submerged leaves, reproduces from rhizomes growing in sediments
Water milfoil, *Myriophyllum heterophyllum*	High branched stem upto 3 m long, vegetative reproduction
Water weed, *Elodea* sp.,*E. canadensis* (most studied type)	Irregular branching stem, vegetative reproduction
Coontail, *Ceratophyllum demersum*	Rootless, branched stem, pinnate leaves, vegetative reproduction
Fanwort, *Cobomba caroliniana*	Highly branched from the base, whorled leaves, vegetative reproduction

Wetland Systems

The lands with water surface near the ground surface for long enough each year to maintain saturated soil conditions alongwith the related vegetation, and being known as wetlands have also been explored for the possible renovation of wastewater.

Swamps, bogs and marshes characterized by dominant trees, mosses and peat and grasses, respectively, have been the types of wetlands used for wastewater treatment. The predominant combinations of involving wetlands and wastewater are as delineated hereunder:

- ✩ Disposal of treated effluent into natural wetlands
- ✩ Use of natural wetlands for further wastewater renovation
- ✩ Use of effluents or partially treated wastewater for enhancement, restoration, or creation of wetlands
- ✩ Use of constructed wetlands as a wastewater treatment process
- ✩ Use of constructed wetlands for treatment of agricultural runoff, combined sewer overflows, landfill leachates, and mine drainage.

Utilization of treated effluents for mitigation, or enhancement, restoration, or creation of wetlands, has been found to be desirable, and environmentally compatible activity, in addition to the cost-effective treatment potential of the constructed wetlands.

Studies indicate that performance of constructed wetland supercedes the natural wetland of equal area, since the bottom is carefully graded and the hydraulic regime in the system is controlled. In addition, process reliability is also improved due to efficient management of vegetation and system components.

Constructed wetlands presently being used mainly comprise free-water surface and subsurface-flow type. Free water surface system has the water surface exposed to atmosphere, the bed of emergent aquatic vegetation, layer of soil to serve as a rooting media, a liner to protect the groundwater. Design of this system with design flow of 1 x 10^6 m^3/d in Egypt has been reported [44].

A subsurface flow wetland consists of excavated basin filled with porous media (mainly gravels), and effluent level is maintained below the top of the gravel. The vegetation being similar to that of free water surface system is present in the upper part of the gravel media. The largest operational system designed for wastewater flow of 13000 m^3/d in USA has been reported [43]. The improved process reliability due to better management of vegetation, and the other system components achieved has also been reported.

Major system components influencing the treatment process in wetlands include the plants, detritus, soils, bacteria, protozoa, and higher animals; the functions and system performances of these biotic components further influenced by water depth, temperature, pH, and dissolved oxygen concentration. Figures 67.3 and 67.4 depict the mechanism that dominate free water surface and submerged bed wetland system [45].

The aquatic macrophytes found in marsh type wetlands are cattails (*Typha* sp.), reeds (*Phragmites communis*), rushes (*Juncus* sp.), bulrushes (*Scirpus* sp.), and sedges (*Carex* sp.). Bulrush and cattails, or a combination of the two, are the dominant species in most of the constructed wetlands. Further details on the characteristics of both emergent and submerged plant species are reported in literature [46,47,48].

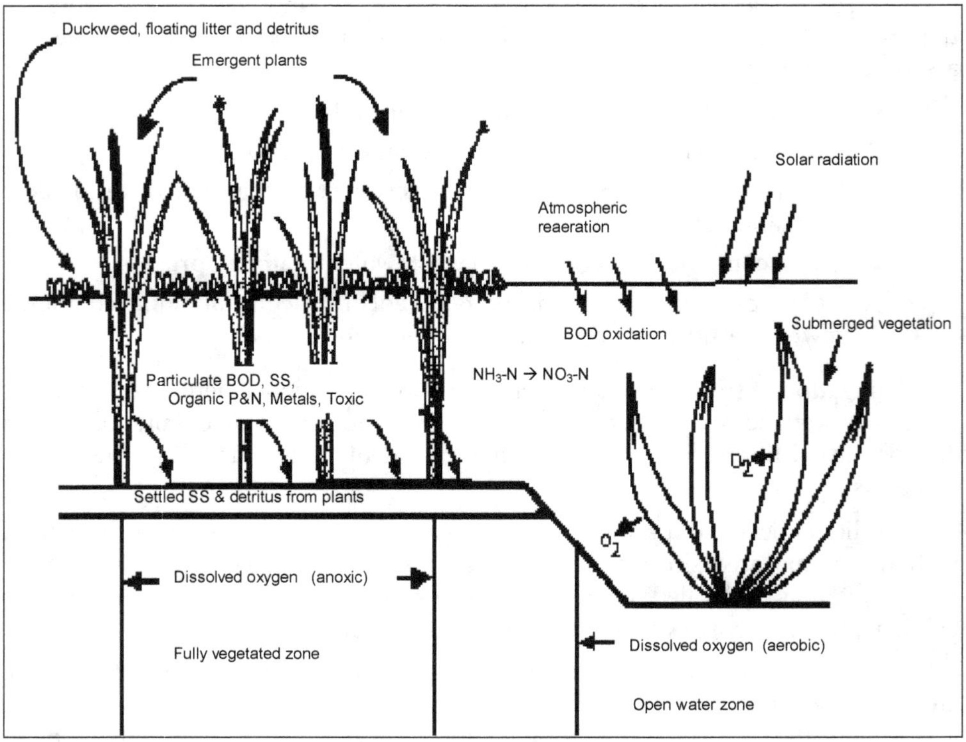

Figure 67.3: Mechanisms dominating Free Water Surface Systems [45].

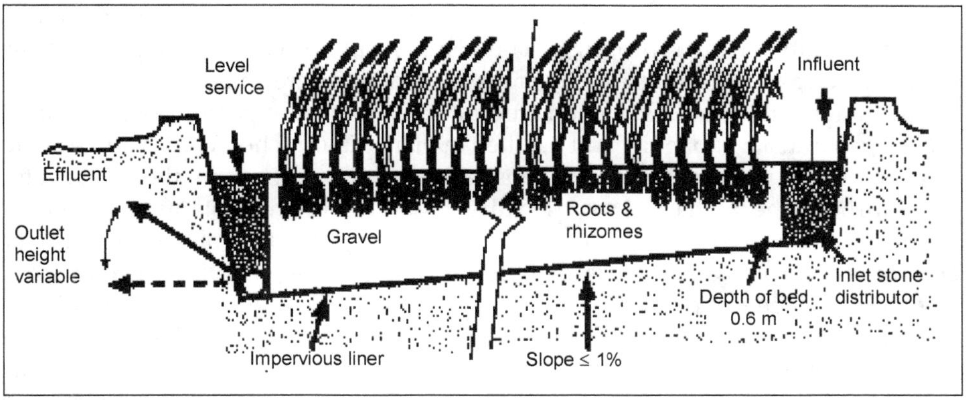

Figure 67.4: Typical Arrangement for a Vegetated Submerged Bed Treatment System[45]**.**

Since natural wetlands contain a wide diversity of plant life, attempts to replicate the same in constructed wetlands designed for wastewater treatment has not been successful. Due to high nutrient content of most wastewaters, the conditions tend to favour growth of cattails, reeds, etc. thus crowding out other less competitive species over time.

Emergent aquatic plants used in wetlands take up nutrients from wastewater as terrestrial plants used in land treatment systems. However, harvesting is not practiced routinely due to problems of access, and relatively high labour costs. Studies [49,50] have shown that harvesting of the plant material from a constructed wetland provides minor nitrogen removal pathway as compared to biological activity in the wetland. It has been indicated that less than 10 per cent of nitrogen removal by the system has been achieved.

Water loss due to evapotranspiration needs to be considered for wetland designs in arid climates during the warm summer months in all locations. Evaporative water losses in the summer months decrease the water volume in the system, and therefore the concentration of pollutants remaining in the system tends to increase even though treatment is very effective on a mass-removal basis.

Due to continuous inundation, the soils, or media in the surface flow wetland are anaerobic, an environment not well suited to support most vegetative species. However, the emergent plant species are all developed with the ability to absorb oxygen and other needed gases from the atmosphere through their leaves and above-water stems, and they have large gas vessels, which conduct those gases to the roots. The roots are therefore, sustained aerobically in an otherwise anaerobic environment. It has been estimated that these plants can transfer between 5 and 45 g of oxygen per day per square meter of wetland surface area, depending on plant density and oxygen stress levels in the root zones [47].

Most of the oxygen is utilized at the plant roots, and availability is limited for support of external microbial activity. However, some of the oxygen is believed to reach the surfaces of the roots and rhizomes, and create aerobic microsites at these points. These aerobic microsites then support aerobic reactions such as nitrification, if other conditions are appropriate. The plants respond with more oxygen as the demand increases at the roots, but the transfer capabilities are limited. Heavy deposits of raw sludge at the head of some constructed wetlands have apparently overwhelmed the oxygen transfer capability and resulted in plant die-off. The oxygen source being of most benefit in the surface flow constructed wetland, wherein, the wastewater flows through the media and comes in direct contact with the roots and rhizomes of the plants. In the free water surface wetland, the wastewater flow above the soil layer and contained roots, and there is no direct contact with the potential oxygen source. The major oxygen source for a free water surface wetland is atmospheric reaeration at the water surface. To maximize the benefit in surface flow, it is essential to encourage root penetration to the full depth of media so as to ensure potential contact points throughout the profile. Removal of ammonia in surface flow wetland has been correlated directly with the depth of root penetration and the availability of oxygen [(52)].

In natural wetlands, most of the nutrients required for plant growth are obtained from the soil by emergent aquatic plants. Soils with some clay content have been found to be very effective for phosphorus removal. In free water surface wetlands, the only contact opportunities available are at the soil surface. During the initial operational period, phosphorus removal achieved excelled due to soil activity and plant development. However, pathways tend to reach equilibrium after prolonged period of operation, and phosphorus removal drop off significantly.

Since soils with clay content or granular media containing clay minerals, have ion-exchange capacity, removal of ammonium (NH_4^+), existing in wastewater in ionic form can be achieved. The capacity, however, rapidly gets exhausted in surface flow and free water surface wetlands, since the contact surfaces continuously under water turn anaerobic.

The basic treatment mechanisms of pollutants removal achieved in wetland systems include sedimentation, chemical precipitation and adsorption, and microbial interactions with BOD and nitrogen uptake by vegetation. Irregular harvesting results in fraction of the decomposing vegetation remaining as refractory organics aiding in the development of peat in wetland systems. The nutrients and other substances associated with the refractory fraction are considered to be permanently removed. Figures 67.5 through 67.8 depict the carbon, nitrogen, phosphorus transformations including generic removal of pollutants in free surface water wetlands [(45)].

Pilot plant studies carried out in Turnkey on constructed wetland for domestic wastewater indicate average removal efficiency of COD–90 per cent, SS–95 per cent, total nitrogen of 61 per cent and phosphate of 39 per cent at 122 gCOD/m^2.d loading conditions [(53)].

Design Kinetics involved in Wetland Systems

Constructed wetlands are considered to be attached growth biological reactors, and their performance can be estimated with first order plug flow kinetics for BOD and nitrogen removal. The basic relationship is given by the following equation [(43)]:

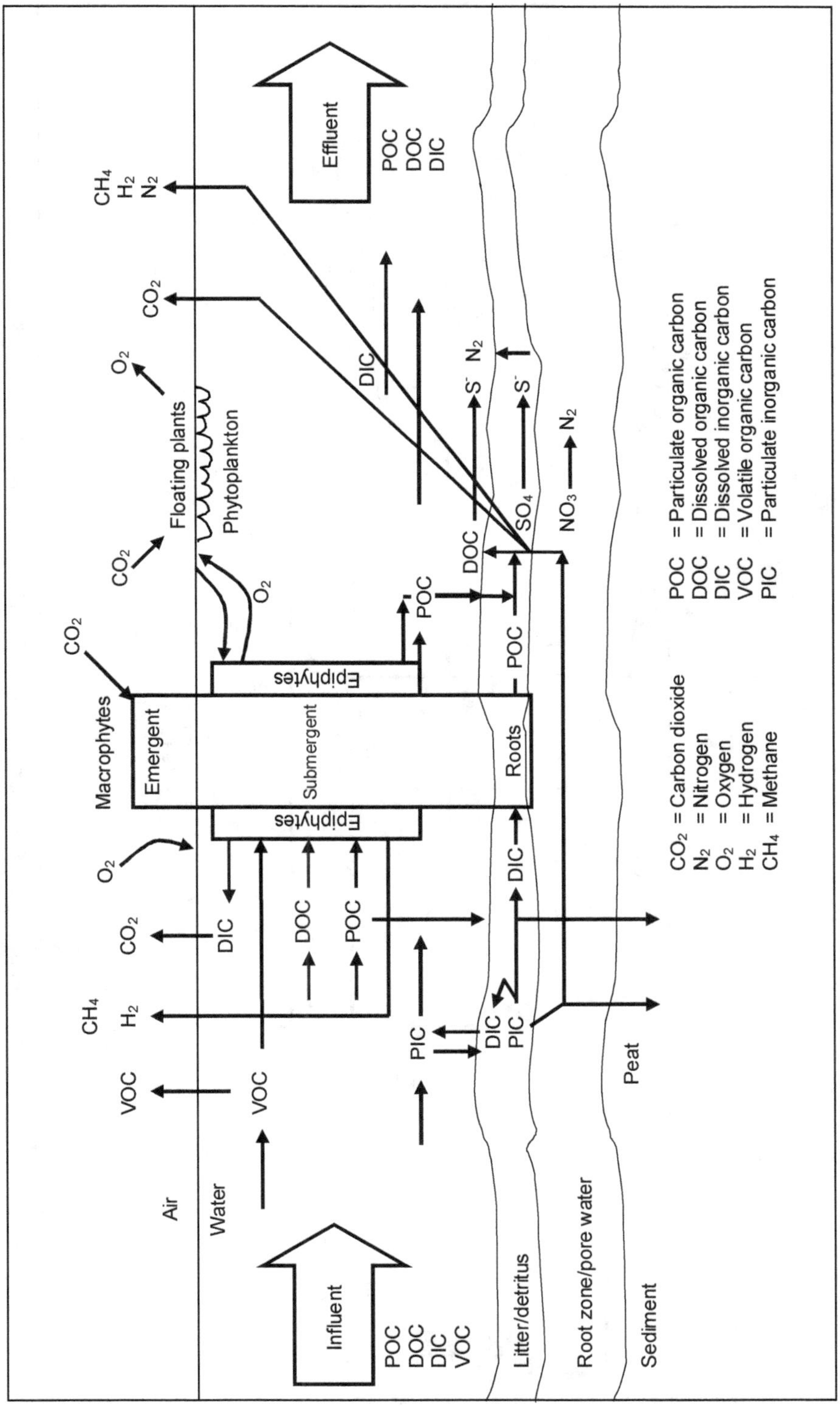

Figure 67.5: Carbon Transformations in Free Water Surface Wetlands [45].

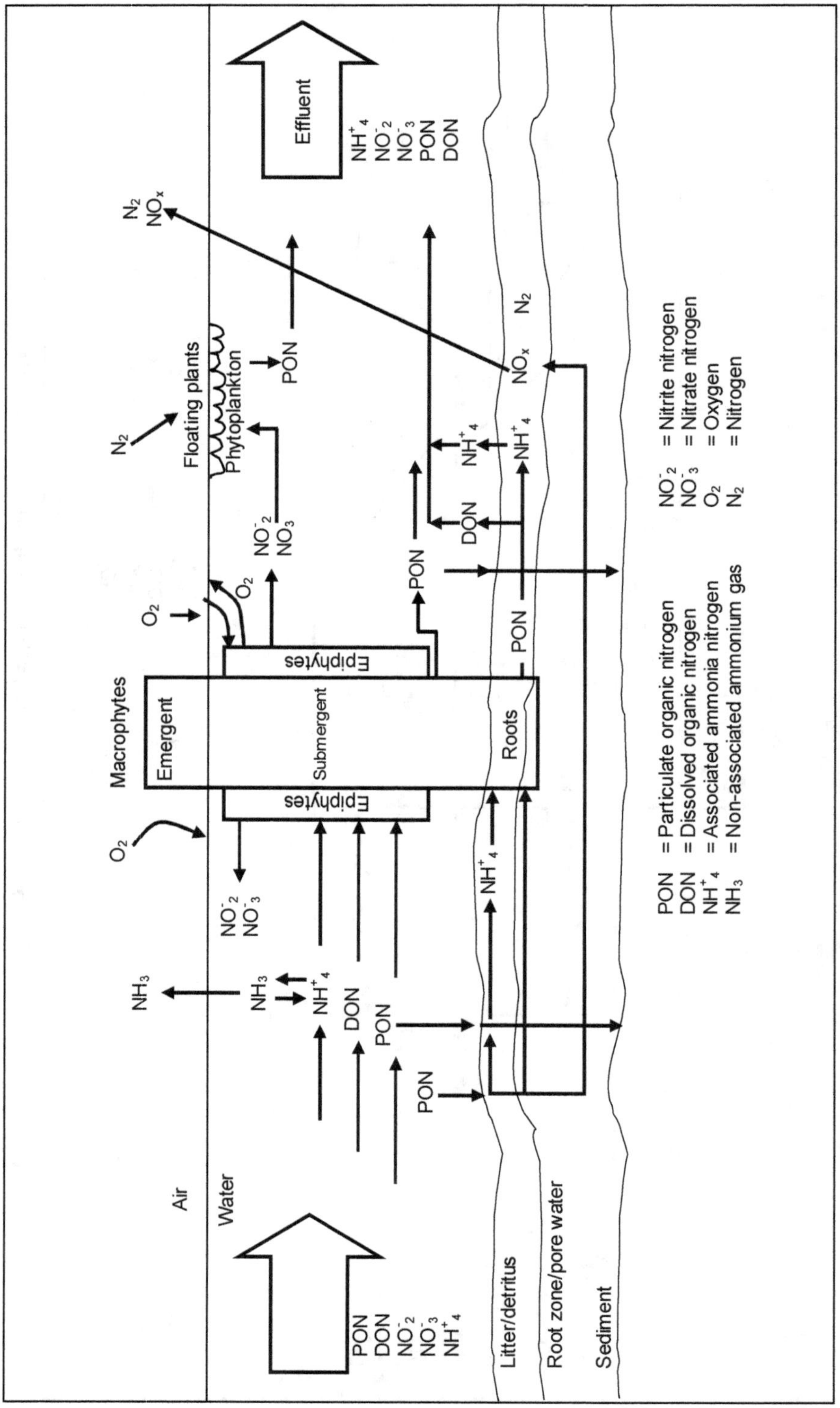

Figure 67.6: Nitrogen Transformations in Free Water Surface Wetlands[45].

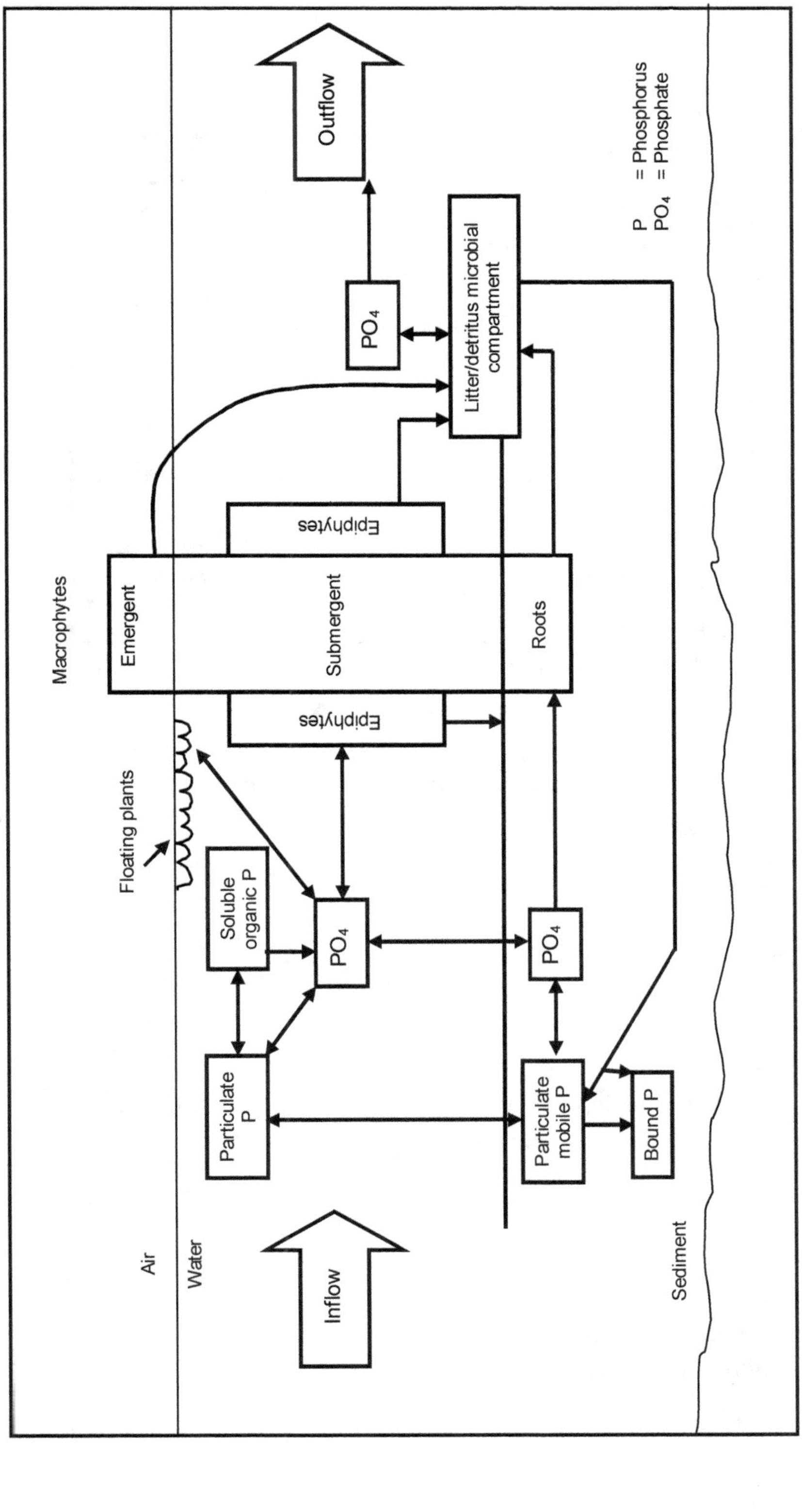

Figure 67.7: Conceptual Model of Phosphorus Cycling in a Hypothetical Wetland [45].

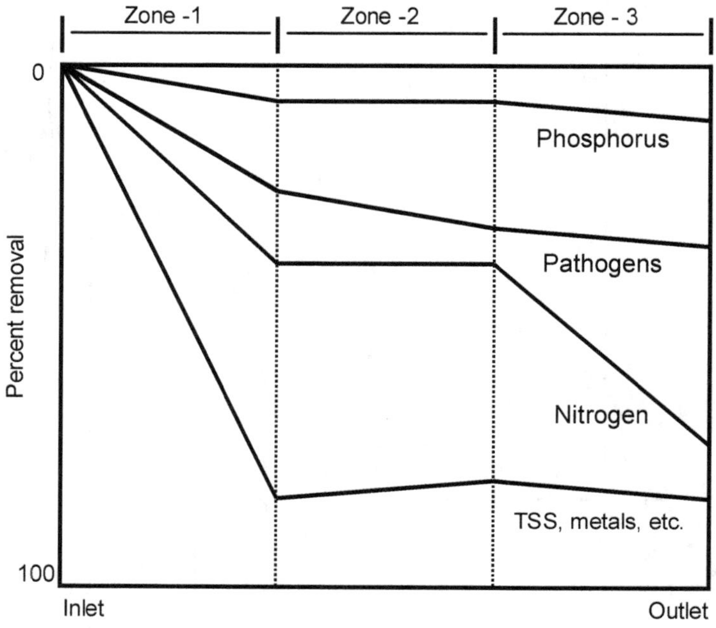

Figure 67.8: Generic Removal of Pollutants in Three-Zone Free Water Surface System[45].

$$\frac{C_e}{C_o} = \exp(-k_T t)$$

where,

C_e = Effluent pollutant concentration, mg/L

C_o = Influent pollutant concentration, mg/L

k_T = Temperature dependent first order reaction rate constant, d^{-1}

t = Hydraulic residence time, d; where

$$t = \frac{LWyn}{Q}$$

where,

L = Length of wetland, m

W = Width of wetland, m

y = Depth of wetland, m

n = Porosity expressed in decimal water to flow through the wetland. Vegetation and litter occupy some space in the free water surface wetland and the media, roots and other solids do the same in surface flow.

Q = Average flow through the wetland, m^3/d; where

$$Q = \frac{Q_{in} + Q_{out}}{Z}$$

Average flow being considered to account for water losses or gain via seepage or precipitation as wastewater flows through the wetland.

The surface area of the wetland can be obtained by combining the above equations:

$$A_s = LW = \frac{Q \ln (C_o / C_e)}{k_T \, yn}$$

where,

A_s = Surface area of wetland, m^2

The aspect ratio (L: W) selected for wetland strongly influences the hydraulic regime and the resistance to flow in the system. In the early design of wetlands aspects ratio of 10: 1 were recommended to ensure plug flow conditions [50]. However, the problem encountered was increased resistance to flow with increase in flow path. Therefore, aspect ratio of less than 1: 1 upto 3: 1 or 4: 1 has been found to be acceptable.

The flow of effluent in free water surface wetlands can be described by Manning's equation. However, in comparison with open channel flow, which assume frictional resistance only at bottom and sides, in these system the resistance to flow is distributed over the entire depth of water column due to emergent plants and litter present in the entire space.

$$v = \frac{1}{n} y^{2/3} s^{1/2}$$

where,

v = Flow velocity, m/s

n = Manning's coefficient

y = Depth of effluent in wetland, n

s = Hydraulic gradient on slope of effluent surface

Since n is dependent on depth of effluent, which further influences density of vegetation, is given by,

$$n = \frac{a}{y^{1/2}}$$

where,

a – Resistance factor

= 0.4 for sparse, low standing vegetation, y > 0.4 m

= 1.6 for moderately dense vegetation with y @ 0.3 m

= 6.4 for very dense vegetation and litter $y \leq 0.3$m

Therefore the modified Manning's equation can be represented as,

$$v = \frac{y^{1/2}}{a} y^{2/3} s^{1/2}$$

$$v = \frac{1}{a} y^{7/6} s^{1/2}$$

Since, $v = \frac{Q}{Wy}$; $W = \frac{A_S}{L}$ *and* $s = \frac{my}{L}$

Length of wetland can be calculated using the following equation [43]:

$$L = \left[\frac{(A_S) y^{8/3} m^{1/2} (86400)}{a\ Q} \right]^{2/3}$$

where,

Q = Flow, m^3/d (average flow $(Q_{in} + Q_{out})/2$

m = Increment of depth serving as head differential, per cent as a decimal (10-30 per cent of potential head available)

Darcy's Law, describes the flow regime in a porous media, and is used for design of subsurface flow wetlands. However the assumptions of Darcy's Law does not hold completely true for the actual subsurface flow wetlands, and, therefore, a reasonable approximation of the hydraulic conditions can only be obtained [43]:

$$v = k_S\ s$$

Since, $v = \frac{Q}{wy}$

$$Q = k_s A_c s$$

where,

Q = Average flow through wetland, m^3/d

= $(Q_{in} + Q_{out})/2$

k_s = Hydraulic conductivity of a unit area of wetland perpendicular to the flow direction, $m^3/m^2.d$

A_c = Total cross-sectional area perpendicular to flow, m^2

s = Hydraulic gradient slope of effluent surface in the flow system, m/m

v = Darcy's velocity, the apparent flow velocity through the entire cross-sectional area of bed, m/d

The minimum acceptable width of wetland can be determined using the following equation:

$$s = \frac{m\ y}{L};\ L = \frac{A_S}{W}\ and\ A_C = W\ y$$

$$W = \frac{1}{y}\left[\frac{Q\ A_S}{m\ k_S}\right]^{0.5}$$

where,

W = Width of wetland, m

A_s = Surface area of wetland, m^2

L = Length of wetland

m = Increment of depth serving as head differential, per cent as decimal

y = Depth of water in wetland, m

The aforereferred equations are valid for laminar flow, wherein Reynold's number (N_R) is less than 10. However, for turbulent flow conditions, the effective hydraulic conductivity will be significantly less than predicted by Dary's Law:

$$N_R = \frac{vD}{\tau}$$

The hydraulic conductivity, k_s varies directly with viscosity of effluent, which depends on temperature,

$$\frac{k_{sT}}{k_{S20}} = \frac{\mu_{20}}{\mu_T}$$

When coarse gravels on rocks are utilised as media, relationship based on Ergun equation [54] can be used to estimate hydraulic conductivity as given below:

$$k_S = (n)^{3.7}$$

where,

n = Porosity of media

Design of subsurface flow wetland is also influenced by organic loading and, therefore, is used for determining width by organic loading (kgBOD/m^2.d) on the cross sectional area at the entry zone of the system to avoid clogging [55].

Since temperature affect both physical and biological activities in wetland system [51, 54], calculations for determining effluent temperature in subsurface flow and free water surface wetlands have been established through extensive research [43].

For subsurface flow wetlands, basic mechanisms involved include conduction to or from the ground, conduction to or from the wastewater, conduction and convection to or from the atmosphere, and radiation to or from the atmosphere. However, simplified models based only on conduction losses to the atmosphere, and developed from basic heat transfer relationships [56,57,58] have been widely accepted.

The available energy in the water flowing through the wetland is given by the following equation:

$$q_G = C_p \delta A_s y n$$

where,

q_G = Energy gain from water, J/° C
C_p = Specific heat capacity of water, 4215 J/kg°C
δ = Density of water
A_s = Surface area of wetland, m^2
y = Depth of water in wetland, m
n = Porosity of wetland, *viz.* space available for water to flow; (the remainder is occupied by the media)

Daily temperature change of effluent can be calculated by substituting A_s/t for A_s in above equation:

$$q_G = C_p \delta \frac{A_s}{t} y n$$

where,

t = Hydraulic residence time, d
q_G = Energy gain during 1 day of flow, J/°C.d

The heat loss from the wetland can be calculated using the following equation:

$$q_L = (T_o - T_{air}) U \sigma A_s t$$

where,

q_L = Energy lost via conduction to atmosphere, J
T_o = Influent temperature to wetland, °C
T_{air} = Average air temperature during study period, °C
U = Heat transfer coefficient at the surface of wetland bed $W/m^2.°C$
s = Time conversion, 86400 s/d
A_s = Surface area of wetland, m^2
t = Hydraulic residence time in wetland, d

The heat conducting capacity of the wetland profile being a combination of the thermal conductivity of each of the major components, can be divided by its thickness as given below:

$$U = \frac{1}{\left(\frac{y_1}{k_1}\right)+\left(\frac{y_2}{k_2}\right)+\left(\frac{y_3}{k_3}\right)+\left(\frac{y_n}{k_n}\right)}$$

where,

$k_{(1\ to\ n)}$ = Thermal conductivity of layers 1 to n, w/m.°C

$y_{(1\ to\ n)}$ = Thickness of layers 1 to n, m

Further refinement can be achieved by incorporating energy gains and losses from solar radiation, and conduction to or from the ground. Estimation of heat input from the solar radiation can be carried out using the following equation[43]:

$$q_{solar} = \phi A_s t s$$

where,

q_{solar} = Energy gain from solar radiation, J

ϕ = Solar radiation for site, $J/m^2.d$

A_s = Surface area of wetland, m^2

t = Hydraulic retention time for wetland, d

s = Percentage of solar radiation energy which reaches the effluent (typically 0.05 or less)

While incorporating the thermal aspects in the free-water surface wetlands, freezing of the surface exposed to atmosphere is of great concern where its installation in colder parts of the country is required.The sequence of analysis include the following:

- Prior to ice formation
- Flow under ice cover
- Thickness of ice formation.

In the freewater surface wetlands, ice formation begins when the bulk temperature in the liquid approaches 3°C due to bulk difference and correction losses at the surface [56,59] and the following equation can be used sequentially with heat transfer coefficient U_s to calculate effluent temperature:

$$T_w = T_{air} + (T_o - T_{air})\exp\left[\frac{-U_s(x - x_o)}{\delta y v C_p}\right]$$

where,

T_w = Effluent temperature at distance x, °C

T_{air} = Av. Air temperature during study period, °C
T_o = Water temperature at distance xo, the entry point for wetland, °C
U_s = Heat transfer coefficient at wetland surface W/m².°C
δ = Density of effluent
y = Depth of effluent
v = Flow velocity in wetland, m/s
C_r = Specific heat, 4215 J/kg.°C

Rearranging the above equation for distance x at which the temperature becomes 3°C

$$(x - x_0) = \frac{-\delta y v C_p}{U_s}\left[\ln\frac{(3 - T_{air})}{(T_o - T_{air})}\right]$$

Once ice cover formation takes place, heat transfer is not influenced by air temperature. The water temperature under an ice cover can be estimated using following equation:

$$T_w = T_m + (T_o - T_m)\exp-\left[\frac{U_i(x - x_0)}{\delta y v C_p}\right]$$

where,

T_w = Effluent temperature at distance, x °C
T_m = Ice melting point, 0°C
T_o = Efficient temperature at distance x_o, °C
U_i = Heat transfer coefficient at ice/water interface, w/m_2. °C

U_i depends on effluent depth beneath the ice and flow velocity,

$$U_i = \phi\frac{v^{0.8}}{y^{0.2}}$$

where,

ϕ = Proportionality coefficient, 1622 J/m$^{2.6}$ s$^{0.2}$. °C
v = Flow velocity, m/s
y = Effluent depth, m

The thickness of ice formation over a day's period can be estimated by the following equation[56]:

$$y = \frac{t\sigma}{\delta\Omega}\left[\frac{(T_m - T_{air})}{y_s/k_s + y_i/k_T + 1/U_s} - U_i(T_w - T_m)\right]$$

where,

y = Thickness of ice formation per day, m/d
t = Time period of concern, d
s = Time conversion factor, 86400 s/d
δ = Density of ice, 917 kg/m^3
W = Latent heat, 334,944 J/kg
T_m = Melting point of ice, 0°C
T_{air} = Average air temperature during time period of concern °C
y_s = Depth of snow cover, m
k_s = Conductivity of snow
y_i = Depth of daily ice formation, m
k_i = Conductivity of ice
U_s = Heat transfer coefficient at surface
U_i = Heat transfer coefficient water to ice
T_w = Average effluent temperature during study period

The above equation has been modified, and the Stefan formulation is given below [60]:

$$y = m\left[(T_m - T_{air})\, t\right]^{1/2}$$

where,

y = depth of ice which will form over time period t, m
T_m = freezing point of ice, 0°C
T_{air} = average air temperature during time period t, °C
t = number of days in the period
m = proportionality coefficient
0.027 for open water zones, no-snow
0.018 for open water zones, with snow
0.010 for wetland with dense vegetation and litter

Model Simulations for Design of Wetland Systems

Since the surface area A_s of wetlands equal LW, the area required to achieve desired level of treatment can be obtained by the equation as hereunder [61,62]:

$$A_s = \frac{Q(\ln C_o - \ln C_e + \ln A)}{k_T\, y\, n}$$

where,

A_s = Surface area of freewater surface wetland, m^2
K_T = $K_{20}\,(1.06)^{(T-20)}$

K_{20} = 0.2779 d^{-1}

n = Porosity of system

A = 0.65-0.75 (lower value for clause and mature vegetation) fraction of BOD not removed as settleable solids near the headworks of the system, a variable depending on water quality (as decimal fraction)

0.052 for primary effluent

0.7-0.85 for secondary effluent

0.9 for tertiary effluent

C_o = Influent BOD, mg/L

C_e = Effluent BOD, mg/L

y = Average effluent depth in the system, m

To obviate the difficulties involved with evaluating A_s, another approach arrived at by evaluating performance data of operational wetland systems as given below:

$$\frac{C_e}{C_o} = e^{-k_T t}$$

where,

K_T = K_{20} $(1.06)^{(T-20)}$

K_{20} = 0.678 d^{-1}

Wetland surface area can be determined by the following equation:

$$A_s = \frac{Q(\ln C_o - \ln C_e)}{k_T\, y\, n}$$

where,

k_T = Rate constant, d^{-1}

y = Design depth of effluent in the system, m

n = Porosity of wetland, 0.65-0.75

However, the limitation of these equations are that the final effluent BOD is influenced by the production of residual BOD within the wetland from decomposition of plant detritus and other naturally occuring organics.

Knight *et al* [63] have proposed the following equation based on regression analysis for both type of wetland systems:

C_e = 0.192 Co + 0.097 HLR

where,

C_e = Effluent BOD, mg/L

C_o = Influent BOD, mg/L

HLR = Hydraulic loading rate, cm/d

Since the basic mechanism for BOD removal in subsurface wetlands are same as those for free water surface, the same models are applicable. However, the responses proceed at higher rate as the subsurface systems have more submerged surface area, and therefore greater potential for development of attached growth organisms. Therefore, the differences include variation in magnitude of porosity and rate constant K_{20}. Porosity varies with type of media proposed for use, and rate constant at 20°C = 1.104 d^{-1}.

The removal of suspended solids in both types of wetlands have been achieved due to physical processes, and influenced only by temperature through the viscosity effects on the flow of effluent. The equations developed for both these wetlands based on their performances for municipal wastewater at various hydraulic loading rates are as given hereunder [43]:

Subsurface wetlands: $C_e = C_o$ [0.1058 + 0.0011 HLR)

Free water surface: $C_e = C_o$ [0.1139 + 0.00213 HLR]

where,

C_e = Effluent total suspended solids, mg/L
C_o = Influent total suspended solids, mg/L
HLR = Hydraulic loading rate, cm/d

In freewater surface wetlands, ammonia removal is assumed to be due to nitrification with no plant uptake, since harvesting is not normally practiced. The models utilized in terms of ammonia concentrations are as follows:

$$\frac{C_e}{C_o} = \exp(-k_T\, t)$$

$$A_s = \frac{Q\ln(C_o/C_e)}{k_T\, y n}$$

where,

A_s = Surface area of wetland, m^2
C_e = Effluent ammonia concentration, mg/L
C_o = Influent TKN concentration, mg/L
K_T = Temperature dependent rate constant, d^{-1}
= 0, d^{-1} at °C
= 0.1367 x $1.15^{(T-10)}$ d^{-1} at 1 to 10°C
= 0.2187 x $1.048^{(T-20)}$, d^{-1} at 10°C and above
n = Porosity of wetland, 0.65-0.75
t = Hydraulic residence time, d
y = Depth of effluent in the wetland, m
Q = Average flow through the wetland, m^3/d

$$Q = \frac{Q_{in} + Q_{out}}{2}$$

While designing the free water surface wetland for both BOD and ammonia removal, larger of the area obtained from the respective models is provided.

Other models published in the literature [61] for ammonia removal include,

$\ln (C_e) = 0.688 \ln C_o + 0.655 \ln (HLR) - 1.107$

$$A_s = \frac{100\,Q}{\exp\left[1.527 \ln C_e - 1.050 \ln C_o + 1.69\right]}$$

where,

HLR = Hydraulic loading rate, cm/d

C_o = Influent ammonia concentration, mg/L

C_e = Effluent ammonia concentration, mg/L

A_s = Surface area of wetland, m^2

Q = Design flow, m^3/d

Equation developed by Hammer and Knight [64] with regression analysis is given as,

$$C_e = \frac{18.31\,C_o Q}{A_s} - 0.16063$$

These models account for conversion of ammonia to nitrate. However, when actual nitrogen removal is considered, denitrification requirement also needs to be incorporated.

The recommended model for etimating nitrate removal via denitrification are as follows [43]:

$$\frac{C_e}{C_o} = \exp(-k_T t)$$

$$A_s = \frac{Q \ln(C_o / C_e)}{k_T\, y n}$$

where,

A_s = Surface area of wetland, m^2

C_e = Effluent initial concentration, mg/L

C_o = Influent initial concentration, mg/L

k_T = Temperature dependent rate constant ($0 d^{-1}$ at 0°C, $1\,(1.15)^{(T-20)}$ at 1°C and above)

n = Porosity of wetland, 0.65-0.75

t = Hydraulic residence time, d
y = Depth of effluent in the wetland, m
Q = Average flow through the wetland, m^3/d

The area requirement for total nitrogen removal can be obtained by applying the following model [61]:

$$C_e = 0.193\, C_o + 1.55 \ln (HLR) - 1.75$$

$$A_S = \frac{100Q}{0.645C_e - 0.125C_o + 1.129}$$

where,

HLR = Hydraulic loading rate, cm/d
C_e = Effluent total nitrogen, mg/L
C_o = Influent total nitrogen, mg/L
A_s = Surface area of wetland, m^2
Q = Design flow, m^3/d

Due to subsurface flow, rate of atmospheric reaeration is significantly less than for free water surface wetland. However, the roots and rhizomes of the vegetation have been found to contain aerobic microsites on their surfaces and, as a result conditions for nitrifiction and denitrification are achieved in the same reactor.

Since major source of oxygen in this type of system has been the plant roots, it is absolutely essential to ensure that root system penetrates to the full design depth of the bed, thus emphasizing the fact that design depth of the bed should not be greater than the potential root depth of the plant intended for use, if oxygen requirement needs to be met for ammonia removal.

Extensive research with subsurface flow systems indicates oxygen availability at the root surfaces to support biological activity ranging between 5 and 45 $gO_2/m^2.d$ of wetland surface area [47, 65]. Studies [66] carried out in one of the system in USA indicate that rate of nitrification is depended on depth of root zone in the bed of wetland, and defined by the following relationship:

$$K_{NH} = 0.01854 + 0.3922\,(rz)^{2.6077}$$

where,

K_{NH} = Nitrification rate constant at 20°C, d^{-1}
rz = Per cent of subsurface flow bed depth occupied by root zone, a decimal fraction (0 to 1)

The K_{NH} value ranged between 0.4107 – 0.01854 with fully developed root zone and no vegetation conditions, respectively. Similar results are also highlighted by EPA studies [52] carried out, and also by Bavor *et al* [67]. Ammonia removal via nitrification can be defined through the following relationship:

$$\frac{C_e}{C_0} = \exp(-k_T t)$$

$$A_s = \frac{Q\ln(C_o / C_e)}{k_T\, y n}$$

where,

A_s = Surface area of wetland, m^2
C_e = Effluent ammonia concentration, mg/L
C_o = Influent TKN concentration, mg/L
k_T = Temperature dependent rate constant, d^{-1}
n = Porosity of the wetland
t = Hydraulic residence time, d
y = Depth of effluent in the wetland, m
Q = Average flow through the wetland, m^3/d

At 0°C, $K_o = 0\ d^{-1}$
At 1-10°C, $K_T = K_{10}\ (1.15)^{(T-20)} d^{-1}$
At 10°C and above, $K_T = K_{NH}\ (1.048)^{(T-20)} d^{-1}$

Similar to free water surface wetland, total nitrogen removal calls for requirement of both nitrification and denitrification. Even though subsurface flow wetlands have more surface area for biological responses, availability of carbon source in the system limits the denitrification rate. The model for estimating nitrate removal via denitrification is as follows:

$$\frac{C_e}{C_0} = \exp(-k_T t)$$

$$A_s = \frac{Q\ln(C_o / C_e)}{k_T\, y n}$$

where,

A_s = Surface area of wetland, m^2
C_e = Effluent nitrate concentration, mg/L
C_o = Influent nitrate concentration, mg/L
k_T = Temperature dependent rate constant, d^{-1}
[$0\ d^{-1}$ at 0°C, $1.000\ (1.15)^{(T-20)}.d^{-1}$ at 1°C and above]
n = Porosity of the wetland
t = Hydraulic residence time, d
y = Depth of water in wetland, m
Q = Average flow through the wetland, m^3/d

Phosphorus removal has not been found to be effective in both types of wetlands. The removals achieved over the long term, however, have been through accumulation of sediment, via particulate sedimentation, chemical precipitation and refractory organics from senescing vegetation mostly as precipitates of iron, aluminium or calcium.

Kadlec [54] has proposed first order rate constant equal to 10m/year for estimating phosphorus removal in constructed wetland systems:

$$\frac{C_e}{C_o} = \exp\left(\frac{-k_p}{HLR}\right)$$

where,

C_e = Effluent phosphorus, mg/L
C_o = Influent phosphorus, mg/L
k_p = 2.73 cm/d
HLR = Average annual hydraulic loading rate, cm/d

$$A_S = \frac{bQ\ln(C_o / C_e)}{kp}$$

where,

A_s = Surface area of wetland, m^2
b = Conversion coefficient (100 cm/m)
Q = Average flow through wetland.

This model though developed for free water surface wetland, can also be used for subsurface flow wetland since it is dependent on areal surface deposition, and not on biological reactions occurring on the specific surface area of the media or detritus within the flow zone.

Design of onsite systems for relatively small facilities has been highlighted by providing several approaches, and are detailed elsewhere [55]. Several simplifying assumptions for design of wetlands for small onsite systems have also been documented in USEPA [52].

Application Status

The application status of constructed wetlands include domestic wastewater, municipal wastewater, commercial and industrial wastewaters, storm water runoff, combined sewer overflows, agricultural runoff, live stock wastewaters and mine drainage.

Considering cost economics in the application of these systems, the major cost factors for free water surface wetlands include land, liner, size of dikes or berms proposed for the system. In addition to these components, gravel media for subsurface wetlands is also a major factor. The cost per unit area for subsurface flow wetland might be about 50 per cent higher, depending on local cost of aggregate. However, subsurface wetland requires smaller land area to treat same volume of wastewater.

The routine operation and maintenance for wetlands are similar to pond systems. Removal of vegetation residues on a long term basis may be desirable, if interference with the flow in free water surface wetland are observed.

Mosquitoes being common inhabitants of wetlands. Mosquito control becomes difficult in polluted effluents with high organic content. Gambusia fish and Pupaecide (Altosid) have been successfully used for mosquito control in free water surface wetland in USA (49). The use of *B. thuringiensis isracliensis* has also been recommended after several trials with different insecticides (68).

Terrestrial Systems

Terrestrial treatment processes include slow rate, overland flow and rapid infiltration methods. The system is controlled application of wastewater to soil to achieve treatment of constituents in the wastewater. The slow rate and rapid infiltration processes utilize the soil matrix for treatment after infiltration of the wastewater, the major difference between the processes being rate at which the wastewater is being loaded. In addition, these processes require presence of vegetation as major treatment component. The overland flow process utilizes the soil surface and vegetation for treatment with the treated effluent collected as runoff. The characteristics of these systems are compared in Table 67.9, while the design features and performances of various land treatment systems are detailed in Table 67.10 (69,70).

Table 67.9: Characteristics of Terrestrial Wastewater Treatment Systems

Characteristics	*System Type*		
	Slow Rate	*Overland Flow*	*Rapid Infiltration*
Application method	Sprinkler or surface	Sprinkler or surface	Usually surface
Minimum pre-application treatment	Primary	Fine screening	Primary
Annual loading rate (m/yr)	0.5-6.0	3-20	6-125
Disposition of applied wastewater	Evapotranspiration and percolation	Surface runoff and evapotranspiration	Percolation

Slow Rate Systems

The system is predominant form of land treatment of municipal and industrial wastewater, and the technology being similar to that of conventional agricultural irrigation (71,72,73). The pretreatment to be provided prior to slow rate land treatment depends on the type of wastewater, type of crop, degree of public access to the site, and the percolate quality requirements (74,75,76,77,78).

The crop selection is significant as it aids in removal of nitrogen and increases wastewater infiltration. In addition, crop characteristics of importance are evapotranspiration, water tolerance, salinity tolerance and revenue potential, including ease of management (minimal cultivation and harvesting requirement).

Table 67.10: Terrestrial Treatment Systems, Design Features, and Anticipated Performance [69,70]

Systems	*Treatment Goals*	*Typical Criteria*				*Effluent Characteristics (mg/L)*
		Climate Requirement	*Vegetation*	*Area (ha)**	*Hydraulic Loading (m/yr)*	
Slow rate	Secondary, advanced treatment	Warmer seasons	Yes	23-280	0.5-6.0	BOD <2 TSS <2 TN <3# TP <0.1 FC** 0
Rapid infiltration	Secondary, advanced, or groundwater recharg	None	No	3-23	6-125	BOD < 5 TSS <2 TN 10 TP < 1$ FC <10
Overland flow	Secondary, nitrogen removal	Warmer seasons	Yes	6-40	3-20	BOD <10 TSS <10@ TN <10
On-site	Secondary to tertiary	None	No	Not applicable for a flow of 3785 m^3/d. Size of bed and performance depend on preliminary treatment level		

*: For design flow of 3785 m^3/d.

#: Nitrogen removal depends on type of crop and management.

**: FC- Fecal coliform, #/100 mL.

$: Measured in immediate vicinity of basin; increased removal with longer travel distance.

@: Total suspended solids depend in part on type of wastewater applied.

The slow rate systems are classified based on operation and effluent utilization potential. These include type I systems which is designed to apply maximum possible amount of wastewater to the minimum possible land area, while type II system is designed to optimize the effluent reuse potential.

The design of slow rate systems are limited by either hydraulic or nitrogen loading rates. For type I system, hydraulic loading rate is based on soil permeability comprising application periods and drying periods. For type II systems, the hydraulic loading rate is based on irrigation requirements reflecting crop evapotranspiration rates, and leaching fraction which is utilized to prevent buildup of salts in soil profile.

Hydraulic loading rate arrived at on the basis of water balance for type I slow rate system is given by the following equation[43]:

$$L_w = E_T - P_r + P_w$$

where,

L_w = Wastewater hydraulic loading rate

E_T = Evapotranspiration rate

P_r = Precipitation rate

P_w = Percolation rate

In the design of this system, where protection of potable groundwater is concerned, nitrogen loading is the limiting factor. The loading rate based on nitrogen balance is given by the following equation:

$$L_n = U + f\,(L_n) + A\,C_p\,P_w$$

where,

L_n = Mass loading of nitrogen, kg/ha.yr

U = Crop uptake, kg/ha.yr

f = Fraction of applied nitrogen lost to denitrification, volatilization and soil storage

A = Conversion factor

C_p = Percolate nitrogen concentration, mg/L

P_w = Percolate flow, cm/hr

In the type II slow rate system, usually located in arid regions, the hydraulic loading is often based on crop irrigation rather than soil permeability. The hydraulic loading rate depending on irrigation requirement and precipitation is given by,

$$L_w = IR - P_r$$

where,

L_w = Hydraulic loading rate

IR = Crop irrigation requirement

P_r = Precipitation

Considering leaching factor and irrigation efficiency, the afore-referred equation is modified as follows:

$$L_w = (E_T - P_r)\,(I + LR)\,(100/E)$$

where,

E_T = Crop evapotranspiration
P_r = Precipitation
LR = Leaching requirement
E = Efficiency of the irrigation system

The land requirement for slow rate systems is significantly high and include field application area, area of roads, buffer zones, storage ponds and pre-application treatment. The field area requirement is given by,

$$A = \frac{Q + V_S}{CL_w}$$

where,

A = Field area, ha
Q = Annual flow, m^3/yr
V_s = Net loss or gain in stored wastewater volume
C = Constant (100)
L_w = Design hydraulic loading rate, cm/yr

For slow rate systems to operate efficiently, the soil and crop needs to be well managed, and the soil conditions to be maintained include infiltration rate, compaction, nutrient status and chemical characteristics.

The storage requirements, application scheduling distribution techniques, control of surface runoff and underdrainage for slow rate systems are detailed in literature [43].

Overland Flow Systems

The overland flow treatment process being land treatment system, the wastewater is treated as it flows through a carefully graded grass-covered slopes, the surface runoff being the design requirement. The desired runoff is achieved by providing soils either slowly permeable, or compacted during construction to limit percolation or by inclusion of an impermeable layer just below the surface.

The process has also been utilized for advanced treatment designed for removal of BOD, SS and nitrogen [79,80]. Modifications to achieve significant phosphorus removal by precipitation with added alum on the overland flow slopes have also been reported [81,82].

Grass is the prime vegetation utilized in this system in the slopes due to its ability to provide a support medium for microorganisms, to minimize erosion, and to take up nitrogen and phosphorus.

Recent research [36, 37] has shown that the performance of overland flow system is related more closely to application rate than to hydraulic loading rate, and the relationship is:

$$L_w = \frac{q\ P(100cm/m)}{Z}$$

where,

L_w = Hydraulic loading rate, cm/d
q = Application rate per unit width of slope, m^3/hr.m
P = Application period, hr/d
Z = Slope length, m

Since the field area requirement for the overland flow system depend on the flow, application rate, slope length and the period of application the following equation can be used for estimation of area:

$$A_s = \frac{Q\ Z}{q\ PC}$$

where,

A_s = Field area required, ha
Q = Wastewater flow rate, m^3/d
Z = Slope length, m
q = Application rate, m^3/hr.m

Incorporating the additional area required for wastewater storage, modified equation arrived at is as follows:

$$A_s = \frac{365\ Q + V_s}{DL_w\ C^1}$$

where,

V_s = Net loss or gain in storage, m^3/yr
D = Number of operating days per year
L_w = Design hydraulic loading, cm/d
C^1 = Conversion factor

Overland system requires minimal management except grass cutting, two to three times, per annum. However, system monitoring include influent and runoff quality and flow, groundwater quality and levels, surface water quality, and soil and vegetation characteristics.

Rapid Infiltration System

The rapid infiltration type of land treatment process utilizes the concept of treatment achieved as the effluent percolates through permeable soil. Vegetation is not usually involved as a component, due to application of too high loading rates for nutrient uptake. However utilization of vegetation in stabilizing surface soils, and maintaining high infiltration rates have been reported [83].

The design objectives of rapid infiltration systems apart from treatment also include the following [84]:

- ✰ Recharge of streams by interception of groundwater.
- ✰ Recovery of water by wells or under drains, with subsequent-reuse or discharge.
- ✰ Groundwater recharge.
- ✰ Temporary storage of renovated water in the aquifer.

Enhanced treatment performance of the rapid infiltration systems with decreasing hydraulic loading rate has been observed [85]. Ammonia removal by nitrification has been accomplished in rapid infiltration system. Nitrification as high as 67 kg/ha.d have been reported [85]. However, for nitrogen removal by denitrification, adequate detention time and organic carbon needs to be provided.

In these systems, limit for organic loading has not been developed. However, experience with winery wastewater indicate that BOD loading rates more than 670 kg/ha.d lead to odour problems [86].

Land requirement in rapid infiltration systems comprises basin area, access roads, pre-application treatment and buffer zone, and is given by the following relationship:

$$A = \frac{C\,Q(365\,d/yr)}{L_w}$$

where,

A = Application area, ha

C = Conversion factor (10^{-4} ha/m^2)

Q = Average wastewater flow, m^3/d

L_w = Annual loading rate, m/yr.

The aforereferred relation does not account for additional area requirement for dikes or berms and access ramps, and therefore, needs to be incorporated if required.

The system management includes scarification, scraping or disking of the basin surface. For monitoring the system, monitoring wells are required to be placed upgardient and downgradient of wastewater application area, the details being available elsewhere [87]. The application rate, groundwater mounding, basin construction and operation of the rapid infiltration system are also detailed elsewhere [45].

Pilot plant studies carried out by Mottier *et al*[88] on infiltration percolation plant treating sewage of Mazagon in South of Spain, indicate that at 0.26 m^3/m^2.d of loading, 90 per cent COD removal and 98 per cent N-NH_4 oxidation have been achieved. However, disappointing disinfection performance has been achieved in comparison with reported removals, *viz.* helminth eggs [89], virus [90,91] and faecal coliform [92,93,94].

Epilogue

Natural systems explored for their potential for renovation of wastewater include aquatic, wetland and terrestrial treatment systems. These systems being adopted as viable wastewater management options, depend mainly on natural processes, with minimal dependence on mechanical elements.

The concept of natural systems emphasizes on environmental ethic of recycle and reuse of resources, wherever possible. Based on extensive research and operational experiences, it has been evaluated that under favourable site conditions, these natural systems could be constructed and operated for less cost, and with less energy requirement and complexity of operation than the conventional mechanical technologies. These processes need to be given priority for planning new systems, and for upgrading or retrofitting existing systems thereby providing an effective wastewater management alternative.

The issues hint at far reaching fruitful consequences for a large number of cities, and medium and small size towns in India, which remain to be provided with safe and hygienic wastewater treatment and disposal facilities [95]. Medium and small size towns with low population require creation and strengthening of economic infrastructural base [96]. It has been revealed in the document of the Ninth Five Year Plan (1997-2002), Volume II, that Integrated Development of Small and Medium Towns (IDSMT) is an extremely valuable developmental instrument for small and medium towns. Under this scheme, there will be development of the infrastructure, primarily the physical components in selected towns with population upto 5 lakhs that have the potential to emerge as regional centers of economic growth and employment, and serve as catchment centers to arrest population migration to large town and cities. These are the towns, in addition to some large cities with adequate land available on the outskirts, were attention should be focused on long term basis for implementation of small and large scale wastewater treatment facilities using any of the natural systems discussed. This will go in a long way in improving environmental quality through appropriate wastewater treatment and management, and subsequent prevention of water and land pollution.

References

1. Bdeient, P.B., Springer, N. K., Baca, E. Bouette, T. C., Hutchins, S.R. and Tomson M. B. (1983), Ground-water Transport from Wastewater Infiltration, *ASCE EED Div. J.*, 109 (2), 485-501.
2. Bouwer, H. (1978), *Groundwater Hydrology*, McGraw Hill, New York.
3. Danel, P. (1953), The Measurement of Ground-Water Flow, in *Proceedings Ankara Symposium on Arid Zone Hydrology*, UNESCO, Paris, 99-107.

4. Roberts, P.V., McCarty, P.L., Reinhard, M., and Schriner J. (1980), Organic Contaminant Bheavior during Groundwater Recharge, *J. Water Pollution Control Fed.*, 52 (1), 161-172.

5. U.S. Department of the Interior, Bureau of Reclamation, *Drainage Manual*, U.S. Government Printing Office, Washington, DC, 1978.

6. Harremoes, P. (1998), Upgrading our Inherited Urban Water Systems, *Wat. Sci. Tech.*, 37 (9), 1-9.

7. Bastian, R. K., and Reed S.C. (eds.), (1979), Aquaculture Systems for Wastewater Treatment, EPA 430/9-80-006, U.S. Environmental Protection Agency, Washington, DC.

8. Middlebrooks, E.J., Middlebrooks, C.H., and Reed, S.C. (1981), Energy requirements for Small Wastewater Treatment Systems, *J. Wat. Polt. Contl. Fed.*, 53 (7), 1172-1197.

9. U.S. Environmental Protection Agency, (1983), Process Design Manual for Municipal Wastewater Stabilization Ponds, EPA 625/1-83-015, Center for Environmental Research Information, Cincinnati, Ohio.

10. Reed, S.C., Bastian, R. Black, S. and Khetty R. K. (1984), *Wetlands for Wastewater Treatment in Cold Climates in Proceedings Water Reuse III Sympsoium*, American Water Works Association, Denver, CO.

11. U.S. Environmental Protection agency (1982), *Fate of Priority Pollutants in Publicly Owned Treatment Works*, EPA 440/1-82-303, EPA, Washington, DC.

12. Hutchins, S.R., Tomsom, M.B., Bedient, P.B. and Ward C.H. (1985), Fate of Trace Organics during Land Application of Municipal Wastewater, *Crit. Rev. Environ. Control*, 15 (4), 355-416.

13. Jenkins, T.F., Leggett, D.C., Parker, L.V., and Oliphant, J.L. (1985), Toxic Organics Removal Kinetics in Overland Flow Land Treatment, *Water Res.*, 19 (6),707-718.

14. Reed, S.C. (1979), Health Aspects of Land Treatment, GPO 1979-657-093/7086, U.S. Environmental Protection agency, Centre for Environmental Research Information, Cincinnati, OH.

15. Gloyna, E. (1971), Waste Stabilisation Ponds, *WHO Monograph Series* No.60, World Health Organization, Geneva, Switzerlandd.

16. Arceivala, S. (1980), *Wastewater Treatment and Disposal*, Marcel Dekker, New York, USA.

17. Toms, I. Owens, M., Hall, J. and Mindenhall, M. (1975), Observations on the Performance of Polishing Lagoons, *J. Inst. Water Poll.* Comt., U.K.

18. Oswald, W. and Gotaas, H. (1957), Photosynthesis in Sewage Treatment, *Trans. ASCE*, 122.

19. Oswald, W. (1972), Complete Waste Treatment Ponds, in Advances in Water Pollution Research, *6th International Conference*, Jerusalem, Israel.

20. Neel, J.J., McDermott and Monday,C. (1961), Experimental Lagooning of Raw Sewage in Fayette, Missouri, J. WPCF, 33.

21. Thirumurti, D. (1974), Design Criteria for Waste Stabilisation Ponds, *J. WPCF*, 46.

22. Gloyna, E. and Espino, E. (1969), Sulphide Production in Stabilisation Ponds, *J. ASCE*, San. Eng. Div.

23. Marais, G. (1974), Fecal bacterial Kinetics in Stabilisation Ponds, *J. ASCE*, Div., 100.

24. Pano, A., and Middlebrooks, E.J. (1982), Ammonia Nitrogen Removal in Facultative Wastewater Stabilization Ponds, *J. Water Pollution Control Fed.*, 54 (4): 344-351.

25. Reed, S.C. (1984), *Nitrogen Removal in Wastewater Ponds*, CRREL Report 84-13, Cold Regions Research and Engineering Laboratory, Hanover, NH.

26. U.S. Environmental Protection Agency: Design Manual (1982), *Sewage Stabilization Ponds*, EPA 625/1-83-010, Centre for Environmental Research Information, Cincinnati, OH.

27. Reed, S.C., Bastian, R. and Jawell, W. (1981), Engineers Assess Aquaculture Systems for Wastewater Treatment, *Civil Eng.*, 64-67.

28. Stephenson, Turner, M.G., Pope, P., Colt, J., Knight, A. and Tchobanoglous, G. (1980), The Use and Potential Aquatic Species for Wastewater Treatment, in Appendix A, The Environmental Requirements of Aquatic Plants, Publication No.65, California State Water Resources Control Board, Sacramento.

29. Wolverton, B.C., and McDonald, R.C. (1978), Nutritional Composition of Water Hyacinths Grown on Domestic Sewage, *Econ. Bot.*, 32 (4), 363-370.

30. Reddy, K.R. and Debusk, W.F. (1985), Nutrient Removal Potential of Selected Aquatic Macrophytes, *J. Environ. Qual.*, 14 (4), 459-462.

31. Wolverton, B.C. (1979), Engineering Design Data for Small Vascular Aquatic Plant Wastewater Treatment Systems, in Proceedings Aquaculture Systems for Wastewater Treatment, EPA 430/9-80-006, MCD 67, U.S. Environmental Protection Agency, Office of Municipal Pollution Control, Washington, DC, 179-192.

32. Weber, A.S., and Tchobanoglous, G. (1985), Nitrification in Water Hyacinth Treatment Systems, *J. Environ. Eng. Div. ASCE*, 11 (5), 699-713.

33. Reddy, K.R. (1985), Nutrient Transformations in Aquatic Macrophyte Filters Used for Water Purification, *Proceedings Water Reuse III*, American Water Works Association Denver, CO, 660-678.

34. Reddy, K.R., and deBusk, W.F. (1985), Nutrient Removal Potential of Selected Aquatic Macrophytes, *J. Environ. Qual.*, 14 (4), 459-462.

35. Gee and Johnson Engineers (1980), Water Hyacinth wastewater Treatment design Manual for NASA/NSTL, West Palm Beach, FL.

36. McNabb, C.D. (1976), The Potential of Submerged Vascular Plants for Reclamation of Wastewater in Temperate Zone Ponds, in Biological Control of Water Pollution, University of Pennsylvania Press, Philadelphia, 123-132.

37. Hillman, W.S., and Culley, D.C. (1978) *The Use of Duckweed*, Am. Sci., 66, 442-451.

38. Leslie, M. (1983), Water Hyacinth Wastewater Treatment Systems: Opportunities and Constraints in Cooler Climate, EPA 600/2-83-075, U.S. Environmental Protection Agency, Washington, DC.

39. Wolverton, B.C., and McDonald, R.C. (1979), Upgrading Facultative Wastewater Lagoons with Vascular Aquatic Plants, *J. Water Pollution Control Fed.*, 51 (2), 305-313.

40. Coleman, M.S. (1974), Aquaculture as a Means to Achieve Effluent standards, in Proceedings Wastewater Use in the Production of food and Fiber, EPA 660/2-74-041, U.S. Environmental Protection Agency, Washingdon DC, 199-214.

41. Dinges, R. (1982), *Natural Systems for Water Pollution Control*, van Nostrand Reinhold, New York.

42. Ryther, J.H. (1979), Treated Sewage Effluent as a Nutrient Source for Marine Polyculture, *Proceedings Aquaculture Systems for Wastewater Treatment*, EPA 430/9-80-006, U.S. Environmental Protection Agency, Washington, DC, 351-376.

43. Reed, S.C., Crites, R.W., and Middelbrooks, E. (1995), *Natural Systems for waste management and treatment*, McGraw Hill Inc.

44. U.N. Developmental Program (1993), Global Environmental Facility: Egyptian Engineered Wetlands, Lake Manzala.

45. Kreissl, J.K. and Brown, D.S. (2000), Use of Constructed Wetland as a Major Wastewater Treatment Process, *Wat. Env. Tech.*, Vol. 12, No. 10, 36-41.

46. Hammer, D.A. (1992), *Creating Freshwater Wetlands*, Lewis Publishers, Chelsea, MI.

47. Lawson, G.J. (1985), Cultivating Reeds (Phragmites australis) for Root Zone Treatment of Sewage, Project Report 965, Institute of Terrestrial Ecology, Cumbria, England.

48. Thornhurst, G.A. (1993), *Wetland Planting Guide for the Northeastern United States*, Environmental Concern, Inc., St. Michaels, MD.

49. Gearheart, R. J., Wilbur, S., Williams, J., Hull, D., Finney, B. and Sundberg, S. (1983), Final Report City of Arcata Marsh Pilot Project Effluent Quality Results – System Design and Management, Project Report C-06-2270, City of Arcata, Department of Public Works, Arcata, CA.

50. Herskowitz, J. (1986), Town of Listowel Artificial Marsh Project Final Report, Project No.128RR, Ontario Ministry of the Environment, Toronto, Ont.

51. Gearhart, R.A., Klopp, A. F. and Allen, G. (1989), Constructed Free Surface Wetlands to Treat and Receive Wastewater: Pilot Project to Full Scale, in D.A. Hammer (ed.), *Constructed Wetlands for Wastewater Treatment*, Lewis Publishers, Chelsea, MI, 121-137.

52. Reed, S.C. (1993), Subsurface Flow Constructed Wetlands for Wastewater Treatment: A Technology Assessment, EPA 832-R-93-008, U.S. Environmental Protection Agency, Washington, DC.

53. Ayaz, S.C. and Akca, I. (2000), Treatment of Wastewater by Constructed Wetland in Small Settlements, *Wat. Sci. Tech.*, Vol. 41 No. 1, 69-72.

54. Kadlec, R.H. (1989), Hydrologic Factors in Wetland Water Treatment, in D.A. Hammer (ed.), *Constructed Wetlands for Wastewater Treatment*, Lewis Publishers, Chelsea, MI, 21-40.

55. Steiner, G. R., and Watson, J. T. (1993), General Design, Construction and Operational Guidelines Constructed Wetlands Wastewater Treatment Systems for Small Users including individual Residences, 2nd ed., TVA/WM-93/10, Tennessee Valley Authority, Chattanooga, TN.

56. Calkins, D. (1993), U.S. Cold Regions Research and Engineering Laboratory, Hanover, NH, Personal communication.

57. Chapman, A.J. (1974), *Heat transfer*, 3rd ed., Macmillan, New York.

58. Ogden, M. (1994), South West Wetlands Group, Santa Fe, NM, Personal communication.

59. Ashton, G. (ed.) (1986), *River and Lake Ice Engineering*, Water Resources Publications, Littleton, CO.

60. Stefan, J. (1891), Theory of Ice Formation, Especially in the Arctic Ocean (in German), Wien Stizunsber. Akad. Wiss. A, 42 (2): 965-983.

61. Reed, S. C. (ed.) (1990), Natural Systems for Wastewater Treatment, MOP FD-16, Water Environment Federation, Alexandria, VA.

62. U. S. Environmental Protection Agency (1988), Design Manual Constructed Wetlands and Aquatic Plant Systems, for Municipal Wastewater Treatment, EPA – 625/1-81-013, Center for Environmental Research Information, Cincinnati, OH.

63. Knight, R., Kadlec,R. and Reed, S. (1993), Database: North American Wetlands for Water Quality Treatment, U.S. Environmental Protection Agency, Risk Reduction Environmental Laboratory, Cincinnati, OH.

64. Hammer, D.A. and Knight, R. L. (1992), Designing Constructed Wetlands for Nitrogen Removal, in Wetland Systems in Water Pollution Control, International Association Water Quality, Sydney, Australia, 3.1-3.37.

65. Boon, A.G. (1985), Report of a visit of Members and Staff of Water Resources Centre to Germany to Investigate the Root Zone Method for Treatment of Wastewaters, Water Research Centre, Stevenage, England.

66. Gersberg, R. M., Elkins, B. V., Lyons, S.R. and Goldman, C.R. (1985), Role of Aquatic Plants in Wastewater Treatment by Artificial Wetlands, *Water Res.*, 20, 363-367.

67. Bavor, H. J., Roser,D. J., Fisher, P.J. and Smalls, I. C. (1986) Joint study on Sewage Treatment Using Shallow Lagoon-Aquatic Plant Systems, Water Research Laboratory, Hawkesbury Agricultural College, Richmond, NSW, Australia.

68. Tennessen, K.J. (1993), Production and Suppression of Mosquitoes in Constructed Wetlands, in Constructed Wetlands for Water Quality Improvement, Lewis Publishers, Chelsea, MI, 591-601.

69. U.S. Environmental Protection Agency (1981), Process Design Manual – Land Treatment of Municipal Wastewater, EPA 625/1-81-013, Center for Environmental Research Information, Cincinnati, OH.

70. U.S. Environmental Protection Agency (1984), Process Design Manual Supplement on Rapid Infiltration and Overland Flow, EPA 625/1-81-013a, Center for Environmental Research Information, Cincinnati, OH.

71. Folsom, C.F. (1876), Seventh Annual Report of the State Board of Health of Massachusetts, Wright and Potter, Boston.

72. Jewell, W. J. and B. L. Searbrook (1979), A History of Land Application as a Treatment Alternative, EPA 430/9-79-012, U.S. Environmental Protection Agency, Office of Water Program Operations, Washington, DC.

73. Seabrook, B.L. (1975), Land Application of Wastewater in Australia, EPA 430/9-75-017, U.S. Environmental Protection Agency, Office of Water Program Operations, Washington, DC.

74. California Fertilizer Association, (1985), *Western Fertilizer Handbook*, 7th ed., Interstate Printers and Publishers, Danville, IL.

75. Loehr, R.C. and Overcash, M. R. (1985), Land Treatment of Wastes: Concenpts and General Design, *J. Environ. Eng. Div. ASCE*, 111 (2), 141-160.

76. Pound, C. E. and Crites, R. W. (1973), Wastewater Treatment and Reuse by Land Application, EPA 660/2-73-006b, U.S. Environmental Protection Agency, Office of Water Program Operations, Washington, DC.

77. U.S. Environmental Protection Agency (1981), Process Design Manual for Land Treatment of Municipal Wastewater, EPA 625/1-81-013, Center for Environmental Research Information, Cincinnati, OH.

78. Westcot, D. W. and Ayers, R. S. (1984), Irrigation Water Quality Criteria, in Irrigation with Reclaimed Municipal Wastewater – A Guidance Manual, California State Water Resources Control Board, Sacramento.

79. Kruzic, A. P. and Schroeder, E. D. (1990), Nitrogen Removal in the Overland Flow Wastewater Treatment Process- Removal Mechanisms, *Res. J. Water Pollution Control Fed.*, 62 (7): 867-876.

80. Thomas, R.E., Jackson, K. and Penrod, L. (1974), Feasibility of Overland Flow for Treatment of Raw Domestic Wastewater, EPA 660/2-74-087, U.S. Environmental Protection Agency, Office of Research and Development, Washington, DC.

81. Thomas, R. E., Bledsoe, B. and Jackson K. (1976), Overland Flow Treatment of Raw Wastewater with Enhanced Phosphorus Removal, EPA 600/2-76-131, U.S. Environmental Protection Agency, Office of Research and Development, Washington, DC.

82. Water Pollution Control Federation (1990), Natural Systems for Wastewater Treatment, Manual of Practice FD-16, Alexandria, VA.

83. Reed, S.C., Crites, R. W. and Waalace, A.T. (1985), Problems with Rapid Infiltration – A Post Mortem Analysis, *J. Water Pollution Control Fed.*, 57 (8): 854-858.

84. Crites, R.W. (1985), Nitrogen Removal in Rapid Infiltration Systems, *J. Environ. Eng. Div. ASCE*, 111 (6), 865-873.

85. US Environmental Protection Agency, Process Design Manual for Land Treatment of Municipal Wastewater (1981), EPA 625/1-81-013, Centre for Environmental Research Information, Chincinnati, OH.

86. Crites, R. W. (1985), Nitrogen Removal in Rapid Infiltration System, *J. Environ. Eng. Div. ASCE*, III (6), 865-873.

87. Singor, D.C. (1985), Groundwater Sampling during Artificial Recharge: Equipment, Techniques and Data Analysis in T. Asano (ed.), *Artificial Recharge of Groundwater*, Buttaworth, Stoneham, MA, 151-202.

88. Mottier, V., Brissaud, F., Nieto P. and Alamy, Z. (2000), Wastewater treatment by infiltration percolation: a case study, *Wat. Sci. Technol.* Vol. 41 No. 1, 77-84.

89. Guessab, M., Bize, J., Schwarzbrod, J., Mau, A., Nivault, N. and Schwarbrod, L. (1993), Wastewater treatment by infiltration-percolation on sand: results in Ben Sergao, Moroco. *Wat. Sci. Tech.*, 27 (a), 91-95.

90. Lance, J. C., Rice, R.C. and Gilbert, R. G. (1980) Renovation of Wastewater by Soil Columns Flooded with Primary Effluent, *Journal WPCF*, 52 (2), 381-388.

91. Ho, G.E., Gibbs, R.A. and Mathew, K. (1991), Bacteria and Virus Removal from Secondary Effluents in Sand and Red Mud Columns, *Wat. Sci. Tech.*, 23 (1-3), 261-270.

92. Lefevre, F. (1988), Epuration Des Eaus Usees Urbaines Par Infiltration Percolation. Etude Experimentale et definition de cprocedes. *Doct. Thesis*. Montpellier II University.

93. Carre, J. and Dufils, J. (1991), Wastewater Treatment by Infiltration Basins: Usefulness and Limits – Sewage Plant in Creance (France). *Wat. Sci. Tech.*, 24 (9), 287-293.

94. Brissaud, F. and Lesavre, J. (1993), Infiltration Percolation in France: 10 years experience. *Wat. Sci. Tech.*, 28 (10), 73-81.

95. Shukla, S.R. (1999), Urban Water Sector –beyond 2000 AD , *J. IWWA*, 31 (3), 167-174.

96. Gopal, B. (2000), Urban Scenario Planning and Development- Policy Preparation, *Employment News*, XXIV (50), 1-3.

Chapter 68

Application of Industrial Ecology Concept in Developing SDP Complex with Particular Emphasis on Wastewater Management

S.K. Goyal

CSIR-NEERI, Nagpur

ABSTRACT

The concept of industrial ecology has been applied in developing an industrial complex affiliated to sugar industry, where the sugar industry generated two of the major wastes; molasses and bagasse are used as main raw material in distillery and paper industry for manufacturing alcohol and paper products respectively. Further, to meet the energy requirement in the complex, the wastewaters generated from these industries are utilized for biogas production through anaerobic digestion. A systematic approach has been presented in this paper to demonstrate the suitability of industrial integrity of sugar, distillery and paper industry (SDP Complex) with particular emphasis on wastewater management for sustainable industrial development.

Introduction

Industrial ecology is a systems framework for managing energy and materials flows for high efficiency and low pollution. It is a systems approach that enables executives, managers and designers to balance financial, technical and environmental objectives. Further, application of industrial ecology opens opportunities for cost savings, new revenues, and definition of new product lines.

Essentially, industrial ecology refers to the exchange of materials between different industrial sectors, where the 'waste' output of one industry becomes the 'feedstock' of another. In this approach number of different industries, usually in close proximity to each other, exchange a variety 'waste' outputs. Implementing industrial ecology involves such things as life cycle analysis, closed loop processing, reusing and recycling, design for environment and waste exchange [1].

Need for an integrated complex of sugar-distillery and paper mills has been emphasized, as the sugar industry generates two major by-products; bagasse and molasses, which can be better utilized. At the same time distillery and paper industry are known to have very high pollution potential. Treatment of high organic load containing streams for biogas generation together shall have twin advantage of pollution load reduction at low cost and energy recovery, reducing fuel (coal) consumption in the industries. Hence, development of integrated industrial ecological complex is the need of the hour to achieve sustainable industrial development.

Techno-economic feasibility analysis of a 2500 tcd sugar mill along with a small paper mill and distillery is presented in this paper, with particular emphasis on wastewater management in the industrial complex. In India, there are number of distilleries associated with sugar industry, and few paper industries with sugar mill, but a complex integrating these three are in the exploratory stage.

Status of Sugar Industry

Globally, sugar is produced in 121 countries and sugar production in the year 2000-01 was about 128 million tons. Approximately 70 per cent of sugar is produced from sugar cane, which is largely grown in the southern hemisphere tropical countries and the remaining 30 per cent from sugar beet in the temperate zones of the north. The five largest exporters (Brazil, the European Union, Australia, Cuba and Thailand) supply 70 per cent of the world free market exports. India is the largest consumer and second largest producer of sugar in the world. It contributes nearly 15 per cent of the total world's production, and exports only about 5.5 per cent of its total sugar production [2].

With over 450 sugar factories located throughout the country, the sugar industry is amongst the largest agro processing industries, with an annual turnover of Rs. 150 billion. It plays a major role in rural development and its importance for India stretches far beyond the role of a sweetener supplier.

The number of operating sugar mills in the country has increased from 29 in the sugar year 1930-31 to 412 by the year 1996-97. The average capacity of the sugar mills has considerably increased from just 644 TPD in the year 1930-31 to 2656 TPD in the year 1996-97. But the growth in the Indian sugar industry was driven by horizontal growth (increase in number of units) compared to the vertical growth witnessed in other countries (increase in average capacity). The average size of Indian sugar mill is very small as compared to international capacity standards. In India, sugarcane-crushing capacity varies from 400 to 8000 tcd. The distribution of sugar mills according to their capacity range is; up to 1250 tcd: 52 per cent; 1251-2500 tcd: 34 per cent and above 2500 tcd: 14 per cent [3,4].

Indian sugar industries are classified under three major categories; farmers own cooperative sector, private sector and public sector. Currently, out of 453 sugar mills in the country, 252 are in the co-operative sector, 134 are in the private sector and 67 are in the public sector. Besides 136 units in the private sector are in various stages of implementation [3].

Sugar produced in India is mainly of granulated type. The demand for sugar has increased from 3.69 million ton in SY1975-76 to 14.8 million ton for SY1997-98 at a CAGR of 6.48 per cent. The per capita consumption of sugar in the country is only about 25 kg/year (including about 9 kg//year for gur and khandsari), as compared to 54 kg/year in Brazil, 49 kg/year in Cuba, 46 kg/year in Australia and 32 kg/year in USA [2].

Sugar Manufacturing Process

In India, sugar is mainly produced only from sugarcane, as the country's climatic conditions are suitable for cane cultivation. Harvested sugarcane is generally crushed within 8 to 10 hours of cutting, to avoid its deterioration and achieve maximum recovery. The usual sugarcane constituents are: water: 69-75 per cent, sucrose: 8-16 per cent, reducing sugars: 0.5-2 per cent, organic substances: 0.5-1 per cent, nitrogenous bodies: 0.5-1 per cent, ash: 0.3-0.8 per cent and fiber: 10-16 per cent of the total weight. The sugar manufacturing process consists of juice extraction, clarification, evaporation, crystallization and centrifugation.

The quantum of sugar produced by a mill is determined by the factors like daily crushing capacity, duration of crushing season and percentage of sugar recovery. The crushing season in the country starts from October and reaches its peak in January before finally ending in March or April of the next year; however, it may also vary depending upon the cane availability in different parts of the country. The recovery rate for sugar mills does vary widely across the country, and in the past it has been in the range 9.25 to 10.5 per cent (overall average 10 per cent) [3].

By-products of Sugar Industry

The sugar mill produces many by-products along with sugar. Bagasse molasses and press mud are the major by-products, which if not utilized/disposed properly, have high pollution potential. In fact, these are very valuable products and are used for different purposes as briefly described below [4,5].

Bagasse

Bagasse is a fibrous residue of cane stalk that is obtained after crushing and extraction of juice. It consists of water, fiber and relatively small quantities of soluble solids. The usual bagasse composition is; moisture: 46-52 per cent, fibre: 43-52 per cent, soluble solids: 2-6 per cent. On an average, a sugar mill generates one-third of bagasse of the total cane crushed. Bagasse is usually used as a fuel (calorific value about 1920 Kcal/kg) in the furnaces for production of steam and motive power required (up to 95 per cent) by the industry. It is also used as raw material for production of paper and as feedstock for cattle.

Molasses

Molasses is the final effluent obtained in the preparation of sugar by repeated crystallization. It is a yellowish or dark brown, thick viscous liquid and contains very high organic matters, and as such, it has very high pollution potential. Typical characteristics of molasses are: ; pH–3.5-4.1; TDS–200-320 g/L; BOD–440 g/L; COD–960 g/L; chlorides–32 g/L and sulphates–15 g/L. The yield of molasses varies between 3.5 to 4.5 per cent of the sugar cane crushed.

Molasses is mainly used for the manufacture of ethyl alcohol (ethanol), yeast and cattle feed. Ethanol is in turn used to produce potable liquor and downstream value added chemicals such as acetone, acetic acid, butonol, acetic anhydride, MEG, etc. Approximately, 80 per cent of molasses produced by organized sector is utilized as raw material for production of industrial (70 per cent) and potable (30 per cent) alcohol. There are around 250 distilleries in the country with annual installed capacity of about 2.4 million KL.

Press Mud

Press mud is produced in the clarification process of cane juice from the vacuum filtration unit. It is commonly known as filter cake and is used as manure by the farmers. It is combustible and also used in brick kilns for brick making. The production of press mud is about 3 per cent of cane crushed in the sulphitation process based mills (90 per cent in India) and 7 per cent in the carbonation-based mills. Press mud constitutes; crude wax and fat: 5-14 per cent; fibre: 15-30 per cent, sugar: 5-15 per cent, CaO: 1-4 per cent, P_2O_5: 1-3 per cent, MgO: 0.5-1.5 per cent. Application of press mud to the soil is reported to improve the physical and chemical properties of the soil. When mixed with the spent wash generated by the distilleries, the resultant product is enriched organic manure, which has the capability to improve yields in farms [6].

Development of an Integrated Industrial Complex for Utilization of By-products

During the manufacturing process of sugar from sugarcanes, sugar industry produces many useful by products. Among which, Bagasse and Molasses are the most important. Most of the molasses is used as raw material in distillery for manufacture of alcohol and various other chemicals, and these industries are generally located near the sugar industry. Bagasse is normally used as fuel in the sugar industry boiler to meet its steam and captive power requirement. Bagasse is also known to have a very good potential for paper manufacture, and several bagasse-based industries are operating in the country along with other non-conventional raw materials.

It is proposed that the concept of industrial ecology or industrial symbiosis may be adopted in case of sugar industry, where the waste of one industry becomes the feedstock to the other industry. An integrated industrial complex (SDP) comprising sugar-distillery and paper industry should be developed in order to achieve maximum utilization of various resources and also for combined treatment of wastewater streams, as these industries are not only water intensive, but also have high pollution potential. The approach has been explained through a case study of 2500 tcd capacity sugar mill, because as per the Govt. of India policy, new sugar factories should have minimum economic capacity of 2500 tcd with no maximum limit on capacity. However in industrially backward areas, co-operative and public sector new units are allowed with an initial capacity of 1750 tcd subject to the condition that the units would expand their capacities to 2500 tcd within a period of 5 years of going into production [3].

A 2500 tcd sugar mill is estimated to produce about 250 MT (@10 per cent recovery) of sugar, and in the process shall generate three major by-products; about 417 MT bone dry bagasse, 112.5 MT (@4.5 per cent) molasses and 80 MT (@3.2 per cent) press mud. Accordingly, three more industries utilizing these valuable waste products can be included in the industrial complex. A schematic diagram for the industrial complex along with biogas generation scheme described here briefly.

Utilization of Bagasse – Paper Mill

A huge quantity of bagasse generated can be used partly for paper manufacture and the rest as fuel in boiler for steam and power generation. Depending upon the financial status and availability of other input resources, a small (30 TPD without chemical recovery) or a large (150 TPD with chemical recovery) paper mill can be set up. In order to make self sufficient complex with respect to energy resources, a small paper mill with 30 TPD capacity manufacturing industrial papers and boards is proposed, a major portion of which can to be utilized in the complex itself.

The total demand for paper in the country is about 3.5 million tons per annual, whereas the actual production is around 2.8 million tons. Thus there exists a huge gap between demand and supply, for which additional paper industries are required to be set up in the country.

For a 30 TPD pulp and paper mill producing industrial papers from bagasse (70 per cent), waste gunny (10-15 per cent) and waste paper (10-15 per cent), the bagasse requirement would be around 67 MTPD (bone dry), which is about 6 per cent of the total bagasse produced. With a pulp yield of about 30-35 per cent, the annual bagasse requirement would be around 25000 MT (equivalent to 2 months of bagasse production). Since, the sugar mill is operative only for about 6 months in a year, therefore, an equal amount (70-75 MTPD) of bagasse shall need to be stored separately for utilization during non-sugar season. Remaining 275 MT bagasse can be used as fuel in boiler for steam and power generation.

The details of manufacturing process for small paper mills, particularly based on non-conventional raw materials are well established [7]. There are two major sections in a paper mill producing industrial papers and boards as; pulp mill and paper machine section. Pulp mill section involving chemical digestion of raw materials (bagasse, gunny etc.) and then pulp washing contribute to about 70-80 per cent of the total pollution load from the paper industry. Black liquor containing very high organic load can be separated through screw press to get most concentrated form for the subsequent biogas recovery.

An Indian paper mill (M/s Pravara Pulp and Paper Mill, Distt. Ahmednagar, Maharashtra) has adopted this approach. About 400 m^3/day black liquor is processed through screw press, of which about 100 m^3/day is used for make up volume in the digester and 300 m^3/day black liquor, thus separated is further mixed with brown stock washer filterate to the volume of 500 m^3/day, and is sent to biogas generation [8].

Utilization of Molasses–Distillery

On an average, about 45 MT of molasses is produced for every 100 MT of sugar cane crushed, thus a 2500 tcd sugar mill shall generate about 112 MT molasses daily, to be used for manufacture of alcohol and other products. To produce one litre of alcohol, 3-10 kg (avg. 6 kg) molasses is required. Hence, a total of 18.75 KL alcohol can be produced daily. Since sugar mill operates only for about 6 months in a year, a distillery of 9 KLPD capacity can be set up in the complex, so that it is operated throughout the year. The requisite molasses for the non-sugar mill operation period can be stored properly.

The details of manufacturing of alcohol from molasses are well established [**9**], and at present about 250 distilleries are operating in the country. For every litre of alcohol produced about 12-15 litre of spent wash, containing very high BOD and COD is generated, which has a very high potential for biogas generation, otherwise leads to high water pollution.

Utilization of Press Mud–Chemical Industry

In India, the average crude wax content of dry sulphitation press mud is approximately 10 per cent. The wax from press mud can be recovered by various solvent extraction methods. Solvents normally used are n-hexane and n-heptane. This wax (crude) is a mixture of hard and soft waxes with other minerals, colouring and resinous matters etc. Hard wax of high melting point representing 70 per cent of crude wax can be separated from crude wax by adopting suitable refining methods. Hard wax can be used for the manufacture of floor polishes, candles, waxed card board, containers and paper cups for electrical insulation etc.

Sugar cane wax promises to be very good source for meeting the demand for wax in the country as well as for export purposes. Moreover, the manurial value of dewaxed press mud is reported to be better than that of the ordinary press mud. It is estimated that for every million ton of sugar production, India can produce approximately 9630 MT of crude wax or 6740 MT of hard wax. A 2500 tcd sugar mill can produce 2.4 MT of crude wax or 1.6 MT of hard wax daily.

Utilization of Waste Streams for Biogas Generation

A huge quantity of water is used in sugar as well paper industry, with moderate quantity in distillery. During the manufacturing process of various products, wastewater containing varied amount of pollution loads are generated. The typical characteristics of various streams in sugar mill [10], distillery [11] and agro-residues based small paper mills [12] are summarized in Table 68.1. Characterization of various wastewater streams from these three industries suggest that the following streams containing high organic content can be utilized for biogas generation through anaerobic digestion:

- ✰ Wastewater from mill house and boiling house of sugar industry
- ✰ Black liquor and washer filtrate from pulp mill section of paper mill
- ✰ Spent wash from distillery

Table 68.1: General Characteristics of Wastewaters from Different Sources in Sugar, Distillery and Paper Industry

Parameter	*pH*	*Suspended Solids (mg/L)*	*COD (mg/L)*	*BOD (mg/L)*
Sugar Industry				
Mill house	4-6	500-550	1000-1500	700-1000
Boiling house	4-6	400-600	2000-3000	1500-2000
Condensates	6-7	5-10	250-300	100-150
Lime house	8-9	3500-4000	200-250	100-150
Boiler blow-down	6-7	50-100	500-550	30-40
Combined	6-7	150-300	1800-3000	1000-1500
Distillery				
Spentwash	4.0-4.7	8600-13000	62000-85000	30000-42000
Vat cleansing	3.8-4.7	15000-150000	61000-148000	25000-78000
Spentlease	3.4-3.9	10-290	10000-18000	5000-8000
Floor washing	4.5-4.8	600-800	11000-15000	5800-6500
Condensate water	5.7-7.2	16	100	25
Paper Industry				
Black liquor	9-10	2340	53600	10200
Pulp wash	9-9.5	1850	8800	2450
Bleaching	6-7	440	1020	290
Paper Machine	6-7	480	580	180
Combined	7-7.5	850	1100	330

Biogas can be produced through any of the anaerobic digestion process. A fixed bed, fluidized bed, a contact process or UASB system can be employed for anaerobic digestion of these streams. Typical wastewater characteristics used for biogas generation for a 2500 tcd sugar mill [3], 40 KLPD distillery [13] and 30 MTPD paper mill [8] are presented given in Table 68.2. Since the COD and BOD content of sugar industry stream is comparatively very low, it would be desirable to treat/recover biogas from this stream separately in an anaerobic reactor (AR I).

The other two streams from paper mill and distillery can be mixed together before feeding to another anaerobic reactor (AR II). Another reason for separately treating sugar industry stream is that it operates only for a period of 6 months in a year, whereas the other are operated for the whole year. Further, the hydraulic retention time (HRT) required for AR I and AR II vary considerably, *i.e.* 6 hrs and 10 days respectively for the UASB type reactor. Hence, sugar industry stream needs to be treated separately.

Table 68.2: Typical Characteristics of Segregated Wastewater Streams and Combined Effluent for Biogas Generation

Parameter	*Sugar Industry (Mill house and boiling house)*	*Paper Mill (Pulp Mill Section)*	*Distillery (Spentwash)*	*Combined Effluent (Paper Mill + Distillery)*
Flow, m^3/day	1060	500	140	640
pH	4-6	7-8	4-4.5	6-7
COD, mg/L	2200	22000	80000	34700
BOD, mg/L	1600	5500	50000	15200
SS, mg/L	540	700	10000	2700

About 65-80 per cent reduction in COD and 85-90 per cent reduction in BOD can be achieved in the process with biogas yield of about 0.5 m^3/kg COD destroyed. The design parameters and anticipated performance of the UASB anaerobic reactor are summarized in Table 68.3, and along with quality of effluent. The total biogas generation would be around 8151 m^3/day with 65-70 per cent methane content, to be used in boiler. This along with the bagasse combustion during the sugar season shall be equivalent to 140 MT of coal per day, and about 9.0 MTPD during non-sugar season. The details are presented in Table 68.4. The total energy available during sugar season shall be sufficient to meet the requirements in all the three industries, whereas during non-sugar season, additional arrangement to the tune of about 50 per cent will have to made through coal fired boiler.

Integrated Wastewater Treatment

The remaining wastewater streams from all the three industries can be treated together separately before final disposal to water bodies/land for irrigation. The characteristics of remaining wastewater streams are presented in Table 68.5 along with combined effluent.

All the remaining streams from sugar mill, paper mill, distillery, and anaerobic reactors can be mixed together in flow equalization tank. The approximate quality of combined effluent has been estimated for the two periods; with and without sugar mill operation.

The wastewater treatment scheme followed shall involve; primary clarifier, anaerobic lagoon, and aerobic lagoon with surface aeration. A schematic diagram of wastewater treatment along with anticipated effluent quality at each stage during sugar season. The finally treated effluent conforming to MINAS with respect to pH, SS and BOD can be used for irrigation in the nearby land. The characteristics of treated effluent at every stage of treatment process are presented in Table 68.6, with corresponding pollution load reductions during sugar and non-sugar seasons.

Table 68.3: Design Parameters and Anticipated Performance of UASB Reactors

Parameter	*Sugar*	*Paper and Distillery*
Reactor Capacity, m^3	280	6400
Retention Time	6 hrs	10 days
COD loading rate, Kg/m^3	8.3	3.5
COD Reduction, per cent	75-85	60-70
BOD Reduction, per cent	85-90	85-90
Biogas Production, Nm^3/kg COD destroyed	0.5	0.5
Total COD destroyed, kg/day	1866	14435
Total biogas production, NM^3/day	933	7218
Final Effluent Quality		
pH	7-7.5	7-7.5
COD, mg/L	440	12145
BOD, mg/L	160	2280

Table 68.4: Energy Generation during Sugar and Non-Sugar Season

Parameter	*Value*
During Sugar Season	
Bagasse available for combustion	275 MT/day
Calorific Value of bagasse	1900 Kcal/kg
Total heat value of bagasse	52.25×10^7 Kcal/day
Total Biogas Production	8151 m^3/day
Calorific value of biogas	5000 Kcal/kg
Total heat value of biogas	4.08×10^7 Kcal/day
Total heat value	56.33×10^7 Kcal/day
Calorific value of coal	4000 Kcal/kg
Equivalent quantity of Coal	140 MT/day
During Non-Sugar Season	
Total Biogas Production	7218 m^3/day
Total heat value of biogas	3.61×10^7 Kcal/day
Equivalent quantity of Coal	9.02 MT/day

The concept of SDP complex may be adopted at least in the new developmental sugar projects, besides wherever it is possible in the existing units, and can be benefited of various benefits of industrial ecology. Some of which are:

- ✫ Better utilization of waste product; bagasse
- ✫ Reduction in requirement of forest based raw material for paper manufacture
- ✫ Reduction in high pollution loads

Table 68.5: Estimated Characteristics of Remaining Wastewater Streams and Combined Effluent

Parameter	*Sugar Industry (A)*	*Paper Mill (Paper Machine Section) (B)*	*AR I (Sugar) (C)*	*AR II (Paper + Distillery) (D)*	*Combined Effluent Sugar Season (A+B+C+D)*	*Combined Effluent Non-Sugar Season (B+D)*
Flow, m^3/day	865	1800	1060	640	4365	2440
pH	6-7	6-6.5	7-7.5	7-7.5	7	7
SS, mg/l	705	750	380	1900	820	1052
COD, mg/l	264	1500	440	12145	2560	4292
BOD, mg/l	57	500	160	2280	590	967

Table 68.6: Estimated Characteristics of Combined Effluent at each Stage of Treatment

Stage	*Parameters*				
	Flow (m^3/day)	*pH*	*SS (mg/L)*	*BOD (mg/L)*	*COD (mg/L)*
During Sugar Season					
After flow equalization	4365	6.5-7.5	820	590	2560
After primary clarifier (DT 4 hrs)	4365	6.5-7.5	246 (70%)	442 (25%)	1536 (40%)
After anaerobic lagoon (DT 16 days)	< 4365	6.5-7.5	49 (80%)	147 (60%)	768 (50%)
After aerated lagoon (DT 4 days)	< 4365	7-7.5	< 20 (70%)	< 30 (85%)	< 500 (40%)
During Non-Sugar Season					
After flow equalization	2440	6.5-7.5	1052	967	4292
After primary clarifier (DT 4 hrs)	2440	6.5-7.5	316 (70%)	725 (25%)	2575 (40%)
After anaerobic lagoon (DT 16 days)	< 2440	6.5-7.5	47 (85%)	290 (60%)	1288 (50%)
After aerated lagoon (DT 4 days)	< 2440	7-7.5	< 20 (70%)	< 50 (85%)	< 800 (40%)

Value in brackets represents expected reductions.

- ☆ Energy generation from high organic matter containing wastewaters
- ☆ Reduction in coal requirement
- ☆ Reduction in freshwater requirement
- ☆ Availability of nutrient rich treated wastewater for irrigation
- ☆ Reduction in various infra-structural requirements of the individual industry

☆ Energy recovery from small paper mill effluent (black liquor) together with distillery spentwash, as a small paper mill is unable to do so individually for economic reasons etc.

To gain maximum benefits from the industrial ecology concept, there need to be continuous research within the industrial society to further look for most beneficial, efficient and cost effective methods to create the best products for the consumer, releasing the least waste (of energy, materials and capital). This in turn shall help in maximizing the profits and achieving the goal of sustainable development.

Conclusion

The concept of industrial ecology has been demonstrated in developing an integrated sugar-distillery and paper (SDP) industrial complex with particular emphasis on wastewater management for a 2500 tcd capacity sugar mill. Further, a 30 TPD small paper mill (without chemical recovery) and 9 KLPD distillery can be setup in the complex.

Biogas equivalent to about 9 MT coal can be generated from distillery and paper mill wastewaters. During sugar season, the total energy requirement can be met from bagasse combustion and biogas generation (equivalent to 140 MT coal). However, there is a need for further study to look critically for most beneficial, efficient and cost effective methods to create the best products with least waste generation and thus help in achieving the goal of sustainable development.

References

1. Garner Andy and Keoleian G.A. (1995). *Industrial Ecology – An Introduction,* National Pollution Prevention for Higher Education, University of Michigan, Ann Arbor, MI, Nov. (Article available on the web).
2. International Facts and Figures about Sugar Industry (Article available on the web).
3. India Infoline Sector Reports – Sugar Industry ((Article available on the web).
4. TIFAC Document (1996). Techno-market Survey on Pollution Control Technologies for Sugar Mills, prepared by Technology Information, Forecasting and Assessment Council (TIFAC), Department of Science and Technology, Delhi.
5. Comprehensive Industry Document for Sugar Industry, prepared by Central Board for Prevention and Control of Water Pollution (CPCB), New Delhi, COINDS/8/1980-81.
6. Jain N.C. and Bajpai A. (1989). Utilization of Filter mud – Present and Future Prospects, Indian Sugar, April.
7. Comprehensive Industry Document for Small Pulp and Paper Mill, prepared by Central Board for Prevention and Control of Water Pollution (CPCB), New Delhi, COINDS/22/1986.
8. Ray S.S., Subramaniyan L.R. and Ghogare B.V. (1998). Treatment of Effluent in Smaller and Medium Capacity Pulp Mill (20-30 TPD) – Agro Pulping – A Case Study of Pravara Pulp and Paper Mill, *IPPTA*, Vol. 10 (1), March, pp. 29-34.

9. Comprehensive Industry Document for Fermentation Industries, prepared by Central Board for Prevention and Control of Water Pollution (CPCB), New Delhi, COINDS/1981-82.

10. Kaul S.N., Pathe P.P. and Nandy T. (1990). An Overview of Waste Management in Sugar Industry, *IAEM*, Vol. 17, pp. 55-78.

11. Kaul S.N., Nandy T. and Trivedy R.K. (1999). *Pollution Control in Distilleries*, published by Enviro-media, Karad, p. 13.

12. Singh S.P. and Rao N.J. (1988). A Quick Estimation of Batch Digester Discharge, *IPPTA*, Vol. 25 (3), pp. 28-33.

13. Goyal S.K., Seth R. and Handa B.K. (1996). Diphasic Fixed Film Bio-methanation of Distillery Spentwash, *Bio-resource Technology*, Vol. 56, pp. 239-244.

Chapter 69

Low Cost Adsorbents for the Removal of Heavy Metals

D.S. Ramteke, A. Swami, R. Malik, M. Mukherjee and S.N. Kaul

CSIR-NEERI, Nagpur

Introduction

Water is the world's most precious natural resource, and due to this realization, it is becoming a prevalent source of concern. More attention and emphasis is placed on its maintenance and purification than ever before. Industry is also very prominent in today's society and its impact on water is increasingly significant. Oil refineries, plastics and paint industries discharge harmful effluents, which have the potential to devastate receiving waters. Metals are a unique class of toxicants. They occur and persist in nature; their chemical form may be changed because of physio-chemical, biological or anthropogenic activities. Their toxicity may be drastically altered as they assume different chemical forms. Most of them have some value to humans because of their uses in industry, agriculture, or medicine. Some are essential elements, required in various biochemical or physiological functions. On the other hand, they may pose health hazards to the public because of their presence in food, water or air and to workers engaged in mining, smelting, and a variety of industrial activities. Heavy metals are of special concern as they are non-degradable and persistent.

There has been increasing concern in recent years over the discharge of heavy metals to the aquatic environment. Water is a key source of heavy metal toxicity and ingestion of toxic metals in food chain. In past decades, so many incidences have been occurred due to metal hazards and its occurrence in food chain. " Mina Mata" bay hazard is one of them. Due to increase in population and industrial growth, there is a great need of metal free potable and usable waters. The growth in demand for clean water has placed a heavy demand on the concept of water reuse. The high concentration of metals in water system is generally caused by the discharge of industrial waste to sewer or infiltration of water that has passed through deposits of metal bearing minerals. The cost of removal of heavy metals from rich industrial effluents, before being discharge into waterways, is quite high. Metal exists in wastewater in many forms, including soluble, insoluble, inorganic, reduced, oxidized, free metal, precipitated, adsorbed and complexed (1). It is very much essential to remove heavy metals from wastewater before its discharge into sewage system or surface water streams.

Numbers of technologies are in progress for the treatment of wastewaters containing heavy Metals. Accordingly, various types of treatment plants have been installed for the purification of water and wastewater to make it usable for drinking and industrial purposes. Current methods for wastewater treatment of metals include precipitation, coagulation/flocculation, sedimentation, flotation, filtration, membrane process, electrochemical process, ion exchange, adsorption, biological process and chemical reaction. These conventional technologies appear to be useful, but often create some problems such as the separation of sludge and the generation of secondary waste.

It is known that adsorption is one of the most efficient methods for removal of heavy metals from waste streams. Activated carbon is the most widely used adsorbent due to its excellent adsorption capability for heavy metals. However, its use is often limited due to high cost, making these methods unfavorable for the needs of developing countries. For removing metals from wastewater, particular attention is given to effective and simple processes.

Metal removal by low-cost adsorbent materials may be advantageous, compared with abovementioned traditional processes. These low-cost adsorbent materials recently studied include activated carbon and modified activated carbon, biosorbents, mineral adsorbent, biopolymer adsorbents and widespread and cheap natural materials or reused wastes such as clay, slag, peat and fly ash. Over the last century considerable research has taken place in the area of activated carbons and to develop other low cost adsorbents for various uses. This chapter mainly highlights the toxicity of metals such as Cadmium and Chromium, existing treatment technologies including adsorption, their efficacy and development of low cost adsorbents *viz.* Natural polymeric substances and activated carbons from waste materials.

Heavy Metal Occurrence and Toxicity

Toxicity is defined as the property of a material, or a grouping of materials, to restrain the metabolic process of cells without completely destroying them in a particular species under certain physical and biological conditions for a specified time and concentration. Cadmium and Chromium are among the metals of major health concern, because of their impact resulting from environmental pollution as well as the serious nature of their toxic effects (2).

Cadmium occurs in nature mainly in lead and zinc ores. It is thus released near the mines and smelters of the metals. It is used as a pigment in ceramics, electroplating, making alloys and alkali storage batteries. Its level in the air is in the range of nanograms per cubic meter, but may amount to several milligrams per cubic meter in certain workplaces. The level in water is very low (~ 1 µg/l) except in contaminated areas. Cadmium is also used in packing industries, plastic industries, ceramic paints etc. Cadmium is an oxyphilic and sulfophilic element. It undergoes multiple hydrolysis at pH values encountered in the environment. It begins to hydrolyze at pH 9. Cadmium forms moderately stable complexes with a variety of organic matter found in natural waters including amino acids, sugars, poly saccharides and humic substances.

Cadmium is less toxic to aquatic plants. Inhibition of growth and photosynthesis occurs at 0.02-1.0 mg/l of cadmium concentration. But chronic effects may occur at concentration as low as 1µg/l. At toxic levels, plants exhibit chlorosis and reduced growth. In case of animals, LC_{50} value for freshwater invertebrates vary from 0.003 to <0.5mg/l. LC_{50} value vary from 0.09-10.5 mg/l for freshwater fishes and 8.8 mg/l for estuarine species. Exposure of fishes with cadmium generally reduces their ability to osmoregulate. Cadmium is highly toxic to humans because of the absence of homeostatic control in human body. It interacts or competes with other metals and induces deficiency symptoms. Cadmium acts as inhibitor to sulphydryl enzymes. It can disrupt pathways of oxidative phosphorylation. The symptoms of cadmium toxicity produced by enzymatic inhibition include hypertension, respiratory disorders, damage to kidney and liver, amioaciduria (urinary excretion of amino acids), hypercalciurea (excretion of excessive calcium), protein urea, (excretion of proteins), osteoporosis (decalcification of skeleton), and formation of kidney stones (2- 4) Carcinogenic and teratogenic effects have also been observed. It increases the risk of prostate cancer in man.

Chromium occurs in ores, mining, smelting and industrial uses which tend to increase its environmental levels. It is used in making stainless steel, various alloys and pigments. Fossil-fuel-powered plants and cement producing plants are also sources of environmental pollution. However, its levels in air, water and food are generally very low. The major human exposure is occupational. The principal chromium emissions into surface waters are from metal- finishing processes such as electroplating, pickling, and bright dipping. Uncontrolled emissions have great potential for contaminating the freshwater with the relatively toxic form, Cr^{6+}. Other smaller discharges of Cr^{6+} are from the additive in circulating cooling waters, laundry chemicals, and animals glue manufacture. Sources of Cr^{3+} contamination include liquid waste discharges from leather, tanning and textile dyeing containing several thousand mg L^{-1} of chromium.

Toxicity of Cr^{3+} and Cr^{6+} to aquatic plants is generally low. Toxicity of plants depends on pH and other factors such as the presence of organic chelators, cations, nutrients and other heavy metals in solutions. Sub lethal/chronic effects of chromium intoxication include decreased/growth and body size, reduction in the rate of reproduction and survival of progeny. Chromium (VI) is found to be more toxic than Chromium (III). Chronic exposure to chromate dust has been correlated with increased incidence of lung cancer. Oral administration of excessive levels (150ppm) is associated with liver and kidney damage and depression. Chromium (VI) causes allergic skin irritation, dermatitis, irritation to mucous membranes, conjunctiva and gastro intestinal ulcers. A higher rate of bronchitis among chromate workers has been reported as a condition called 'chronic chromate lung' which may be carcinogenic.

Techniques for Metal Removal

Numerous reviews (5-17) are available regarding the techniques employed for heavy metals removal include precipitation, ion-exchange adsorption, cementation, evaporation, reverse osmosis, electrolysis and electrodialysis apart from liquid-liquid extraction, floatation and freeze precipitation. Some of these techniques are summarized in Table 69.1.

All the processes discussed in Table 69.1 have one or the other disadvantages as compared to adsorption. In wastewater treatment processes, primary and secondary treatment is not sufficient unless tertiary treatment is applied. In the primary treatment, only suspended solids and degradable compounds are reduced while in chemical treatment, parts of dissolved materials are reduced. The treatment system becomes full phase only when adsorption treatment is applied. In this treatment, pollutants, which are not removed by the earlier two methods, are adsorbed by passing the liquid through column containing adsorbent. In this way, activated carbon plays an important role in the purification system for the removal of residual organics and inorganics, left in primary/secondary treatment system. Nevertheless, it is clear that the problems of dependability, sensitivity and cost are no worse and often less than those encountered with more conventional treatment technologies, particularly biological. In some cases, it is apparent that physico-chemical treatment with carbon offers potential benefits in stability, dependability and even cost, which outweighed the risks inherent in the use of new technology. The applicability of activated carbon in the tertiary wastewater treatment system depends mainly on the nature of that particular wastewater stream.

The main disadvantage of this technology is the cost of Adsorbent. Inexpensive and more easily available adsorbents would make the removal of pollutants an economically viable alternative. The solid/industrial waste materials being carbonaceous in nature could be used as alternate raw materials in place of costly raw materials for the preparation of activated carbon. Lot of research work is going on to produce quality adsorbent in less cost. In this context, this chapter mainly imparts focus on the preparation and evaluation of activated carbons from waste materials like paper mill sludge, coke fine waste, water hyacinth, walnut and ground nut shells along with modified natural polymeric substance like tree barks and use of these adsorbents for removal of heavy metals.

Potential of Waste Material Generation

(i) Paper Mill Sludge

Pulp and paper manufacturing is one of the largest industries in the world and has shown rapid and steady growth. In India, at present, there are total 594 mills for manufacturing of paper (18). The current productions of all the mills are around 22000 TPA (approx) as against the installed capacity of 8.500 million ton per annum (19). During the manufacturing processes, large amount of primary and secondary

Table 69.1: General Techniques for the Treatment of Heavy Metals Containing Wastewater

Sl.No.	*Techniques*	*Mechanism*	*Advantages*	*Disadvantages*
1.	Precipitation	Heavy metals are removed from effluent by precipitating them as insoluble hydroxides, carbonates and sulphides using NaOH, Na_2CO_3, and Na_2SO_4 respectively.	It is an effective technique for removing metals	Theoretically, minimum solubilities for different metals occur at different pH, which require multistage precipitation. pH should be controlled for maximum removal. Consequent post treatment of effluent is done to bring the pH within permissible limit.
2.	Cementation	It is a process of spontaneous electro-chemical reduction to the elemental state and consequent oxidation of a sacrifice metal solution	The advantages are simple control equipments and low energy utilization	The major disadvantage is the process optimization and thermodynamic limitation.
3.	Evaporation	It is a process of concentration of solute through evaporative treatment for the recovery of product.	Advantage of evaporation is that it permits recovery of a wide variety of process chemicals	Major disadvantage is that it consumes large amount of energy.
4.	Ion Exchange	Ion exchange is a process in which ions are held by electrostatic force to charge functional group on the surfaces of a solid and exchanged for ions of similar charge in solution in which the resin is immersed.	It is highly efficient at recovering metal containing chemicals. If effectively removes Al, As, Cd, Cr (VI, III) Cu, and Pb	Effluents characteristics affect the chemistry of ion exchanger. Handling and Cost of treatment is higher as compared to other treatment techniques
5.	Reverse osmosis	Osmosis is a process of passage of a liquid through a semi permeable membrane from a dilute to a more concentrated solution. Reverse osmosis is an operation where in pressure is applied to the more concentrated solution, forcing the permeate to diffuse through the membrane and into the more dilute solution. The filtering action produces a concentrate and a permeate on opposite sides of the membrane.	The concentrate can then be further treated or returned to the original form for reuse	Major limitation is that membrane materials are fairly limited, and narrow acceptable temperature range for satisfactory operation. Higher cost.

Contd...

Table 69.1–*Contd...*

Sl.No.	*Techniques*	*Mechanism*	*Advantages*	*Disadvantages*
6.	Electrolysis	In this process, there is electrochemical reduction of metal ions at the cathode of elemental metal and evaluation of O_2 at anode.	This process is used to recover different metals from the waste containing metals. Efficiency of this process has been tremendously increased by innovative designs such as eco-cell involving rotating electrodes and extended surface electrolysis	Costly technique for the recovery of Metals.
7.	Electrodialysis	In this process, dissolved colloidal species are exchanged between two liquids through selective semi permeable membranes under the influence of electric potential. Dissolved species separate according to their charges. Semi-permeable membranes allow passage of certain charged species while rejecting the oppositely charged species.	It is highly efficient at recovering metal containing chemicals. It effectively removes Al, As, Cd, Cr (VI, III), Cu, and Pb.	Cost of treatment is higher as compared to other metals.
8.	Adsorption	This treatment process is also called as polishing treatment. It comes after primary and secondary treatment. In this treatment materials, which are not removed by the two steps, are adsorbed by passing the liquid through activated carbon column	**High Efficiency**: ALM series of adsorbents developed in Japan containing S and N functional group has reported to reduce heavy metal levels in mining and industrial wastewater. **Low Cost**: Adsorbent like activated carbon can be prepared from low cost materials like peat, moss and even waste, *e.g.* paper mill waste and Agricultural waste. Adsorption can remove inorganics, pesticides and organics like hydrocarbons etc. from wastewater **Regeneration**: Now a days regeneration of adsorbent is also possible	Cost of good adsorbent makes this method costlier than others

sludge is generated (20) accounting 0.5 tons of dry sludge per ton of paper (21). The present practice is dewatering of sludge followed by land disposal or incineration and is not cost effective. The sludge is fibrous and cellulosic in nature.

(ii) Coke Fines

Coke fines are waste product generated during coal (caking type) carbonization for making coke, which is used in the steel plant. Quantity of coke fines produced at coal carbonization industry is roughly about 40 MTPD. Since the base material is coal, it constitutes a good carbonaceous material containing high percentage of carbon (80-90 per cent). The present disposal method is to use as fuel or to dispose off on roadsides. It can be better used as a raw material for the preparation of activated carbon.

(iii) Water Hyacinth

Water Hyacinth is widespread in the tropical countries, considered as one of the most obnoxious weed creating long time problem for eradication. It is a tropical plant of exquisite beauty possessing good potential for reduction of pollution load of sewage, industrial effluents and also a source of energy, food and livestock food. In India, it grows profusely in the northeastern region and regarded as a menace in agricultural production. Although the present practice is to use it as local manure and biogas production, the majority of these are wasted. It is carbonaceous in nature containing 60-70 per cent of carbon on dry basis.

(iv) Walnut Shell

Walnut shell is a solid waste material produced from dry fruit processing industries creating lots of disposal problems in Gulf countries. The present production rate in India and worldwide is 14,000 MTPA and 676,260 MTPA respectively. The present disposal practice is to dump in a wasteland site creating nuisance conditions due to microbial degradation or to use as a fuel. It is also a carbonaceous material containing 60-70 per cent carbon and could be used as an alternate raw material for the production of better product.

(v) Groundnut Shell

India is the second largest producer of Groundnut throughout the world, which has 8-milion hectares area of cultivation producing 8004,000 metric tons/yr. Nutshell is a carbonaceous, fibrous solid waste which encounters disposal problem and is generally used for its fuel value. Therefore, it was of interest to prepare a higher value product such as activated carbon from Groundnut shell. These byproducts (22) offer the advantage of greater percentage of non-carbon constituents in their composition compared to coal or peat and therefore afford a greater chance of retaining functional groups, especially oxygenated groups in the carbonized product. However, the lower carbon content of by- products translates to lower activated carbon yields, but the low cost of waste cancel out the lower yields.

Preparation of Low Cost Adsorbent

Modified Bark

Several studies have been carried out for the utilization of natural polymeric substances for the removal of pollutants. Some workers described the use of various tree barks and various agricultural byproducts (23-26) such as peanut skin, onion skin, paddy husk, paddy straw, sugar cane baggase, green embilica leaves, garlic skin etc. for the removal of toxic and hazardous substances from mining and industrial wastewaters. Some workers have described the use of barks (particularly coastal red wood barks) for selective removal of toxic heavy metal cations from mining and industrial wastewaters. The metals were bound to the bark substrates by ion exchange with hydrogen, presumably from the phenolic groups in tannin compounds. Impregnation of the bark with dilute formaldehyde in acid medium prevented the bleeding of colors into the liquid, perhaps by polymerization and precipitation of the colored compounds. Formaldehyde treatment enhances the capacity of barks to remove heavy metal ions from solution. Formaldehyde treated western Hemlock had a binding capacity of about 0.068 gm of lead per grams of the bark from solution of 18 ppm Pb^{+2}, with essentially Zen bleed of colour into solution. Studies on adsorption of heavy metal ions such as copper, cadmium and chromium on various pinewoods barks shows that adsorption capacity of the barks depends on their tannin contents (27). Other Natural Polymeric Substances are cotton, wool, peat, rice husk etc. These substances are the basic material for the unconventional methods of wastewater treatment.

The ion exchange and complexing property of cotton rags is due to ethylenimine network polymer. The ion exchange capacity of the polymer was found to be relatively low as compared to most of the granular ion-exchange resins but it has high capacity for complexing certain heavy metals. The potential use of the impregnated fabric in removing heavy metals from polluted water was also reported (28). Studies are also done on the removal of the mercury from aqueous solution by nitrogen-containing chemically modified cotton. Cotton cellulose modified with 2 diethylaminoethyl or 2-amino ethyl substitutes were found to be more effective at varying concentration above 100 ppb. The sorbed mercury could be readily removed from the chemically modified cottons for recovery of mercury and reuse of the substrate (29). Studies are also carried out on modified barks for scavenging toxic heavy metal ions such as Cu, Pb, Zn and Cd. The binding of mercuric chloride and other metal salts to bark, activated sludge (miloganite), chitosan, poly (P-amino styrene) and other natural and synthetic materials are also reported. Wide range of natural materials are potentially useful for the removal of toxic and precious metals (such as Hg, Pb, Zn, Cd, Cu, Ni, Co, Fe, Mn, Ag, Pt, Pd, and Au), that may be present in industrial effluents, mine water or other water supplies (30-31). The use of water hyacinth (32-38) for adsorption of remarkably high rate of heavy metals, minerals, nutrients, organic substrates etc. has been demonstrated by several workers in recent years.

It has been found that the binding of heavy metal ions on the bark samples is pH dependent, probably partly because tannins in the bark act as weak acid ion exchangers, exchanging two hydrogen ions into the solution for each divalent metal ion bound to the solid. The initial pH of solution has been selected on the basis of separate experiments at different pH values. Hardwickia binata bark (39) has also been found to have good sorption capacity for mercury (II). Studies indicated that the sorption of Hg (II) increased as the pH increases and a contact time 2 hr. was found to be optimum. Some workers (40) also removed Cr (VI) from wastewater by using *Ablesmoschus esculentus* (Ladyfinger plant). The removal of Cr (VI) was in general most effective at low pH values. They found that the uptake is highly dependent on the initial concentration of the metal ions in the solution. The higher uptake at low concentration can be attributed to the availability of more parking spaces (*i.e.* active centers) on the surface of the adsorbent for lesser number of adsorbate ions.

From the above review, it is clear that some natural polymeric substances including barks are capable of binding appreciable quantities of heavy metal ions. For present study, two Indian barks *viz. Acacia arabica* (Babul) and *Artocarpus heterophyllus* (Jack fruit) are considered to bind toxic heavy metal ions such as cadmium and chromium.

Process Development for Modified Bark

The *Acacia arabica* and *Artocarpus heterophyllus* barks were collected locally and dried at 100°C for 24 hours. Both barks were pulverized in order to increase surface area and to achieve a particle size that would facilitate penetration of the solution in the bark. The dried powder was further treated with 20 parts of 0.02 N H_2SO_4 and 5 parts of 39 per cent formaldehyde, kept in an oven at 100° C for 6 hours and stirred. Both barks were washed with double distilled water several times and dried at 60° C. The dried and sieved powder of both the barks was used in the experiments.

Activated Carbon

Commercially available activated carbons are generally prepared from coal (Anthracite and Bituminous), wood and coconut shell. In India, good quality activated carbons are generally imported from the other countries. It is a main cause for it's high cost. Lot of research work has been done to assess the potential of different type of solid wastes for the preparation of activated carbons. Gallop R.A. (41) studied the pyrolysis of straw grains, potatoes, pits, waste oils, cardboards and newspapers for the preparation of charcoal and activated carbon. He observed that charcoal and activated carbons play a key role in 'Polishing' effluents to a very high purity, which could be regenerated and recycled.

P.R. Bell and Virjarandi (42) described a large-scale pilot plant for pyrolysis of municipal bulk wastes and thereby, producing useful products in the form of combustible gases, oils and activated carbons. The pilot plant converts solid wastes into usable by-product on a continuous basis by thermal degradation in an oxygen deficient atmosphere using heating method.

C.D. Barett (43) pyrolysed a cellulosic type organic waste at different temperatures, *i.e.* 400°C, 520°C and 600°C respectively. He reported that, the quality of the product depends on the primary and secondary pyrolytic reactions related to temperature, time and size of the raw material.

In Canada, (44) pilot plant studies were carried out for the preparation of low cost activated carbon from pulping wastes liquors by pyrolysis. The process is economical for obtaining high quality activated carbon.

Backmans et. al. (45) pyrolysed a filtered sludge from municipal activated sludge in a continuous feed multiple hearth furnace. The carbon content of pyrolyzate under optimum conditions was found to be 14 per cent. The pyrolyzate was tested for its utility to adsorb COD and colour. Parallel adsorption tests were also made with fly ash and activated carbon. The pyrolyzate was found to be intermediate in its adsorption capacity between fly ash and activated carbon, suggesting that it could prove useful as a tertiary treatment adsorbent in wastewater renovation.

Many common organic wastes (46) and also straw (47), coconut shell (48), Sawdust (49), peat mass (50), nutshells (51), rice husks (52) and other fibrous plant residues, wood board and newspapers have been used as substrate for the production of activated carbon by pyrolysis.

Schlesinger, M.S. et. al. (53) developed a process to convert a typical waste material including municipal refuse, wood waste, battery scraps and sludge manure into gas, liquid and solid fuels. He found that the process is thermally self sufficient, and energy in the feed is almost all recovered in the products. Godd, A.A. (54) suggested a new gasification process to convert carbonaceous waste material into activated carbon. He found an improved yield of activated carbon by gasification process.

In India, different raw materials have been used time to time for production of activated carbon. S. Banerjee and R. Haque (55-56) have prepared activated carbon from coconut shell and coal respectively by using different activators. An attempt was made by Ramteke, D.S. (57) to develop a technology for waste utilization and set-up a pilot-plant for the use of industrial solid waste by converting it into a valuable product known as 'Activated Pyrochar'.

In this context, waste materials *viz.* paper mill sludge, coke fine waste, water hyacinth, walnut and groundnut shell were utilized for the preparation of activated carbons. The products were characterized for their physico-chemical and adsorptive properties and also evaluated for their application in heavy metals removal.

Process Developed for Activated Carbons from Waste Materials

The waste materials selected are carbonaceous in nature. Carbonization of such waste materials under controlled conditions may prove to be an economical method of disposal without creating any environmental problem. Pyrolysis usually refers to chemical decomposition of organic materials heated in an environment having insufficient oxygen for combustion (58). Basically, there are two different processes for the preparation of activated carbon: physical and chemical activation (59). Physical activation involves carbonization of carbonaceous precursor followed by activation of resulting char in the presence of activating agents such as carbon dioxide or steam. Chemical activation on the other hand, involves the carbonization of precursor in the presence of chemical agents (*e.g.* H_3PO_4, $ZnCl_2$). In physical activation, elimination of large amount of internal carbon mass is necessary to obtain a well developed porous structure, whereas in chemical activation process, chemical agents used are dehydrating agents which influence pyrolytic decomposition and inhibit the formation of tar, thus enhancing yield of carbon. Chemical activation has more advantages (60) over physical activation with respect to higher yield, more surface area and better development of porous structure in carbon. It also helps to develop oxygenated surface complexes on the surface of activated carbon.

Operating conditions for pyrolysis depends on the ratio of flow and feed material. It is carried out in two different sequences known as Low Temperature Carbonization (LTC) and High Temperature Carbonization (HTC) in the range 250-600°C and 600-1000°C respectively. Low temperature carbonisation produces more char and less gaseous products, while high temperature carbonization produces less residue and large quantity of fuel gases. A few intermediate products (liquid) are also obtained in both the processes.

Pyrolysis Unit

The pyrolysis unit (Figure 69.1) consists of a stainless steel cylindrical reactor having diameter of 10.2 cm and 91.5 cm in length. It is wounded accurately with heating material like nicrome wire and insulated by asbestos powder. For temperature measurement, furnace is provided with two thermo couple Ni-Cr/Ni-Al as a multiple temperature indicator. The reactor was connected at the top through heat resistant tubing to a water condenser. A tar and oil trap was kept beneath the condenser to collect the condensate. Further, it was joined with three glass tubes connected in series and kept in freezing mixture to collect the incondensable products. These are further connected with the water scrubber to wash the generated gases and then measured with the gas meter attached to the system. The complete unit was kept airtight during experiment. Different raw waste based activated carbons were prepared on laboratory scale. The process details are as follows:

Activated Carbon from Paper Mill Sludge

For pre-treatment, ground and sieved dry sludge was mixed thoroughly with 0.1 per cent zinc chloride solution. The impregnated material was kept for soaking for 24

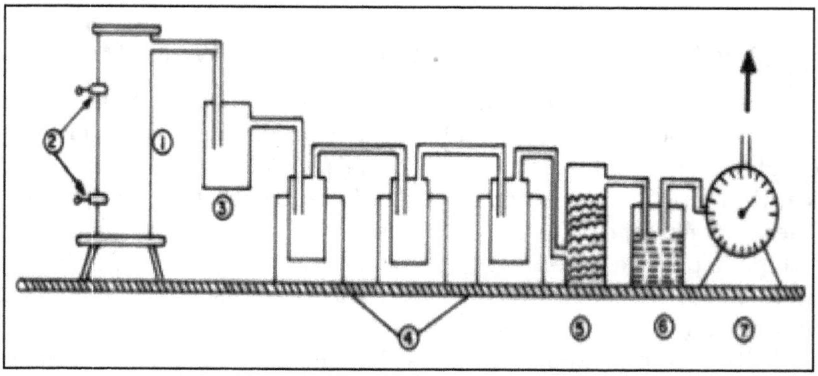

Figure 69.1: Pyrolysis Unit.

1. Reactor; 2. Thermocouple; 3. Tar Trap; 4. Receivers Kept in Freezing Mixture; 5. Scrubber; 6. Water Trap; 7. Gas Meter.

hours with regular mixing and then dried in a hot air oven at 110°C for 24 hours. The dried impregnated material was then fed in to a closed airtight reactor of the pyrolysis unit and heated to 700°C. The reactor temperature was maintained at 700°C for one hour. After cooling, the residue was removed from the reactor and the liquid obtained was thoroughly mixed and subjected to the fractional distillation to obtain different organic fractions. The gases escaping out of the wet test gas meter were collected and analysed for the various constituents.

The product (residue) was further washed with water to remove the excess of zinc chloride used during pre-treatment. The wash water was collected for the recovery of zinc chloride for reuse. It was further washed with hot hydrochloric acid (1+1) for the removal of traces of zinc chloride and the mineral ingredients. After the acid treatment, the material was again washed with hot distilled waster until the acidity and the chloride content was completely removed. The washed product was then dried in hot air oven at 110°C for 24 hours. This was further sieved through 325-mesh sieve and stored in a plastic bottle for further use. The flow sheet diagram for preparation of activated carbon is presented in Figure 69.2.

Activated Carbon from Coke Fines

Coke fines based activated carbon is prepared by adopting single carbonisation process followed by chemical activation. The schedule of experiments carried out for optimizing conditions for activation of coke fines were variation of impregnation time, ratio of activating agent and carbonaceous material, carbonization time and temperature.

Coke fines (1 kg) were impregnated with zinc chloride by maintaining a ratio of Lewis acid and coke fines as 1: 2 for an optimum time of seven days. The impregnated carbon was dried at 100–120°C and subsequently carbonized in pyrolysis unit at 650°C for 15 minutes. Carbonisation results in removal of volatile substances and dehydration by stripping the non-carbon elements like hydrogen and oxygen. The acid treated coke fines were kept in contact with HCl for a minimum period of 24 hours. The acid stream containing nearly 90 per cent of Lewis acid can be concentrated for recovery or this stream can be reused for activation purpose. The acid washed activated carbon was thoroughly washed with warm distilled water to remove chloride for complete neutralization or acidity of the activated carbon. The schematic presentation of the process developed is depicted in Figure 69.3.

Activated Carbon from Water Hyacinth

The stems of water hyacinth were separated from the plants, washed thoroughly with water to remove mud debris etc. and the washed biomass was air dried and finely ground. The dried material (1kg.) was then carbonized in a specially designed reactor (Pyrolysis unit) in absence of air at 400°C for 30 minutes. The carbonized material was then impregnated with zinc chloride with an optimised ratio of 1.33 for zinc chloride to carbonised char. The impregnated material was dried at 100°C and subsequently carbonized at 700°C for 30 minutes. The product was treated with hydrochloric acid (1+1) for the removal of excess of zinc chloride. The acid stream containing zinc chloride can be reused for activation purpose. The acid washed

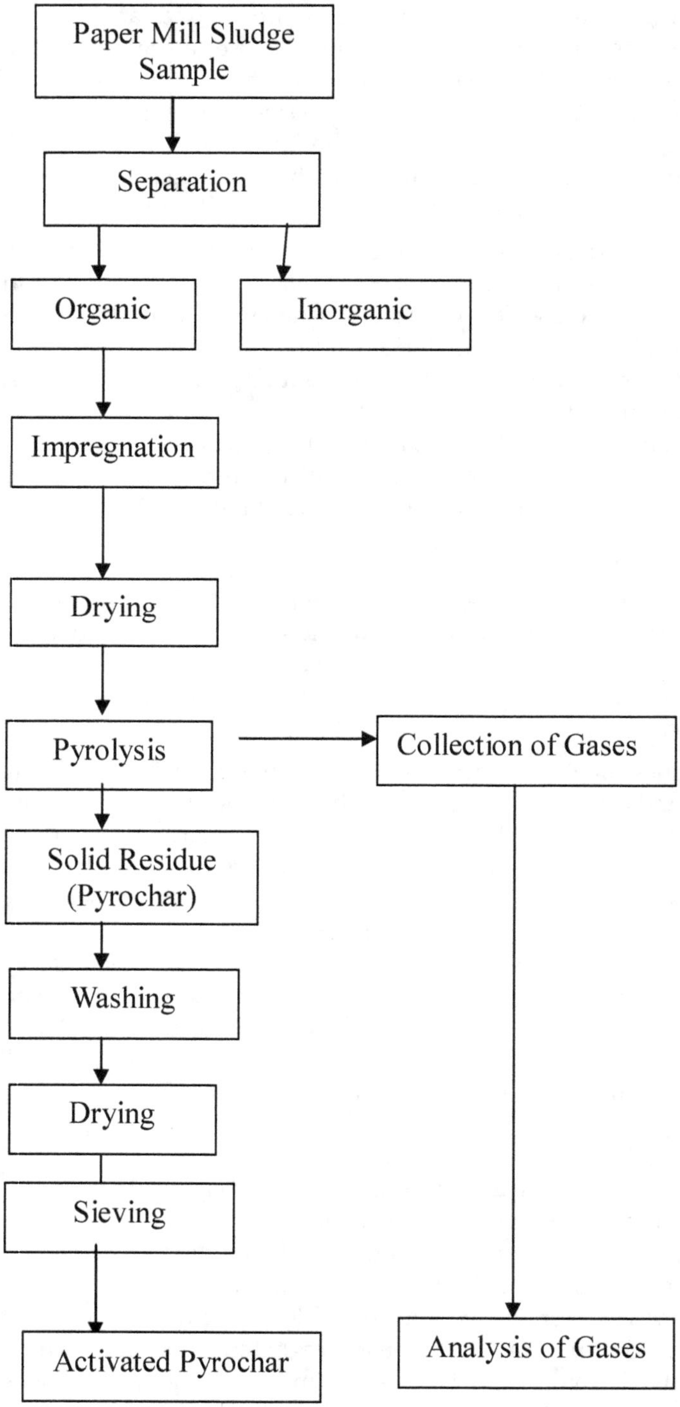

Figure 69.2: Flow Sheet Diagram for Preparation of Paper Mill Sludge Based Activated Pyrochar.

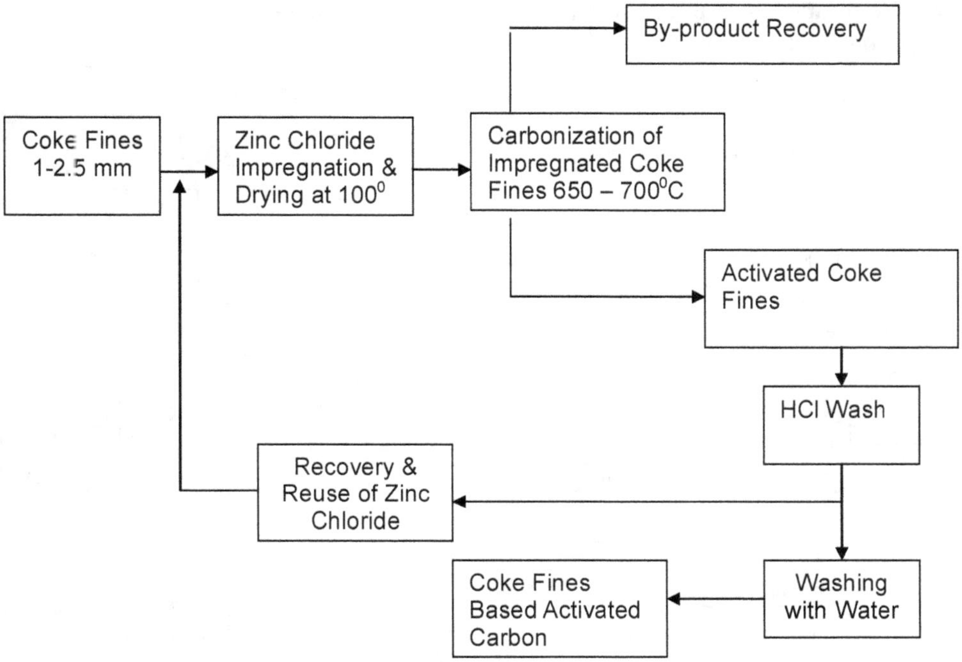

Figure 69.3: Schematic Presentation for Preparation of Granular Activated carbon from Coke Fines.

product was thoroughly washed with hot distilled water to remove acidity and chlorides.

Walnut Shell Based Activated Carbon (Figure 69.4)

The process developed for the preparation of activated carbon is based on double carbonisation technique. The shredded and sized walnut shell was carbonized at 400°C for 30 minutes in a pyrolysis unit to prepare char. Experiments were conducted in the activation process by optimizing different conditions *viz.* temperature, time and $ZnCl_2$/char ratio for final product and impregnated char was carbonized at 700°C for 30 minutes, acid treated and washed to obtain granular activated carbon.

Groundnut Shell Based Activated Carbon (Figure 69.4)

Groundnut shell was collected from local oil mill and washed thoroughly with water to remove soil particles etc. and dried in sunlight. The dried Groundnut shell was shredded to 2 mm size for carbonization purpose. Groundnut shell was filled in stainless steel vessels with a tight lid and it was placed in electric muffle furnace for charring at 400°C for 30 min. It was impregnated with $ZnCl_2$ for 24 hrs and dried at 100° C. This dried and impregnated char was further carbonized at 650° C for 15 minutes. After cooling, the excess $ZnCl_2$ on activated carbon particles was leached out by treating it with 1: 1 HCl for 24 hrs. The product was repeatedly washed with hot distilled water for the removal of zinc chloride. The final product was oven dried at 100°C temperature and further used for the treatment of heavy metals. The scheme for preparation of Activated carbon is shown in Figure 69.4.

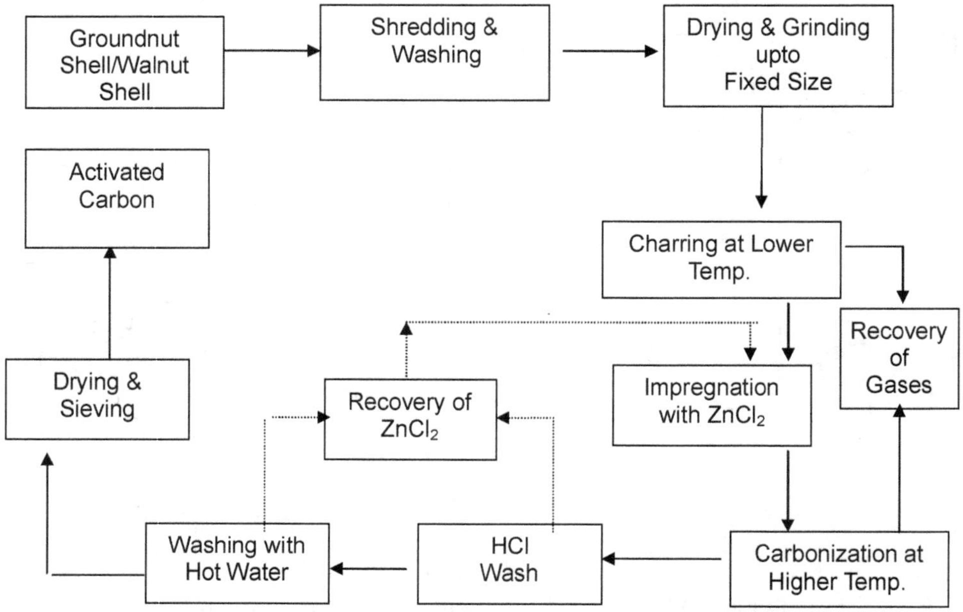

Figure 69.4: Scheme for Preparation of Activated Carbon from Groundnut and Walnut Shell.

Characterization of Adsorbents

The Physical characteristics of the adsorbents *viz.* tree bark and activated carbons were determined by estimating their particle size (61), bulk density (62) and surface area (63) followed by proximate (64) and ultimate analysis (65). The chemical characteristics were determined for their mineral ingredients content along with pH as per standard methods (66). Determination of heavy metals was carried out by AAS model no. GBC 904AA, using air-acetylene flame. All the products were characterized for their adsorptive characteristics in terms of quality control parameters, *i.e.*, Phenol value, Iodine value and Methylene blue number (61-62).

The physico-chemical and adsorptive nature of the adsorbent has profound effect on overall adsorbability. The comparative physical characteristics of modified bark and activated carbons are shown in Table 69.2. The bulk density is a measure of the regenerability of the adsorbent and is used to determine the adsorbent quantities necessary to use in the treatment system. It was found to be 0.50 and 0.33 gm/cm^3 for modified bark *Acacia arabica* and *Artocarpus heterophyllus* bark respectively similarly for paper mill sludge (PSPAC), water hyacinth (WHPAC), coke fines (CFGAC), walnut (WSGAC) and groundnut shell (GSPAC) based activated carbon, it was found to be 0.32, 0.33, 0.66, 0.64 and 0.23 gm/cm^3 respectively.

Table 69.2: Physical Characteristics of Modified Barks and Activated Carbons

Sl.No.	*Parameter*	*Modified Bark*		*Activated Carbons*				
		Acacia arabica	*Artocarpus hetero-phyllus*	*PS PAC*	*CF GAC*	*WH PAC*	*WS GAC*	*GS PAC*
1.	Bulk Density (g/cm^3)	0.50	0.33	0.32	0.66	0.33	0.64	0.23
2.	Particle size (mm)	2–5	2 -5	0.044	1-2	0.044	0.80	0.075
3.	Surface area (m^2/g)	378	563	478	526	1247	750	1200

In relation to know the adsorptive behaviour of adsorbents, it is often desirable to have information about the other properties through proximate, ultimate and chemical analysis that influences the utility of the carbon (Table 69.3). Proximate analysis indicates that activated carbons along with tree bark are low in ash content except PSPAC and CFGAC, *i.e.*, 58.2 and 29.7 per cent respectively. Low ash content denotes good adsorption phenomenon (67). However, it's also reported that if ash contains specific mineral ingredients like alkali and alkaline earth metals besides iron, positive adsorption could be achieved (62). Ultimate analysis indicates that all the carbons except PSPAC are high in carbon content but low in hydrogen, nitrogen and sulphur content. This indicates the availability of more pores developed during the activation process.

Chemical characteristics (Table 69.4) indicate mineral ingredients and heavy metals content in all the solid waste based activated carbons and modified tree barks. High availability of mineral ingredients and heavy metals in carbon directly affect the quality of the treated water when it is applied in treatment system. The results indicate that all the products can be successfully applied to the wastewater treatment, as there are negligible chances to leach out any heavy metal or mineral ingredients from the product to the treated water because of very low concentration of minerals and heavy metals in water solubles.

The adsorptive properties of a particular adsorbent is expressed in terms of its adsorption quality control parameters and presented in Table 69.5. Phenol values are found to be 32 and 21 in case of modified barks similarly for activated carbons PSPAC, WHPAC, CFGAC, WSGAC and GSPAC it was found 55, 17, 25, 48 and 20 respectively. On the other hand, the iodine values were observed as 265 and 465 in case of modified Acacia arabica and Artocarpus heterophyllus bark, similarly for solid waste based activated carbons PSPAC, WHPAC, CFGAC, WSGAC and GSPAC it was found 658, 785, 620, 664 and 1114 respectively. Methylene blue numbers of all the products are comparable. This indicates the better adsorption efficiency of adsorbents for removal of taste, odour and color, low and high molecular organic substances when applied in the water and wastewater treatment system.

Evaluation of Low Cost Adsorbents in Wastewater Treatment

Indigenously prepared low cost adsorbents were further applied for the treatment of heavy metal containing wastewater. Batch sorption tests were performed at room

Table 69.3: Proximate and Ultimate Analysis of Modified Barks and Activated Carbons

Carbon	*Proximate Analysis (per cent)*				*Ultimate Analysis (per cent)*			
	Moisture	*Ash*	*V.M.*	*Fixed Carbon*	*Carbon*	*Hydrogen*	*Nitrogen*	*Sulphur*
Acacia arabica	8.0	9.9	58.0	24.1	–	–	–	–
Artocarpus heterophyllus	4.8	16.6	63.4	15.2	–	–	–	–
PSPAC	1.0	58.2	9.3	31.3	29.92	1.15	Traces	Traces
CFGAC	8.4	29.7	4.3	57.6	54.80	0.86	0.98	0.23
WHPAC	2.6	8.8	2.1	86.5	79.80	2.80	3.60	2.40
WSGAC	8.0	5.5	15.9	70.6	79.20	4.20	0.16	0.54
GSPAC	9.0	8.2	11.4	71.4	75.8	2.95	1.53	0.32

Table 69.4: Characteristics of Water and Acid Soluble of Modified Barks and Activated Carbons

Sl.No.	Parameter	Wator Extract							Acid Extract						
		Acacia arabica	*Arto-carpus hetero-phyllus*	*PS PAC*	*CF GAC*	*WH PAC*	*WS GAC*	*GS PAC*	*Acacia arabica*	*Arto-carpus hetero-phyllus*	*PS PAC*	*CF GAC*	*WH PAC*	*WS GAC*	*GS PAC*
1.	pH	4.70	6.84	7.8	7.2	7.3	6.82	7.2	–	–	–	–	–	–	–
2.	Sodium (per cent)	0.0625	0.07	0.12	0.08	0.03	0.026	0.003	0.1375	0.1475	0.012	0.092	0.0004	0.036	0.005
3.	Potassium (per cent)	0.26	0.23	0.04	0.01	ND	0.054	0.0003	0.620	0.45	0.043	0.008	Traces	0.070	0.004
4.	Sulphate (per cent)	0.32	ND	ND	0.815	1.31	0.024	0.045	ND	ND	ND	0.95	0.0313	0.005	0.055
5.	Phosphate (per cent)	ND	ND	0.038	0.024	0.003	0.048	ND	ND	ND	0.0392	0.0248	0.00008	0.052	0.006
6.	Nitrate (per cent)	–	–	0.0023	0.0032	0.0068	0.004	ND	–	–	–	–	–	ND	0.005
7.	Calcium (per cent)	0.0048	0.0112	0.583	0.106	0.28	0.089	0.033	0.83	1.468	0.7216	0.182	0.0096	0.110	0.094
8.	Magnesium (per cent)	0.432	0.00576	0.417	0.28	0.268	0.078	0.021	0.72	0.3224	0.4616	0.2462	0.0024	0.120	0.203
9.	Heavy Metals (per cent)														
	Cadmium	ND	ND	ND	ND	ND	ND	ND	0.011	0.003	0.002	0.001	ND	ND	ND
	Chromium	ND	ND	ND	ND	ND	0.0002	ND	0.043	0.040	ND	ND	ND	0.0004	0.009
	Lead	ND	ND	ND	0.342	0.001	ND	ND	0.175	0.198	ND	ND	0.076	ND	ND
	Zinc	ND	ND	0.272	ND	0.081	0.0016	ND	0.389	0.227	0.368	0.436	0.2121	0.0022	0.201
	Copper	0.009	0.009	0.002	0.084	ND	0.00007	ND	0.330	0.467	0.004	0.001	0.002	0.00009	ND
	Iron	0.019	ND	0.072	0.006	0.074	0.00051	ND	58.90	22.70	0.088	0.092	0.076	0.0036	0.0136
	Manganese	0.062	0.062	0.002	0.003	0.003	0.00046	ND	0.837	0.101	0.005	0.008	0.004	0.0015	0.0005

ND: Not Detectable.

temperature. For each experiment, 100 ml synthetic solutions of cadmium and chromium metals were taken in series in different BOD bottles and fixed doses of modified bark and activated carbons were added in each bottle. pH was adjusted for each from 3 to 10 using 0.1 N NaOH and 0.1 N HCl. Bottles containing solutions were shaken at electric shaker and taken out after 0.5 Hrs and solutions from each bottles were filtered through whatman No. 42 filter paper and filtrates were used for analysis on AAS. After finding the optimum pH and time for maximum removal in each experiment, the procedure was repeated at fixed pH and time for finding optimum dose by using different doses of both the modified barks (*Acacia arabica* and *Artocarpus heterophyllus*) and activated carbons prepared from waste materials.

Table 69.5: Adsorption Characteristics of Modified Barks and Activated Carbons

Sl.No.	*Parameter*	*Modified Bark*		*Activated Carbons*				
		Acacia arabica	*Artocarpus hetero-phyllus*	*PS PAC*	*CF GAC*	*WH PAC*	*WS GAC*	*GS PAC*
1.	Iodine Number	265	465	658	620	785	664	1114
2.	Phenol Number	32	21	55	25	17	48	25
3.	Methylene Blue Number	184	192	242	248	278	285	238

Adsorption Mechanism

The adsorption mechanism of metal ions bound to adsorbents is a rather complex one. Besides ion exchange processes, there are several other phenomena. As ion exchange plays a very important role in the above-mentioned binding, the influence of pH on the chromium and cadmium removal by modified barks and activated carbons must be taken into consideration. Actually, a prominent effect can be predicted as H^+ ions complex with metallic ions, present in the solution according to the following equilibrium:

$$\text{H n Adsorbents} + M^+ \rightleftharpoons \text{M Adsorbents} + n\,H^+$$

Many researchers have reported the relationship between the amount of metal adsorption and the protons released into the solution. The release of hydrated protons exchange the pH of the solution and thereby affects the solubility of the ingredients present on the surface of modified bark and activated carbons. Second mechanism can be explained in terms of the chemical bonding. There are two major chemical bondings responsible for adsorption of heavy metals on the surface of barks and activated carbons, one is surface covalent bonding and another is hydrogen bonding. The covalent bonding results from the sharing of free electron pair between the surface oxygen atoms and metal atoms or the formation of an O-M bonding. The hydrogen bonding is due to the long-range forces between surface oxygen atoms and hydrogen atoms of the hydrated metal ions. This particular type of bonding is related to the free energy. The covalent bonding forms the outer surface complexes and hydrogen bonding forms inner surface complexes, which is mainly related to the available surface oxygen on the barks and activated carbons (67-68).

The schematic representation of the different types of bonding with metal are shown as follows:

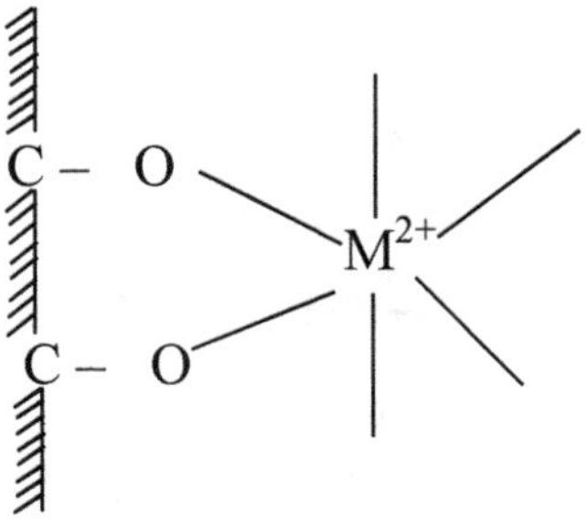

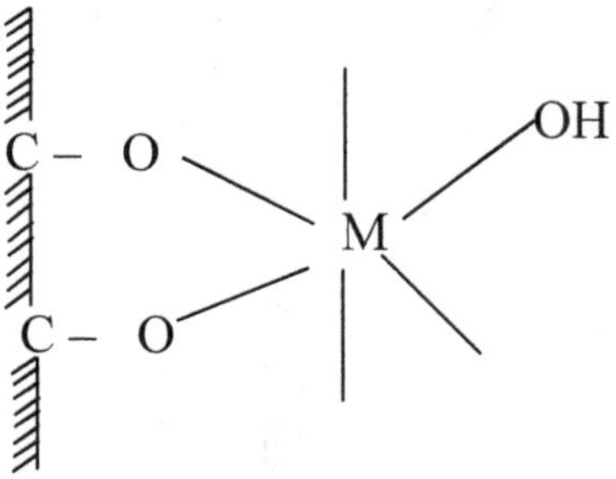

(I) **(II)**

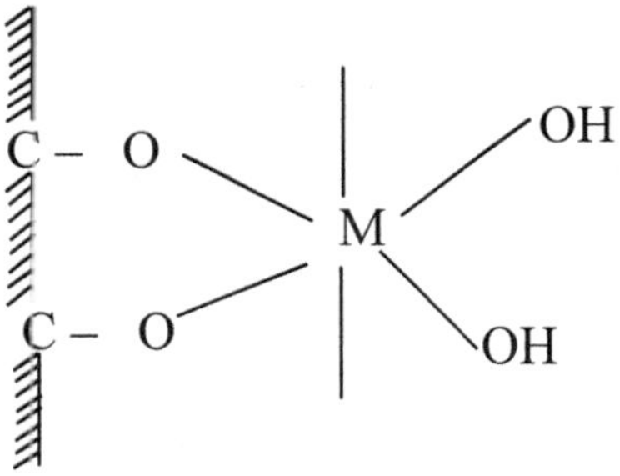

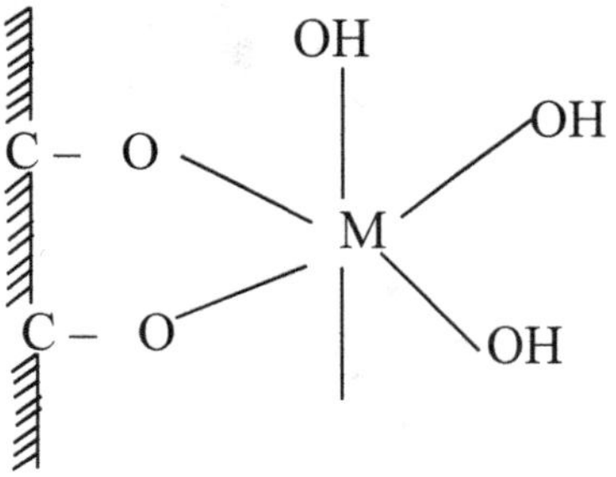

(III) **(IV)**

In order to form an inner sphere complex, it is necessary for the adsorbent to share the same hydration sphere, that is the metal ions must depart from their own hydration sphere and enter into the hydration spare of adsorbent. It is known that pH of the solution at which adsorption occurs, for one or more number of reasons, it influences the extent of removal of pollutants by adsorption on adsorbent surface. This takes place when H^+ and OH^- ions are likely to be adsorbed strongly, thereby affecting the adsorption of metals. Generally adsorption of inorganics increases at lower pH. Besides ionization of such compounds, ionic species and the charge they carry are also important. If the ionic species are charged negatively, then adsorption at lower pH could result for neutralization on the bark and activated carbons surface by higher H^+ ions concentration, which enhances adsorption.

The third mechanism can be explained in terms of the surface groups available on the surface of the adsorbents. These surface groups form complexes with the metal ions at particular pH and chemical adsorption takes place. Such type of adsorption is due to formation of strong bonds on the surface of the adsorbent and cannot be removed easily. The fourth mechanism can be explained in terms of physical phenomena. *i.e.* adsorption takes place according to the surface area and porosity. More the surface area and porosity, more will be the adsorption.

Efficiency of Low Cost Adsorbents for Heavy Metal Removal

The adsorption efficiency of modified barks and activated carbons prepared from waste materials for the removal of heavy metals are presented in Figure 69.5. It was observed that all the adsorbents prepared from waste materials are having better adsorption efficiency and could be used successfully in the wastewater treatment system.

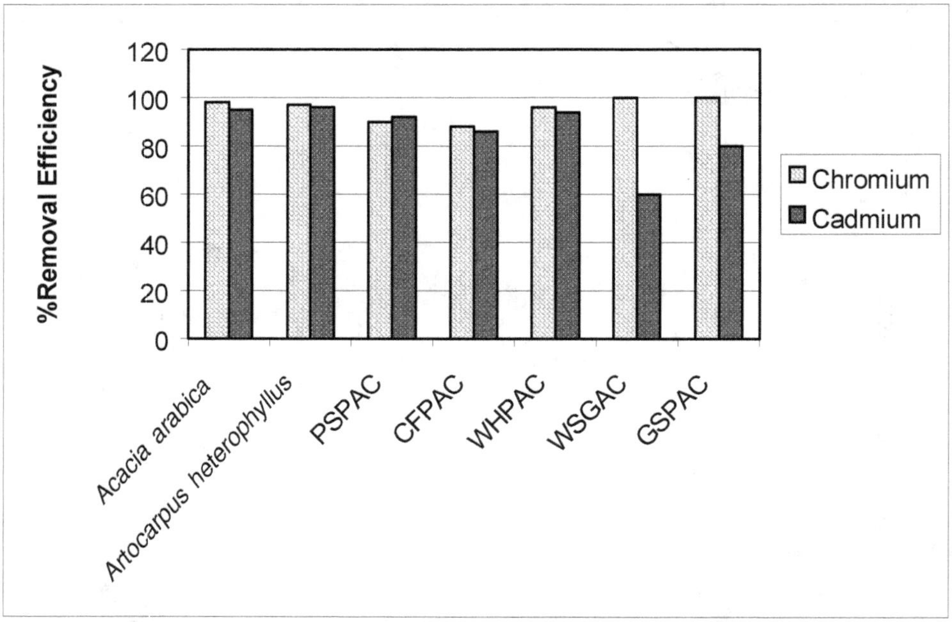

Figure 69.5: A Comparative Removal Efficiency of different Solid Waste based Activated Carbons (at heavy metal conc. of 25 ppm).

The adsorption in a solid-liquid system results in removal of solute from the solution and their concentration at the surface of the solid, till such a time that the concentrations of the solute remaining in solution is in a dynamic equilibrium with that at the surface. As the adsorption proceeds, equal amount of solute is eventually adsorbed and desorbed simultaneously. Consequently, the rates of adsorption and desorption will attain an equilibrium state called adsorption equilibrium. At equilibrium, no change can be observed in the concentration of the solute on the solid surface or in the bulk solution. The position of equilibrium is characteristics of the entire system, *i.e.* the solute, adsorbent, solvent, temperature and pH etc. Adsorbed quantities at equilibrium usually increase with an increase in the solute concentration. The presentation of the amount of solute adsorbed per unit of adsorbent or a function of the equilibrium concentration in bulk solution, at constant temperature is termed as adsorption isotherm. The shape of the adsorption isotherm gives qualitative information about the adsorption process and the extent of the surface coverage by the adsorbate. Several models can be used for the description of the adsorption data. For the present study, only Freundlich model was used. This model shows the efficiency of adsorbent to adsorb the adsorbate from the solution. The Freundlich equation is basically empirical but often useful as a means of data description. The mathematical expression of Freundlich model is as follows:

$$\frac{x}{m} = KCe^{1/n} \quad (1)$$

where, K and n are constants and Ce is the concentration of solute in solution at equilibrium. X is the amount of solute adsorbed (mg) and m is the weight of adsorbent (g). The results are usually fitted to the logarithmic form of the equation.

$$\log\frac{x}{m} = \log K + \frac{1}{n}\log Ce \quad (2)$$

This type of distribution is expressed as the quantity 'X/M' is the amount of solute adsorbed per gram of the adsorbent whereas the quantity 'Ce' is the amount of solute remaining in solution of equilibrium. A plot of log (X/M) against log Ce gives a straight line, the slope and intercept of which correspond to 1/n and log K, respectively. The data obtained from the adsorption system were fitted to the equation and the isotherms were obtained. The Freundlich adsorption parameters are depicted in Table 69.6.

Table 69.6: The Freundlich Adsorption Isotherm Parameters (FAIP) for the performance Evaluation of Modified barks and Activated Carbons

Heavy Metals	*Type*	*Adsorbent*	*FAIP*		
			K	*1/n*	*R*
	Chromium (25 mg/l)	*Acacia arabica*	0.004	0.169	0.8851
		Artocarpus heterophyllus	0.08	0.135	0.8569
		PSPAC	10.965	0.576	0.869
		CFGAC	0.169	0.479	0.829
		WHPAC	0.319	0.214	0.937
		WSGAC	0.278	0.133	0.896
		GSPAC	1.45	0.310	0.953
	Cadmium (25 mg/l)	*Acacia arabica*	2.7144	0.10	0.9755
		Artocarpus heterophyllus	0.2276	0.30	0.9178
		PSPAC	3.879	0.476	0.999
		CFGAC	0.826	0.368	0.856
		WHPAC	2.168	0.634	0.987
		WSGAC	0.006	0.500	0.898
		GSPAC	0.325	0.764	0.876

Summary

The main aim of this study was to produce low cost adsorbents for the removal of heavy metals from wastewater. In this context, two types of adsorbents were produced from waste materials. One is being the well-researched area of activated carbons, the other being that of modified natural polymeric substances like tree bark. The results showed that favorable isotherms were obtained for Cadmium and Chromium adsorption on the both type of adsorbents. Freundlich equation was fitted to the isotherms, resulting in the Freundlich equation being a closer fit to the data, based on the linear regression coefficients. The adsorption correlation coefficient (R) of all the adsorbents indicates better correlation between the adsorbent and adsorbate system. The main objective of the study was to use the waste products as a raw material under waste management practices for the production of valuable product and to evaluate its efficiency as an adsorbent by removing heavy metals. This leads to an economic method of recycling and reuse of the wastes generated from various sources, by preparing activated carbon and using this product as a polishing material in wastewater treatment system.

References

1. Toshiro M., Sidney A.H., Jesse M.C; "Metal Removal by Physical and Chemical Processes"; *J. Water Pollut. Contr. Fed.*; 47; 962-975; (1975).
2. Moore, James W. and Ramamoorthy, S.; "*Heavy Metals in Natural* Waters – *Applied Monitoring and Impact Assessment*"; Robert S. Desanto, Series Ed.; Springer Verlag Publ.; New York (1984).
3. Fribeg, L., Piscator, M., Nordberg, G.F. and Kjellstrom, T.; "*Cadmium in the Environment*"; 2nd Ed.; CRC Press; Cleveland, Ohio (1974).
4. Hulton, M.; "Cadmium"; Chapter in "*Lead, Mercury, Cadmium and Arsenic in the Environment – Scope 32*"; Hutchinson, T.C. and Meema K.M. Edn.; John Wiley and Sons Publ. (1987).
5. Fujie K., Nakagone H., Hu H. Y. and Urano K. Mizu Shori Gijustu, 34 (12): 629-33 (Spain), *Chem. Abstr.*, 120: 278856 m (1994).
6. Campbell L. Cardavelli E., Cordatore M. and Mazzei F., *Inquinamento*, 37 (3): 66-70. (1995).
7. Kapoor A., Tanjore S. and Viraraghvan T. *Int. J. Environ. Stud.*, 49 (2): 137-47 (1995).
8. Tyagi R. D., Vembu K. (Eds.); *Wastewater Treatment by Immobilized cells*, CRC Press, Boca Raton Ann. Arbor Boston, (1990).
9. Cheremisinoff P. N., *Hand book of Water and Wastewater treatment Technology*, Marcel Dekker, New York, NY (1994).
10. Mackenzie L. Davis, David A. Cornwell; *Introduction to Environmental Engineering*; McGraw- Hill, Inc., Singapore (1991).

11. Sharma S. K., Rajshree R., " Wastewater Treatment Technology for Heavy Metal Pollution"; ed. Goel P.K., *Advances in Wastewater treatment*; ABD Publishers Jaipur India (2003).

12. Ali Mohammad, Mohamed Najar P. A., Iraqi E.; "Treatment Techniques for the Minimization of heavy Metal Contents in Water and wastewater"; ed. Goel P.K., *Advances in Wastewater treatment*; ABD Publishers Jaipur India (2003).

13. Kelsall G. H., " *Electrochemical Engineering Course*" New Castle Upon Tyne 1994.

14. Corapocioglu, M.O. and Huang, C.P.; "The Adsorption of Heavy Metals on to Hydrons Activated Carbon"; *Water Res.*, 21 (9), (1987).

15. Weber, W.J. (Jr.); " *Physico-chemical Process for Water Quality Control*"; New York, Wiley, Interscience, 230 (1972).

16. Ford D. L., "Wastewater Characteristics and Treatment"; *Activated Carbon Adsorption for Wastewater Treatment*, ed. J.R. Perrich, CRC Press Inc, Florida, (1981).

17. Yenkie, M. K. N., Natarajan, G.S., "Adsorption Equilibrium Studies of Some Aqueous Aromatic Pollutants on Granular Activated Carbon Samples", *Separation Science and Technology*, 26 (5), (1991).

18. Wanumanulu, V. and Subramanyam, P.V.R.; "Planning Of Water Pollution Control Measures For An Integrated Kraft Pulp And Paper Mill", *IPPTA Souvenir* 138-152 (1975).

19. Directorate General of Technical Development (DGTD); Pulp and Paper Directorate, Govt. of India, New Delhi (1978).

20. Subramanyam, P.V.R., Parekh, R.C. and Mohanrao, G.H.; "Low cost methods for treating pulp and paper mill effluents", *Indian Pulp and Paper*, 9,6 (1972).

21. Mohandas Rao, K.P., Chhabaria, N.D.; "Pollution control in pulp and paper mill", *IPPTA*, 4,11 (1972).

22. Christopher A.Toles, Wayne E. Marshall and Mitchell M. Johns, "Activated carbons from agricultural by-products for adsorption of organics and metals, CAER-University of Kentucky, Center for Applied Energy Research, Energeia, Vol. 9, No. 3, (1998).

23. Randall, J.M.; Garrett V.; Bermann R.L.; Waiss Jr. A.C. "Use of bark to remove heavy metal ions from waste solution". *Forest Prod. J.* 24-9-80-84 (1974).

24. Randall, J.M.; Hautala E.; Waiss A.C. Jr. "Removal and recycling of heavy metals ions from mining and industrial waste streams with agricultural byproducts". *Proc. 4th Mineral Waste Utilization Symposium III.*, May 7-8, 329-34 (1974).

25. Randall, J.M. and Hautala, E. "Removal of heavy metal ions from waste solution by contact with agricultural byproducts". *Proc. 30th Industrial Waste Conf.* Purdue University. May 6-8, 412-22 (1975).

26. Randall, J.M.; Hautala E.; Waiss A.C. Jr. and Tschernitz, J.C.; "Modified barks as scavengers for heavy metal ions". *Forest Products J.*, 26-8, 46-50 (1976).

27. Jodai, Susumu, O'nishi, Hirolsugus, Uehara Toru, Goto, Terus. "Adsorption of heavy metal by barks (I) – Adsorption properties of various barks". Shimane Daigaku Nogakuba Kenkya Hokoku. 12, 102-8 (1978).

28. Robert, E.J. and Rowland, S.P. "Ion-exchange and complexing properties of cotton fabric and containing ethyleamine network polymer formed in situ". *Textile Research J.*, 41, 10-864 (1971).

29. Robert, E.J. and Rowland, S.P. "Removal of mercury from aqueous solutions by nitrogen containing chemically modified cotton". *Environ. Sci. and Technol.*, 7-552 June (1973).

30. Pawankumar, Dara, S.S., "Use of modified barks for scavenging heavy metal ions from industrial wastewaters". *Indian Journal of Environmental Health*, 22, 3-196-202, (1980).

31. Masri, M.S.; Reuter F.W.; Friedman, M. "Binding of metal cation by natural substances". *J. Applied Polym. Sci.*, 18-675-81 (1974).

32. Varshney, C.K. and Rzoskg, J., Jemk, W. (Eds). " *Aqautic weeds in south east Asia*", B.V. Publications, the Hague (1976).

33. Making aquatic weeds useful, some perspectives for development coutries, National Academy of sciences, Washington D.C. (1976).

34. Moharao, G.J. *Indian Journal of Environmental Health*, 15 (3), 223-225 (1973).

35. Sheffield, C.W., "Hycinth Control"' *Journal* 4-5, (1975).

36. Wolverton, B.C., "Water Hyacinths for upgrading scanvage lagoons to meet advanced wastewater treatment standards. Part II, NASA, Technical memorandum, T.M.X., 72730, Washington D.C. (1976).

37. Volverton, B.C. and McDonald, R.C., "Water hyacinth sorption rates of lead, mercury and cadmium NASA/ERL report Mp. 170, Washington D.C. (1978).

38. Wolverton, B.C. and McDonald, R.C., "Water hyacinth productivity and harvesting studies", NASA/ERL report Mp. 171 and 173, Washington D.C. (1978).

39. Deshkar, A.M., Bokade, S.S. and Dara, S.S.' "Modified hardwickia binata bark for adsorption of mercury (II) from Water", *J. Water Research*, Vol. 24, No. 8, pp 1011-1016, (1990).

40. Jasuja, K., Parwana, H.K. and Rao, A.L.J., "Removal of Cr (VI) from wastewater by *Ablesmoschus esculentus*", *Indian Journal Environ. Hlth*. Vol.39, No 2, 103-108 (1997).

41. Gallop, R.A.; "Pollution Solution – Recycling of Wastes"; *AICE Rev.*; 25167 (1970).

42. Bell P.R., Varjarandi, J.J.; "*Pyrolysis-Resource, Recovery from Solid Waste*"; Ann. Arbor Mich.; Ann Arbor Science, Publ. (1973).

43. Barret, C.D.; "*Pyrolysis of Organic Wastes, Waste Management, Control, Recovery and Rense*"; Ed. by M.R. Kirov; Ann Arbor Science Publ. (1975).

44. Resenzweig, Mark D.; "Low Cost Activated Carbon from Pulping Waste"; *Chem. Engg*. New York; 76 (27); 46-48 (1969).

45. Backmans, John M.; "Pyrolysis of Sewage Sludge, Its Production and Possible Utility"; *Environmental Science and Technology*; Washington D.C.; 20 (1); 69-71 (1971).

46. "Activated Carbon from Wastes"; *Canadian Symposium on Water Pollution Research; 6th Proc*. Held in Toronto Ont. Can.; 2-3 (1971).

47. Larsen V.J.; "The Use of Straw for Removal of Heavy Metals from Wastewater"; *J. Environ. Qual.*; 10; 188-192; (1981).

48. Alaerts, G.J.; Jitjalrant, V. and Kelderman, P.; "Used of Coconut Shell Based Activated Carbon for Chromium (VI) Removal"; *Wat. Sci. Tech.*; 21; 1901 (1989).

49. Flynn, C.M., Carnahan, T.G.; "Adsorption of Heavy Metal Ions by Zanthated Saw Dust"; Rep. Inverst. U.S., Bur/Mines, R.I.; 8427 (1980).

50. Coupel B.; "The Treatment of Wastewater with Peat Mass"; *Wat. Res.*; 10; 1071-1076 (1976).

51. Rodriguez Reinoso, F, Martin-Martinez J.M., Molina-Sobia M., Perez-Lledo I and Prado-Burgerete C.; "A Comparison of the Porous Texture of Two CO_2 Activated Botanic Materials"; *Carbon*; 23; 19 (1985).

52. Srinivasan, K., Balasubramaniam, N. and Ramkrishna, T.V.; "Studies on Chromium Remvoal by Rice Husk Carbon"; *Ind. J. Enviorn. Hlth.*; 30; 376 (1988).

53. Schlesinger, M.D., Sonner, W.S.; Waltson, D.E., "Pyrolysis of Waste Material from Urban and Rural Sources"; *Mineral Waste Utilization Symposium, 3rd Proc.* held in Chicago (1972).

54. Godd, A.A.; "Manufacture of Activated Carbon by Gasification in Fluidized Bed"; *Int. Conf. on Fluidised Bed Combustion; 2nd Proc*. and held in College Corner; Ohio (1972).

55. Banerjee, S.; "Active Carbon from Coconut Shell"; *Ind. J. of Technology*; 14 (1); 45-46 (1976).

56. Haque, R., Chakraborti, R.K., Dutta, M.L. and Lyengar, M.S.; Production of Carbon Black from Assam Coal"; *Ind. Engg. Chem. Prod. Res. Dur.;* 19 (4) (1971).

57. Ramteke, D.S.; "Physico-chemical Studies of Pyrolysed Product from Industrial Wastes"; A *Thesis* submitted to Nagpur University (1989).

58. NTIS Report, Cincinnati, Ohio (1973).

59. Ahmadpour A., Do D.D.; " The Preparation of Active Carbons from Coal by Chemical and Physical Activation", *Carbon*, 34 (4), 471-479 (1996).

60. Lillo- Rodenas M.A., Carzrola-Ameros D., Linares – Solano A.; "Understanding Chemical Reactions between Carbons and NaOH and KOH –An Insight into Chemical Activation Mechanisms", *Carbon*, 41, 265- 267 (2003).

61. " Process Design manual for Carbon Adsorption"; U. S. Environmental Protection Agency, Technology Transfer EPA; 625; 1071-0029.

62. Hassler, J.W.; (1974); " *Purification with Activated Carbon"; 3rd Ed.*; Published by Chemical Publishing Company, New York.

63. Giles, C. H. and Trivedi, A. S.; (1969); " Specific Surface Area by Methylene Blue Method"; *J. Chem. Ind.*; 40; 1426.

64. "Proximate Analysis, IS- 1350 (Part-1)"; (1969); Publication of Indian Standard Institute, New Delhi.

65. "Ultimate Analysis of Coal and Cokes", IS- 1351-59; (1969); Publications of Indian Standard institute, New Delhi.

66. *"Standard Methods for the examination of water and wastewater"*; 17th Ed. APHA, AWWA and WPCF; (1989); Published by American Public Health Association; New York.

67. Snoeynik, V.L. and Weber, W.J. (Jr.); (1967)"The Surface Chemistry of Active Carbon–A Discussion of Structure and Surface Functional Groups"; *Environ. Science Technol.* 1,228.

68. Swami, A.; "Physico-chemical and Adsorption Studies of Natural Polymeric Substances"; A dissertation submitted to Devi Ahilya University, Indore; (1999).

Chapter 70

Biogas Generation Potential of Municipal Wastewater: A Techno-Economic Evaluation

P. Nema, S.N. Kaul, N.C. Kankal, C.G. Mehta and S.K. Goyal

CSIR-NEERI, Nagpur

ABSTRACT

Anaerobic digestion of wastewater is considered to be most appropriate and economical method of energy generation. Moreover, biogas production is an important parameter for evaluation of the performance of a treatment system. Bench scale studies were conducted using UASB reactors for assessment of treatability and biogas generation potential of two types of wastewaters collected from different parts of a city. These wastewaters were characterized as low strength (COD: 200-300 mg/L) and high strength (COD > 700 mg/L).

The system was operated at varying organic loading rates and hydraulic retention time and their efficacies in terms of removal of COD, BOD, SS and other pollutants as also the biogas generation potential was evaluated. Quantitative and qualitative (composition) assessment of the biogas were also made simultaneously. Based on the data collected from the laboratory scale experiments, two UASB based anaerobic systems of 200 MLD capacity each were designed. In addition, the anticipated revenue earnings through recovery of biogas and its equivalent energy cost were also computed. The exercise demonstrated that a substantial cost recovery towards capital investment (system cost) can be achieved. The paper summarizes the salient findings of the study.

Introduction

A huge quantity of wastewater is generated from domestic sources. A typical daily per capita water consumption varies from 180-300 litres, of which about 80-85 per cent returns as wastewater. On an average, a city with 1 million population generates about 150 MLD wastewater. Presently in India, most of the sewage treatment plants installed particularly in large cities are based on conventional biological treatment systems, which in general aim disposal of treated effluent in the water bodies/on land without any resource recovery. The treated effluent is usually let out into natural water bodies, which get polluted if the effluent quality does not conform to regulatory effluent quality requirements. Recognizing the importance of anaerobic treatment of wastewater, which of late has gained momentum due to its associated advantages, a bench scale study was carried out to assess system's performance together with biogas generation potential.

Experimental Study

The experimental methodology consisted of characterization of wastewater streams followed by bench scale treatability studies. Wastewater samples were collected from the inlet of sewage treatment plants in the city at an interval of 2 hrs, and were composited for 24 hrs. The composited samples were then analyzed for various physico-chemical parameters including nutrients and heavy metals, as per the prescribed standard procedures. Further, bench scale studies were conducted for assessment of treatability as well as biogas generation potential for both the types of wastewaters.

The bench scale system consisted of two 2 litre capacity cylindrical reactors one each for low and high strength wastewater. The system was energised with active anaerobic sludge taken from a sewage treatment plant. A mixture of sludge and wastewater were kept in the laboratory reactor, which was monitored continuously. The reactor was operated at an HRT of about 8-10 hrs at an organic loading rate of 0.7-1.0 (average 0.8) kg COD/m^3.d for low strength wastewater and between 2.1 and 3.0 (average 2.7) kg COD/m^3.d for high strength wastewater. The reactors were operated for a total period of about 60 days. Temperature of the reactor during the study period was recorded around 31-33 °C, whereas pH of treated effluent was in the range of 6.8-7.6. Various parameters representative of the performance of the system were analyzed as per the standard procedures.

Results and Discussion

Wastewater Characteristics

Characteristics of the two types of wastewater streams are summarized in Table 70.1.

The observed concentration of COD, BOD and SS in municipal wastewater was in the range of 246-281, 84-92 and 108-128 mg/L respectively, whereas the concentration of these parameters in the mixed wastewater was considerably high, *i.e.* COD: 633-892; BOD: 170-340 and SS: 396-472 mg/L. Concentrations of sulphate and sodium were also higher in mixed wastewater than municipal waste as expected due to industrial waste component.

Table 70.1: Wastewater Characteristics (24 hrs composite samples collected at an interval of 2 hrs)

Sl.No	*Parameter*	*Municipal*		*Municipal and Industrial Mix*	
		Range	*Average*	*Range*	*Average*
1.	PH	7.4-7.9	—	7.1-7.8	—
2.	COD	246-281	259	663-892	755
3.	BOD	84-92	87	170-340	230
4.	SS	108-128	117	396-472	438
5.	VSS	50-64	55	180-264	221
6.	TDS	1664-1840	1745	1866-1934	1910
7.	Total Alkalinity	450-560	510	560-650	590
8.	Chlorides	500-560	527	400-450	427
9.	Sulfates	62-100	78	80-430	217
10.	Sulfides	ND	—	ND	—
11.	Oil and Grease	10-15	12	12-17	15
12.	COD/BOD (avg)	—	2.98	—	3.28
13.	Phosphates	5-6	6	6-8	7
14.	NH3-N	18-21	19	20-24	22
15.	TKN	39-47	43	50-51	50
16.	Calcium	32-56	44	32-52	44
17.	Magnesium	10-53	30	7-36	24
18.	Sodium	460-540	500	680-760	713
19.	Potassium	7-21	13	18-32	23

All values except pH are in mg/L.

Treatability vis a vis Biogas Generation Study

The performance data of the bench scale system with reference to removal of different polutants as well as biogas generation potential are presented in Tables 70.2 and 70.3 respectively for low and high strength wastewaters.

A perusal of Table 70.2 indicates that in low strength wastewater, average reduction in concentration of COD, BOD and SS was about 78 per cent, 80 per cent and 60 per cent respectively. Average gas yield was about 0.48 m^3/kg COD removed, with methane and carbon dioxide content of 62-68 per cent (average 65 per cent) and 32-36 per cent (average 33.6 per cent) respectively.

In high strength wastewater, average reduction in concentration of COD, BOD and SS was about 82 per cent, 84.5 per cent and 60.8 per cent respectively (Table 70.3). Average gas yield was about 0.45 m^3/kg COD removed, with methane and carbon dioxide content of 62-69 per cent (average 65 per cent) and 30-35 per cent (average 33.5 per cent) respectively.

Table 70.2: Treatability of Low Strength Wastewater: Performance Characteristics

Parameter	*Influent (mg/L)*		*Effluent (mg/L)*		*Reduction (per cent)*	
	Range	*Average*	*Range*	*Average*	*Range*	*Average*
COD	238-317	263	40-83	58	67-84	77.9
BOD	86-95	91	14-25	18	73-85	80.2
SS	74-188	139	40-80	56	46-72	59.7
TVA	51-137	95	17-53	42	44-72	55.8
Total Alkalinity	540-680	578	600-1040	748	—	—
NH_3-N	14-22	17	13-19	15	6-14	11.8
TKN	36-56	43	33-49	39	7-12	9.3
PO_4-P	5-9	7	2-6	4	25-66	42.9

Biogas Generation	*Range*	*Average*
Gas Yield, m^3/kg CODr	0.38-0.56	0.48
Methane, per cent	62.8-67.6	65.1
Carbon dioxide, per cent	31.7-35.8	33.6

Table 70.3: Treatability of High Strength Wastewater: Performance Characteristics

Parameter	*Influent (mg/L)*		*Effluent (mg/L)*		*Reduction (per cent)*	
	Range	*Average*	*Range*	*Average*	*Range*	*Average*
COD	757-971	882	119-208	164	78-84	81.4
BOD	200-405	295	37-64	48	75-87	83.7
SS	272-380	314	92-172	123	37-71	60.8
TVA	163-206	177	43-80	60	53-74	66.1
Total Alkalinity	540-720	630	740-1380	970	—	—
NH_3-N	24-36	28	21-30	25	8-16	10.7
TKN	54-86	72	50-78	64	9-13	11.1
PO_4-P	6-14	9	2-12	6	18-78	33.3

Biogas Generation	*Range*	*Average*
Gas Yield, m^3/kg CODr	0.3-0.64	0.45
Methane, per cent	62.8-69.2	65.0
Carbon dioxide, per cent	29.6-35.2	33.5

The difference between removal of COD, BOD and SS in the two wastewaters could be attributed to the initial concentration of these parameter in municipal wastewater in comparison to the wastewater mixed with industrial waste. As regards removal of nutrients NH_3-N was around 10 per cent, which was in line with the known efficacy of anaerobic system, whereas PO_4 showed 33-42 per cent removal.

Proposed Treatment System (Table 70.4)

From the above study, it is concluded that the low as well as high strength wastewaters can be treated successfully through anaerobic digestion involving biogas recovery system, for which an appropriate anaerobic system (UASB, Fixed film etc.) may be adopted. Usefulness of UASB reactor system has been demonstrated successfully for industrial as well as municipal wastewaters by various researchers (available on web). A number of full scale UASB plants, in the capacity range of 5 to 50 MLD are already in operation since early nineties, and several others with higher capacity (up to 200 MLD) are under different stages of completion. Some of the advantages achieved in UASB system include; lower capital and O and M costs, lower energy consumption, generation of biogas, small land area requirement and low surplus sludge production. The treated effluent from UASB being devoid of oxygen will need some aerobic treatment prior to discharge in the natural environment in accordance with the regulatory requirements.

Table 70.4: Design Data for Methane Reactors and Biogas Generation

Sl.No.	*Item*	*UASB System for*	
		Low Strength Wastewater	*High Strength Wastewater*
Reactor Details			
1.	Design flow, MLD	200	200
2.	Detention time, hrs	8	8
3.	Loading Rate, kgCOD/m^3.d	0.8	2.7
4.	No. of units	20	20
5.	Reactor diameter, m	30	30
6.	Liquid depth, m	5.2	5.2
7.	Total depth of reactor	8.0	8.0
8.	COD removal efficiency, per cent	67-84	78-84
9.	BOD removal efficiency, per cent	73-85	75-87
Biogas Generation			
10.	Biogas production, m^3/d (at CODr–75 per cent and Gas yield – 0.45 m^3/kg CODr)	17552	61425
11.	**Methane content, per cent**	65	65
12.	**Coal Equivalent, MT/d**	25.4	87.8
13.	Total Cost (@ Rs. 2200/- MT) Rs. Lacs/day	0.56	1.93
14.	Annual Cost of Biogas produced, million Rs.	**20.4**	**70.4**

The typical arrangement of a UASB type treatment plant shall include; (a) initial pumping, (b) screening and degritting, (c) main UASB reactor, (d) gas collection system, (e) sludge drying bed, and (f) post treatment facility (depending upon final discharge standards). The typical scheme is depicted in Figure 70.1.

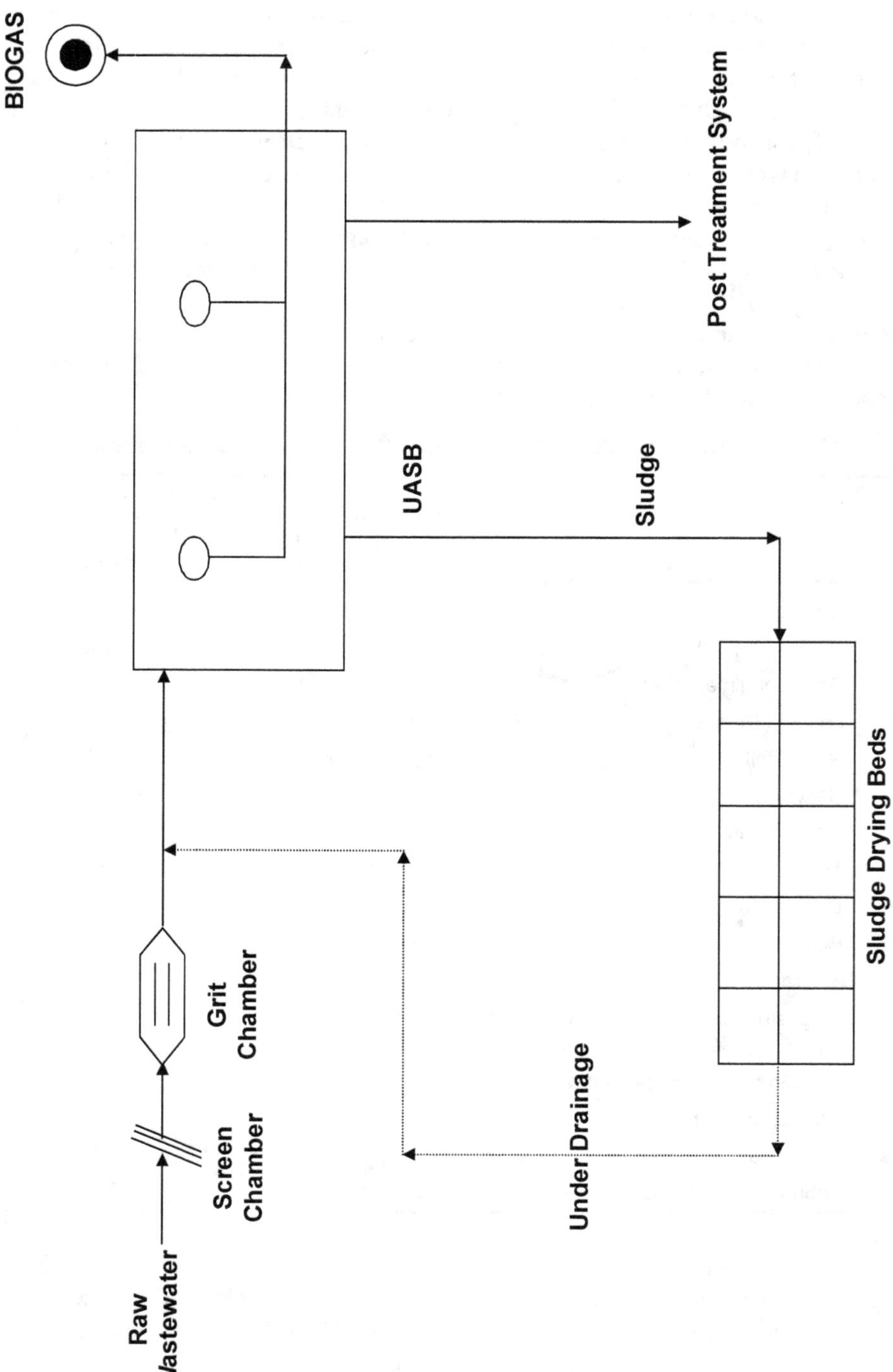

Figure 70.1: Schematics of Wastewater Treatment Through UASB Reactor.

The design parameters worked out based on the bench scale study for a 200 MLD capacity plant (with 10 per cent cushion for the existing load) are given in Table 70.4 along with the biogas generation details.

An HRT of 8 hrs shall be adequate with organic loading rate of about 0.8 and 2.7 Kg COD/day respectively for low and high strength wastewater. A total of 20 UASB reactors with 30 m diameter each handling 10 MLD wastewater shall be required. The biogas generation potential from the system is estimated to be about 17750 and 61425 m^3 respectively from low and high strength wastewater with 65 per cent methane content. The daily revenue generation could be of the order of Rs. 0.56 Lacs and Rs. 1.93 Lacs respectively from the treatment of low and high strength wastewater.

Anticipated quality of UASB treated effluent is given in Table 70.5 for both the types of wastewaters.

Table 70.5: Anticipated Quality of UASB Treated Effluent (HRT–8 hrs)

Parameter	*Influent (mg/L)*	*Anticipated Reduction (per cent)*	*Effluent (mg/L)*
	Low Strength Wastewater		
COD	263	75	66
BOD	91	80	18
SS	139	60	56
	High Strength Wastewater		
COD	755	75	189
BOD	322	80	64
SS	314	60	126

In both the cases further treatment by any of the aerobic processes *viz.* cascade aeration, oxidation/maturation ponds or aerated lagoons with short detention time (depending on the requirements) would be required prior to discharge of effluent into the water body or on land.

Cost-Benefit Analysis

The total quantity of biogas generation from wastewater is estimated to be 47250 m^3/d with 65 per cent methane content, amounting to heat energy worth Rs. 54 million annually. Based on the calorific value of these fuels, equivalent replacement quantity of different fuels, *viz.* LPG, NG, coal, wood, kerosene, diesel etc. is estimated as presented in Figure 70.2. Typically 1 m^3 of biogas is equivalent to 1.43 kg coal (ash content 40 per cent); 1.6 kg firewood, 1.9 kg rice husk, 0.7 litre diesel, 0.25 m^3 of LPG and 0.67 m^3 of natural gas. Various possibilities exist for use of the biogas as a fuel or other energy source and a decision in this regard could be arrived at depending on site-specific requirements as also techno-economic analysis. The revenue generated from the biogas shall offset the input cost of the proposed system to a large extent.

Further, the treated wastewater is also a resource, which can be used for irrigation as it contains a good amount of nutrients *viz.* nitrogen, phosphorus and other trace elements and is a good substitute for chemical fertilizers.

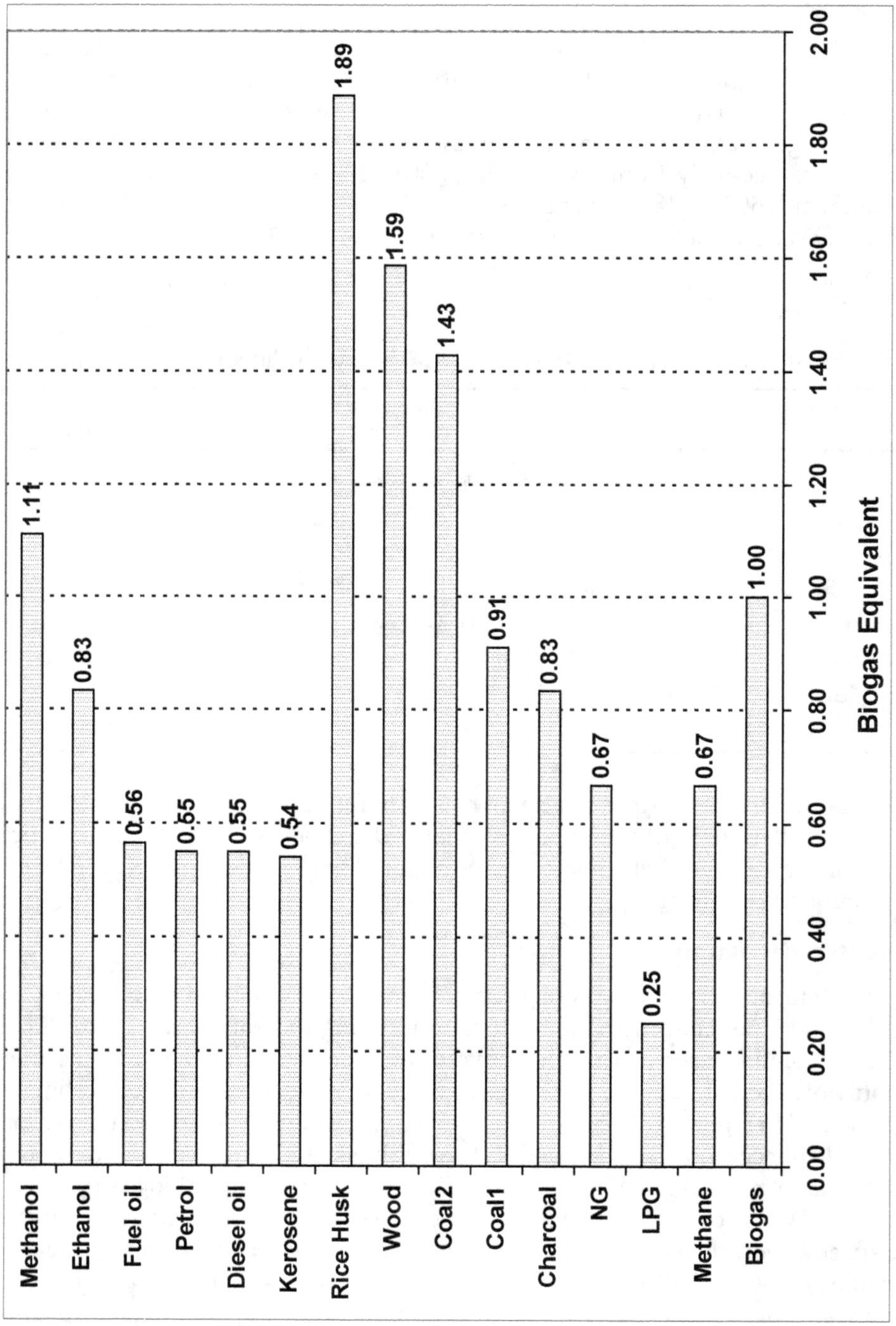

Figure 70.2: Biogas Equivalent of Different Fuels (Based on Calorific Value).

Conclusion

Techno-economic feasibility of municipal wastewater treatment through UASB system along with its biogas generation potential has been evaluated. Bench scale treatability studies indicate the feasibility of a single stage UASB system for the treatment of these wastewaters with the additional benefit of biogas generation. The biogas together with treated effluent containing nutrients is seen as useful resources, which if utilized could offset the input cost to a large extent.

References

1. Arceivala Soli J. (1995). Experience with UASBs for Sewage Treatment in India. Proceedings, 3rd International Conference on Appropriate Waste Management Technologies for Developing Countries, NEERI, Nagpur, Feb. 25-28, 1995.
2. '*Wastewater Treatment for Pollution Control*' by Arceivala Soli J. 2nd edition published by Tata McGraw-Hill Publishing Company Limited, New Delhi, 1998.
3. '*Thermo-chemical Characterization of Biomass Residues for Gasification*', Vol. 1, Edited by P.D. Grover, IIT-Delhi, 1989, p. 56.

Author Index

Subject Index

www.ingramcontent.com/pod-product-compliance
Lightning Source LLC
LaVergne TN
LVHW020836091026
849796LV00003B/32
* 9 7 8 9 3 5 1 3 0 9 1 4 7 *